T0222617

Technologies of Power

Technologies of Power
Essays in Honor of Thomas Parke Hughes and Agatha Chipley Hughes

edited by Michael Thad Allen and Gabrielle Hecht

The MIT Press
Cambridge, Massachusetts
London, England

The MIT Press is pleased to keep this title available in print by manufacturing single copies, on demand, via digital printing technology.

© 2001 Massachusetts Institute of Technology

All rights reserved. No part of this book may be reproduced in any form by any electronic or mechanical means (including photocopying, recording, or information storage and retrieval) without permission in writing from the publisher.

Set in Sabon by The MIT Press.

Library of Congress Cataloging-in-Publication Data

Technologies of power: essays in honor of Thomas Parke Hughes and Agatha Chipley Hughes/edited by Michael Thad Allen and Gabrielle Hecht
p.   cm.
Includes bibliographical references and index.
ISBN 978-0-262-01184-6 (hc: alk. paper)—978-0-262-51124-7 (pbk.: alk. paper)
1. Technology—History. 2. Technology and state—History. I. Hughes, Thomas Parke. II. Hughes, Agatha C. III. Allen, Michael Thad. IV. Hecht, Gabrielle.
T19.T36 2001
609—dc21   00-048041

Contents

Acknowledgments

We would like to acknowledge the numerous individuals who aided us in preparation of this book. We completed the introductory essay while Michael Thad Allen was on leave at the Zentralinstitut für die Geschichte der Technik at the Deutsches Museum in Munich. An informal seminar organized by Matthias Heymann provided a lively forum for the discussion of the guiding ideas around which this book took shape. The authors are grateful for the comments of Alexander Gall, Jörg Hermann, Jeff Lewis, Stephan Lindner, Falk Selinger, Mats Fridlund, and Nina Lerman. For their insightful reviews, we thank David Nye, Ronald Bayor, Brian Balogh, Douglas Flamming, Kenneth Ledford, Richard Kuisel, Geoff Eley, and Karen Sawislak. Thanks also to Larry Cohen of The MIT Press for his support and patience. Our deepest thanks, of course, are reserved for Tom and Agatha.

Gabrielle Hecht, Ann Arbor
Michael Thad Allen, Munich

Disciplined Imagination: The Life and Work of Thomas and Agatha Hughes

John M. Staudenmaier, S.J.

> Throughout almost fifty years she and I loved deeply and did history together.
> —T. Hughes, *Rescuing Prometheus* (Pantheon, 1998), p. 309

Tom and Agatha Hughes, living together for nearly five decades, modeled intellectual creativity for those close to them. They make a strong case for intimacy as the heart of insight and mutuality as an essential condition for sustained attention to the capricious historical record. They loved to talk over what each saw in their several worlds—the academy, art, architecture, the political order, their family, and their church. Having been party to some of those conversations, I understood them to live with the expectation that whatever they perceived, questioned, were annoyed at, or wondered about would not devolve into random fragmentary impressions. Daily listening wove a continuous fabric of interpretation, with each an interlocutor for the other. The editors and contributors to this volume have dedicated it to Agatha and Tom together. When I write about the evolution of Tom's thought and of his place within the history of technology, the roots of sustained insight that I witnessed in their relationship are never far from my mind.

Tom Hughes sought theoretical understanding in densely worked historical evidence, interrogating it and listening for surprises, sticking with his material until it suggested conceptual themes that made sense of how things happened. It was not his style to begin with theory so that he could bring it to bear on evidence—and with good reason. During his early years, it could be said that the closest thing to a theory of technological change—a morally idealistic engineering creed combined with a postwar version of American manifest destiny—was so pervasive that its influence passed virtually unnoticed. The Cold War ideology, rightly emphasized by Gabrielle Hecht and Michael Thad Allen in their introduction to this volume, embodied the belief that science and technology, when well funded and unfettered by local political interventions, led in a clean line toward democratic societies; that science, technology, and democratic society together, uniquely in human history, transcended the superstition, passion, and vested interests of the traditional political order. It made for thin soup as a theoretical basis for historical interpretation.

The history of technology emerged as a (barely) visible discipline in the late 1950s with the founding of the Society for the History of Technology (SHOT) and its journal *Technology and Culture*.[1] The founding members— Tom was one—understood the journal's title as a deliberate step away from the reigning definition of the field seen in the multi-volume internalist histories then appearing in Britain, France, and the Soviet Union.[2] If the series edited by Singer, Holmyard, Hall, and Williams (Oxford University Press, 1954–1958) was called *History of Technology*, the new journal's title would emphasize context—technology *and* culture. By the end of its first 20 years, the field showed solid intellectual promise as an emerging community of discourse. The quality of publications in articles and monographs grew measurably stronger, so that SHOT entered the 1980s as a small but healthy and growing subdiscipline.

We can identify two major generalizations that had become broadly accepted in SHOT by 1980. First, a host of case studies had established the bedrock principle that every technology must be understood in terms of the particularities of its context of origin. Historical actors, understood in terms of their motives, world views, and resources, make a difference in the outcome of technological design decisions; so do the ambient social order, its political and economic character, and its world view.[3] Second, these technological actors have a distinctive cognitive style, a blend of theoretical expertise and experience-based pragmatic judgment. Historians of technology repeatedly explored the relationship of technological practice with science, manifesting near unanimity on the core premise that technology is *not* "applied science." Technological thinkers, these studies indicate, use science as one sort of epistemological equipment among several. Technically creative actors, principally engineers and inventors, know things that the larger society needs to know about the nature of technological creativity, especially the tension between precision and the constraints of the working world.

By 1980, then, historians of technology had arrived at a consensus that appeared to flatly contradict the postwar paradigm of unfettered science and technology driving worldwide development inevitably toward representative democracy. They argued that there is no such thing as autonomous technological progress, resulting from the application of an equally autonomous science, operating free from the constraints and complexities of the human context. What came to be known as "contextualism" had become mainstream.[4]

During these same years, however, a number of mostly unstated working assumptions revealed the latent power of the Cold War model of linear progress. When we look for patterns among the research topics chosen by historians of technology during the period and when we consider topics seldom studied and questions rarely asked of the evidentiary base, we find a shadow consensus that reveals limitations in SHOT's early contextualism.

Consider the most salient of these assumptions: that the term "technology," as described by the patterns of research topic choices, was exclusively Western and was neatly framed by the chronology and geography of Western Civilization courses taught as a core history requirement in most universities at the time. Histories of technology concentrated mostly on successful strategic actors: the engineers, inventors, investors, laboratory research teams, and managers whose work was understood as creating and finalizing the design of technologies and moving them into the world of ordinary use. Once designed and produced, technologies became less historically interesting, perhaps because they were assumed to maintain a stable form until rendered obsolete by newer technologies. Technological success stories appeared much more frequently than failure stories. Paradigmatically, technological cognition meant defining a goal, marshaling resources to achieve the goal, and responding to obstacles as they turned up along the way. Other actors—wage workers, product users, non-Westerners, women—appeared in these accounts, if at all, as deep background, their voices and influence muted. Finally, historians of technology were typically male, should probably have had some engineering background, and lived in the United States or Great Britain.

These are, to be sure, excessively delineated assertions with exceptions at every point. Nonetheless, they reveal the profile of the field's inchoate identity as simultaneously creative and narrow. All the assumptions taken together fit nicely into the Cold War ideology noted briefly above. The thin clean line of an apolitical and ideology-free Western science and technology leading directly toward Western-style democratic societies legitimized a mentality that would brush aside the perspective of anyone who did not hold a seat at the technological design table and treat influences emerging from outside the design plan as interruptions requiring deft management rather than as voices requiring a change of concept. The West emerged from World War II with a fistful of technological and scientific trump cards for the game of global dominance that followed. Wartime research and development experience, especially as seen in the United States, seemed to argue for giving expert strategic actors as free a hand as possible to design seemingly impossible complex systems to serve domestic markets and national defense. In view of their field's origins in the engineering education community, it is hardly surprising that most historians of technology in the founding generation wrote narratives congruent with the spirit of the time.

Still, the field's near-unanimous focus on the designers and proprietors of Western technologies carried powerfully subversive seeds embedded in its central methodological assumption. When SHOT's founders chose *Technology and Culture* as the title of the society's journal, they signaled a break with the predominantly internalist consensus of extant histories of technology. Their new contextualism required that they situate every technology within the particularities of its historical context. Opening the contextual door to the vagaries of the human endeavor meant in principle that linear explanations of design outcomes were subject to interrogation and that technological narrative generally lay open to theoretical interpretation.

Tom participated in SHOT, this emerging community of discourse, as a local citizen, even as his contextual work began to generate a series of exceptionally influential conceptual innovations. Tom first appears in SHOT's public records as chairman of the Program Committee in 1961. In the mid 1960s, and again in the early 1970s, he served on the executive council. During the 1970s, he also chaired the nominating committee and was elected vice president and then (in 1979) president. In the fall of 1973, Tom joined the History and Sociology of Science (H&SS) department at the University of Pennsylvania. He brought with him a growing reputation. His first book, *Elmer Sperry: Inventor and Engineer* (Johns Hopkins University Press, 1971), had won the Texas Institute of Letters' Best Book in the Fields of General Knowledge prize in 1971 and SHOT's Dexter Prize in 1972. Tom settled in at Penn, where over more than 20 years he mentored a steady stream of graduate students, many of them contributors to this volume. He introduced historians of technology into the mix of scholars who graced H&SS's prestigious Monday afternoon colloquium, and he was proactive in early history of technology programming for British and American television. His second book, *Networks of Power: Electrification in Western Society, 1880–1930* (Johns Hopkins University Press, 1983) appeared at the end of SHOT's maturing period—a period in which the field at large, and Tom for the most part, concentrated on the United States and the United Kingdom.[5]

What were Tom's principal thematic contributions during these 20 years? Three stand out. The first, his "technological momentum" metaphor, appeared in a short classic titled "Technological Momentum: Hydrogenation in Germany 1900–1933" (*Past and Present*, August 1969, pp. 106–132). At first glance, momentum looks to be the straightforward observation of an obvious pattern: once a technology takes on institutional life, it becomes hard to stop. But embedded in this simple metaphor were two ideas that have characterized Tom's thinking ever since and have exerted lasting influence on subsequent scholarship. On the one hand, a "technology" that acquires momentum reveals itself to be a subtle complex of linked human and nonhuman agents—some group of experts, research and production facilities, investment habits, communication networks, cognitive paradigms, and so on. Twenty years later, particularly in the "social construction" movement that Tom championed early on, these same notions would appear under headings such as "heterogeneous engineering" and "actor-networks." On the other hand, by locating his story of technological momentum in the Weimar Republic, Tom focused attention on the deeply ambiguous role of the German research and engineering establishment in the rise of National Socialism. Are technological experts morally responsible for the uses to which their creations are put? Can there be any engineering, sponsored in any context, that is morally neutral?

In *Elmer Sperry*, Tom took a different tack, looking in the life of a single prolific inventor for models that would explain how creative individuals negotiate the world of market realities. As with German hydrogenation, Tom found his narrative thread in a specific class of technological problems. Sperry's inventions and business initiatives were all variations on his core insight into feedback control. But Tom also saw Sperry's career as exemplifying an essential technological theme of the twentieth century: that technological creativity is more than individual insight and must be understood in terms of the institutional dynamics that call it into existence and sustain the design work it requires.[6]

Tom's 1985 Dexter Prize citation, for *Networks of Power*, introduces him as a professor at Penn and the Torsten Althin Professor of the History of Technology and Society at the Royal Institute of Technology in Stockholm, the latter a chair he would first occupy in 1986. His involvement with Sweden and his year (1983) at the Wissenschaftskolleg zu Berlin serve as career markers for widening geographical and cultural horizons. *Networks of Power* reflects the expansion of his transatlantic perspective and is a culmination of his long-standing interest in Germany, Britain, and the United States, seen here for the first time as elements within a single interpretative frame. In contrast with the Sperry biography, the central historical figures in *Networks* are systems, which, like any historical agent, must be interpreted in terms of the culturally conditioned complexities of their contexts. Electrical systems cannot be understood without attention to the technical constraints governing power generation and distribution and the economic constraints of their host societies. However, for a thorough understanding of the technological and economic dimensions of a system, we must also study the ideological and even the aesthetic predilections of its key actors. Looking at large technological systems from these several perspectives shows them to be "evolving cultural artifacts rather than isolated technologies."[7] The systems approach, as articulated in *Networks of Power*, remains one of the handful of foundational theories available to scholars in the field.

It is not surprising that Tom encountered the group of European sociologists who studied technologies as socially constructed before many of his American counterparts. Frequent transatlantic travel made him familiar with emerging trends in Europe. Nor is it surprising that the "social con-

struction" school, with its insistence on the essential heterogeneity of technological innovation and design, would commend itself to the author of
Sperry and *Networks*, or that his work would incline social constructivists
to him. Tom's relationship with the group probably helped them catch the
attention of many who could, already by 1980, be called "traditional" historians of technology.[8]

European relationships also provided a seed bed for Tom's next two
major works, *American Genesis* and *Rescuing Prometheus*. *American
Genesis: A Century of Invention and Technological Enthusiasm* (Penguin,
1989) uses Tom's earlier published work and material he developed for
courses in American and European history of technology over the previous
30 years to reinterpret American history as technologically centered.[9] He
tested the *American Genesis* argument by using it as subject matter of courses he taught to graduate students in engineering at the Royal Institute of
Technology in Stockholm beginning in 1986. To scholars who followed
Tom's thinking, *Genesis* broke significant new ground primarily in its treatment of European responses to American industrial prowess and American
modernist aesthetics, particularly in architecture and painting. Tom's principal innovation, however, and again he proved to be an early harbinger of
a later SHOT trend, was to redefine his audience and recalibrate his rhetoric
accordingly. *American Genesis*, his first book to be published by a non-
academic press (first Viking, then Penguin), addresses the larger reading
public.[10] His success in this new venue has been widely acknowledged,
nowhere as notably as in its inclusion in the Pulitzer committee's short list
of finalists for the 1989 Prize.

Rescuing Prometheus (Pantheon, 1998) was completed a week before
Agatha's unexpected death in the summer of 1997. That book, more than
any before it, shows the importance of Tom's close relationships with
practicing engineers, especially at MIT and Stanford. It was also stimulated by a series of international conferences on large technological systems begun in Berlin at the Wissenschaftszentrum and continuing for
nearly 10 years in Germany, the United States, Australia, Sweden, and
France. The sources on which *Rescuing Prometheus* depends include
interviews with a host of players in the Cold War world of big national
defense projects (SAGE, Atlas, ARPANET) and an equally impressive
array of designers, architects, and political leaders involved in Boston's

Central Artery/Tunnel project. Tom's choice of topics shows his growing conviction that complex systems hold the key to understanding the United States in the twentieth century.

Rescuing Prometheus resembles *Networks of Power* more than *American Genesis* in its concentration on technological matters. Despite the international influence of the large-technological-systems conferences, *Rescuing Prometheus* is notably less transatlantic than either *Networks of Power* or *American Genesis*. *Rescuing Prometheus* concentrates on the East Coast and the West Coast as seats of the military-industrial-engineering complex. Project SAGE (centered at MIT) and the Atlas missile effort (centered in Southern California) provide case studies of very complex managerial projects requiring enhanced versions of the systems-engineering and operations-research techniques that emerged from Allied military practice during World War II. These projects were led by engineers who had come to respect the importance of heterogeneity and flexibility in situations where hundreds of contractors had to work together with ample funding but very tight time constraints.

Then, in a hinge chapter, these same big-system innovators run aground during the Johnson administration's attempt to apply systems approaches to America's urban inequities and its war in Southeast Asia. They learned, to their intense frustration, that the absence of an anti-Soviet political consensus rendered Atlas-style project management futile. What they faced here was less problem solving than "managing mess."[11] SAGE and Atlas, with their tight lines of control and their need to continually push the current state of the art to deliver a specified product of enormous complexity, were high-water marks for modernist engineering. In sharp contrast, Boston's "Big Dig" and the ARPANET, the topics of the last two major chapters, represent what Tom calls a postmodern style in which design negotiations include constituents who are not members of the expert community. It is no longer sufficient to push the state of the art. ARPANET's nonhierarchical management structure can still be detected in the Internet's decentralized network design. Boston's Central Artery/Tunnel Project is comparable to the Atlas project in complexity and cost, but not in its participatory character. The CA/T Project is "a messily complex embracing of contradictions."[12] Thus, *Rescuing Prometheus* pulls Tom's earlier work inside out. Instead of seeing complexity as a necessary interpretative element for the study of creativity, he now treats creativity as an important interpretative component for studying the management of complexity.

What generalizations can we draw from the evolution of Tom's work? Do his core publications show an engineer-historian who loves the engineering act, its respect for precision, its supple ability to bring order to the cantankerous vagaries of reality? Or do we see here a scholar acutely aware of the Janus face of the Western engineering tradition, beautiful with elegant design while its monstrous complexities open it to moral critique and ironic teasing? Both, I think. Tom's ironies, he might argue, should not be construed as license to overlook the sheer human achievement we find whenever we look closely at technological endeavors. Nor should respect for that level of achievement distract us from the potential for structured violence and cultural crudity that we find when systems take on the aura of preternatural inevitability.

I find it provocative to wonder how much Agatha's intellectual, aesthetic, and moral scope influenced the growing complexity of Tom's interpretative frames of reference. At academic conferences Agatha routinely considered contextual factors when assessing the content of presentations. Over the years, I found myself seeking Agatha out at such events, both from long friendship and because I always learned from her perception of the proceedings. She could be counted on to interpret the academy with wit and insight, not dismissing the importance of the formal argument but situating it in its present context—the physical arrangement of the room, the social dynamics at work in the audience, even the time of day. She was a master of integration—that is, of what Tom, in his later books, called "managing messy complexity."

I will conclude with an anecdote that says a great deal about Tom and Agatha in another role. Both were committed to the intergenerational responsibilities involved in mentoring graduate students. Their hospitality blended superb meals, living-room warmth, and an understanding ear for personal troubles with seminar intensity and tough editorial assessment. When, after some years of research and taxonomy building, I finally presented Tom with a draft of the first chapter of my dissertation, he read it and scheduled a conference with me. I noted, as I entered his office, that he did not rise and take a seat alongside me, as was his usual practice. He faced me across the large desk and began: "This will be much more difficult than I had thought, and we will be fortunate if we are still on speaking terms when we are finished." Over the better part of the next hour, he dissected my draft, arguing with compelling logic that it was intellectual and rhetorical rubbish. Then he shifted focus, telling me that he did not know what would be required for me to complete the dissertation. Perhaps I needed to leave Philadelphia; I almost certainly needed to stop my counseling practice, because I was "not emotionally engaged" in the project at hand. I left stunned, but with time I saw the wisdom of his diagnosis. I terminated my counseling practice and all other working commitments and settled down for a year's hard writing. Tom gave each chapter meticulous attention, frequently sending me back to work on some still-unclear section. When the task was complete, I headed off into my professional life. About 15 years later, over coffee somewhere in Manhattan, Tom reminisced about that terrible encounter across his desktop. "Agatha and I," he recalled, "agonized for days about what to do with what you had turned in. Clearly, it was terrible, and nothing at all like your earlier seminar papers. But it was not easy to know how to proceed. You were approaching forty rather than thirty, and were a close friend. Finally, we decided that I had to take a very hard line with you. But we didn't find the decision easy at all." Such a lovely moment, to learn that as a graduate student one was the object of agonized and careful attention by these two soul friends. As they treated me, so I think they treated the authors of the essays in this volume and so, most of all, they treated one another. All of us writing in these pages are very much in their debt.

Notes

1. Mel Kranzberg took on the lion's share of the work involved in launching SHOT. For several decades he served both as secretary for the society and as editor-in-chief of *Technology and Culture*. It is the near unanimous opinion of SHOT members that he set the tone of the society. He died in December of 1995. In its July 1996 issue, *Technology and Culture* published an extensive set of recollections in his memory.

2. *A History of Technology*, ed. C. Singer, E. Holmyard, A. Hall, and T. Williams, 5 volumes (Oxford University Press, 1954–1958); *Histoire Generale des Techniques*, ed. M. Daumas, 3 volumes (Presses Universitaires de France); A. A. Zvorikine et al., *Geschichte der Technik*, 2 volumes (Russian edition: 1962; German translation: Leipzig, 1964).

3. For this last notion, Tom used *Zeitgeist* in a 1962 article ("British Electrical Industry Lag: 1882–1888," *Technology and Culture* 3, no. 1, p. 39): "The [British electrical] lag came out of a confluence of the legislative, technological, and economic—and something more: the *Zeitgeist*. . . . The British 'spirit of the times' as manifest in the electrical industry lag was characterized by prudence. American 'go-aheadness' set off this circumspection and caution in Britain, and the Electric Lighting Act of 1882—for one thing—was a symptom of it."

4. For detailed arguments supporting these assertions about contextualism circa 1980, see my book *Technology's Storytellers* (MIT Press, 1985), especially chapter 2 (on the context of origin of new technologies) and chapter 3 (on the relationship between science and technology).

5. Tom was not as preoccupied with the US and the UK as was the field generally. He published his study of technological momentum in Germany 2 years before the Sperry book. *Networks of Power* would give equal weight to electrification in Germany, Britain, and the United States.

6. In the words of the Dexter Prize citation (*Technology and Culture* 14, no. 2, 1973, p. 422): "Professor Hughes has been able to show how, through the life's work of an outstanding individual, great historical changes can be seen at work: the transformation of the self-made single-handed inventor into the research development team, the institutionalization of the process of invention, the effects of war in coupling state and private business in a co-operative research effort."

7. Quoted in the Dexter Prize citation (*Technology and Culture* 27, no. 3 (1986), pp. 566–567) in the following context: "Perhaps the greatest service rendered by Hughes's scholarship is the emphasis he places upon the human dimensions of technological history and the considerations that must severely qualify any attempts to impose theories involving technological autonomy or economic determinism upon the stubborn facts of cultural and other forms of diversity. It is clear from a reading of his work that the personal idiosyncrasies of an Oskar von Miller or a Charles Merz could and did make a difference in the manner in which highly significant electrical power transmission and distribution systems were conceived and developed. . . . Such sensitivities contribute measurably to the quality of Hughes's insight and confirm us in our consciousness that the history of technology is, after all, a humanist discipline."

8. "SCOT" (Social Construction of Technology) is often, and erroneously, used as a generic title for a more complex group of social scientists. Within the larger community of the sociology of technology, social construction is distinguished from actor-network theory. Together, however, this group of scholars have become a remarkably influential source of theoretical models in the history of technology. For an early collection of work as well as an indication of Tom's involvement, see *The Social Construction of Technological Systems*, ed. W. Bijker, T. Hughes, and T. Pinch (MIT Press, 1987).

9. "By 1900 they [Americans] had reached the promised land of the technological world, the world as artifact. In so doing they had acquired traits that have become characteristically American. A nation of machine makers and system builders, they

became imbued with a drive for order, and control. . . . Perceptive foreigners are not so prone to sentimentalize America's founding fathers, frontiersmen, and business moguls. . . . Foreigners have made the second discovery of America, not as nature's but as technology's nation." (*American Genesis*, pp. 1–2)

10. "This book, despite its emphasis on invention, development, and technological-system building, is not a history of technology, a work of specialization outside the mainstream of American history. To the contrary, it is mainstream American history, an exploration of the American nation involved in its most characteristic activity." (*American Genesis*, p. 2)

11. *Rescuing Prometheus*, p. 194, quoting Tom's interview with Russell Ackoff, November 4, 1993.

12. *Rescuing Prometheus*, p. 304.

Technologies of Power

Introduction

Authority, Political Machines, and Technology's History

Gabrielle Hecht and Michael Thad Allen

We have understood for centuries that technology is an instrument of power. Scholarly investigations of how technologies reflect, strengthen, perform, or change power relationships are, of course, much more recent. Yet calls for them issued from the best-known historians and social scientists of the twentieth century. Marc Bloch, soon to face execution at the hands of his Nazi captors, declared in *The Historian's Craft*:

Successive technological revolutions have immeasurably widened the psychological gap between generations. With some reason, perhaps, the man of the age of electricity and of the airplane feels himself far removed from his ancestors. With less wisdom, he has been disposed to conclude that they have ceased to influence him. . . . [To those with a] machine-dominated mentality, it is easy to think that an analysis of their antecedents is just as useless for the understanding and solving of the great human problems of the moment. Without fully recognizing it, the historians too are caught in this modernist climate.[1]

Paradoxically, the pace of technological change seemed, then as now, to render the past more distant while rendering history more essential. Bloch's contemporary Helmuth Plessner argued that unprecedented and seemingly unending changes in the modes of work, communication, and transportation had made the historian indispensable. He or she had to pick through the scraps and remainders of the past in order to weave a coherent whole cloth of national identity. Later generations of historians have been less enthusiastic about forging national identity but still have engaged in the task of presenting the past as a composite picture of that which makes us what we are.[2]

One need only strike "electricity" from Bloch's sentences and replace it with "information technology" to give his text a present-day ring.[3]

Constant dynamic industrial change will and must occasion disputes about cultural, political, and social change. The malleability of technological power has prompted myriad attempts to make technological changes conform to utopianism, or dogma, or virtue, or equality. Though no particular set of values, political institutions, or cultural expressions necessarily accompanies technological change, history has shown time and again that those caught up in that change will always attempt to fix its cultural or political dimensions. Understanding these persistent yet ever-changing links between technology and cultural, political, and social power requires strong voices informed by history.

Among such voices, few have been as influential Thomas Parke Hughes, whose works were greatly enriched by the contributions of his late wife, Agatha Chipley Hughes. The corpus of their scholarship has always sought to explain how political and institutional relationships construct the development, and the power, of technology. From his early work on the inventor Elmer Sperry through his investigation of electrification in Europe and the United States, his study of American technological enthusiasm, and his latest work on postwar, postmodern technological systems, Hughes has repeatedly answered the question "Is technology an autonomous force in society?" with a decisive "No." In developing the notion of "technological systems" as an analytic tool, he has demonstrated how technological change both shapes and is shaped by social change. Hence the double meaning of *Networks of Power*: electricity drives machines, light bulbs, and tramways, but at the same time its constant flux in networks reflects and makes tangible the political life of nation-states. Thus electrical networks are "charged" with corruption in Chicago, with localism in London, and with centralized Social Democracy in Berlin. While for many years scholars tended to focus either on the social construction of the technological or on the technological construction of the social, Hughes always emphasized both these domains simultaneously.

Hughes has argued convincingly that technology (and those who control it) *can* shape history; this shaping, however, must always be understood in social as well as technological terms. Thus, the endurance of large technological systems—their "momentum"—stems both from their embeddedness in social values and from their materiality. Social choices shape technological development. But the resulting physical, financial, and institutional durability of systems means that, once developed, they—and the values they uphold—cannot be changed easily. As material manifestations of human choices, systems acquire momentum. In so doing, they embody, reinforce, and enact social and political power. Thus, human power rides upon the history of things. The theoretical force of Hughes's approach to large technological systems, coupled with his insistence that technology is not the sole determinant of historical destiny, has led historians (and other social scientists) to return over and over again to his work.

Hughes's scholarship on technological systems was instrumental to establishing the history of technology as a field. Yet, never content with disciplinary confines, he has continually endeavored to show how insights into the nature of technological change matter to scholarship—and practice—outside the field. For Hughes, the history of technology is human history. In the metal blades of turbines, the cement, glass, and steel of modern architecture, and the graphs of electric current loads, he—and Agatha perhaps even more so—found human passion where so many "general" historians saw only inert objects. For Hughes, *American Genesis* is not just a history of technology but also a history of the United States.

This concern with broad historical questions is the dimension of Hughes's work that inspired the present volume. Technology is central to human history; everywhere it shapes and is shaped by political, cultural, social, and economic change. Yet outside our field few historians have taken up the challenge of understanding technological change in complex historical terms. This neglect has gotten worse rather than better in recent decades. Ironically, the professionalization of our field may have contributed to its marginalization by luring scholars interested in technology away from larger historical meetings and journals into more specialized terrain.[4] But the theoretical and empirical insights that have given our field its intellectual identity should now compel us to move toward other historians. In 1991 Leo Marx asked "Does it make sense, once historians have abandoned the internalist method, to segregate the history of technology from the rest of history?"[5] Recently our profession has begun to discuss this question anew, asking how we can make our scholarship more visible outside the domain of science and technology studies (STS). By addressing broad historiographic concerns with empirical research on technological change, the contributors to this volume hope to suggest strategies for how the history of technology can contribute to—and reformulate—some of the most pressing questions faced by historians today.

We do this in the spirit that marks the many publications on which Tom and Agatha Hughes labored throughout their lives. Tremendous as their theoretical contributions (not just to the history of technology, but also to STS more generally) have been, they have never indulged in theory for its own sake. The very strength of their theoretical insights has derived from their meticulous empirical grounding. We strive here to emulate this rigor. Rather than limit ourselves to general programmatic pronouncements, we have sought to sketch out how our research illuminates specific historiographic problems: the relationships between expertise and authority, or those between knowledge and the distribution of labor; the development of political ideologies; the changing meanings of modernity; the construction of national identity; the nature of political practices in the Cold War; and more. We further hope to honor Tom and Agatha by organizing the contributions in a manner that evokes the evolution of their scholarship (which John Staudenmaier has outlined above).[6] The essays are thus organized topically and in rough chronological and geographical order. They begin with the late-nineteenth-century United States: W. Bernard Carlson links telephone systems and political ideologies, and Eric Schatzberg explores streetcars, urban politics, and culture. Amy Slaton and Janet Abbate then offer a comparison of US labor standards in the building industry in the early twentieth century with standards in the development of computer network protocols in the late twentieth century. In two essays on Germany, Edmund Todd discusses electrification and national politics in the 1930s Weimar period and Michael Thad Allen explores the meanings of modernity during the Third Reich. Erik Rau, in an essay on Great Britain during World War II, analyzes the power dynamics of expertise based on operations research. Next we move on to the Cold War period, with Gabrielle Hecht's essay on the relationship between nuclear power and national identity in France. Finally, Hans Weinberger contrasts Sweden's official neutrality policy with its purchase of military technology from the United States. The wide range of topics notwithstanding, all these essays address how technology, power, and authority are mutually constituted.

The mutual shaping of technology, power, and authority can seem so obvious that we sometimes lose sight of how politicians, cultural critics, policy makers, and other public figures have thought about these relationships at different historical moments. This is even more true of other social scientists who study technology. STS scholars can become so focused on debunking the myths of technological determinism that they ignore or dismiss ways in which public conceptualizations of technology themselves serve political or cultural functions. How public figures articulate the relationships between technology and society matters on at least two fronts. First and most important, it matters to the development of technology itself. Technology may not drive history, but the fact that influential people believe that it does has real consequences. Several of the authors in this volume trace this historical relationship between ideas and practice, and we will discuss their insights toward the end of this introductory essay. Second, the history of these conceptualizations can help to contextualize and explain why scholarship on technology has become isolated from other historical endeavors. In order to weave the history of technology back into history, we must understand more about the broader context of its isolation.

Indeed, we argue that the Cold War gave a distinctive cast to the multiple ways in which public figures and scholars thought about technology. Public discourse about the glories of technological progress existed well before World War II—such discourse, for example, permeated the twin processes of industrialization and colonialism in the nineteenth century. To an unprecedented degree, however, ideas about technological development became central to state ideologies and policies during the Cold War. Dramatic increases in the funding of scientific and technological research ensued, particularly at universities. It is no surprise, therefore, that the history of technology became increasingly vigorous in this context, acquiring its own urgency. Ironically, however, the context of the Cold War can also help explain the relative isolation of the history of technology from non-STS fields—an isolation supported by the common public conviction that technology and politics, or technology and culture, must be separate entities. The next part of this essay, therefore, explores public conceptualizations of technology during the Cold War in order to elucidate the political meanings of some of the more important epistemological insights of the history of technology. This history, we hope, can help point us forward. Whereas the Cold War may have justified separate historical treatment of technology, the end of the Cold War demands that we draw connections between our work and that of other historians.

A few caveats are in order. First, the following discussion aims to be suggestive rather than comprehensive. This is not the place to cover the full range of public discourse about technology during the Cold War. Second, we focus primarily on the United States. This is because we seek to provide a context for how our field is positioned relative to the American historical profession. We hope that these limitations will, if anything, provoke further scholarly inquiry into the issues we explore below.

After World War II there could be little doubt that the Allies had triumphed because of their superior industrial capacity, especially the ability of the United States to turn its economy into the "arsenal of democracy."[7] Shortly thereafter, as Hughes has argued in *Rescuing Prometheus*, large-scale, accelerated, high-technology research projects gained unparalleled prestige—especially those advanced by proponents of Systems Engineering and Operations Research. This prestige quickly leapt the bounds of "purely technical" efficacy. For example, Karl Popper, C. P. Snow, David Landes, and W. W. Rostow celebrated the supposed connection between the rationality of science, technology, and industrialization, on the one hand, and the values of democracy on the other. These intellectuals—a philosopher, a novelist, a historian, and an economist—had little in common except their faith in the immanent rationality of scientific endeavor and technological development.

During the early days of the Cold War, C. P. Snow claimed that scientists had "the future in their bones" and warned Western democracies to include them in political decision making at the highest level in order to maintain resilience in the face of inevitable scientific and technological change.[8] But it may have been the sociologist Robert Merton who most explicitly asserted that scientific-technological rationality provided the language of democratic society. Merton did not just celebrate the symbiosis of science and democracy; he argued that science might serve as an inoculation against totalitarianism. Authoritarian dogma, he wrote, contradicted the assumptions of modern science to such an extent that true scientists were compelled to resist nondemocratic regimes. Further, Merton viewed technology as science's "respectable sibling" and therefore assumed that it too might buttress democracy:

> The increasing comforts and conveniences deriving from technology and ultimately from science invite the social support of scientific research. They also testify to the integrity of the scientist, since abstract and difficult theories which cannot be understood or evaluated by the laity are presumably proved in a fashion which can be understood by all, that is, through their technological applications.[9]

Technology, here conceived as applied science, thus embodied a kind of "pure reason" for the masses; it justified the ways of science to the hoi polloi. What could connect a metalworker turning out fasteners in Detroit with a nuclear physicist or a field biologist? According to Merton and many others, the answer was an overarching mental state: the habit of rational thinking born of science. Distinguishing this state from superstitious, premodern world views, they equated it with the habits of freedom.[10]

A large and diffuse literature on science and technology coalesced around two themes. First, the history of science and technology could reveal crucial knowledge about the nature of the good society—that is, the liberal, democratic, capitalist nation-state. Second, the epistemology of scientists and engineers could and should guide the thinking of democratic citizens. Thus, David Landes argued in 1969 that "he who is rational in one area [i.e., the rational industrial enterprise] is more likely to be rational in others"—that is, in politics.[11] This intense concern with uncovering and justifying the core habits of thought necessary for Western democracy pervaded the American historical profession.[12]

Yet, as numerous scholars have pointed out, the equation of scientific and technological thinking with democracy during the Cold War rested upon the assumption that rationality was "value neutral."[13] This assumption that science and technics were value neutral *per se* was not new to the Cold War—it went back at least as far as Francis Bacon or the first Royal Society. Nor were links between science and democracy new to the Cold War. In the United States, faith in the democratizing spirit of science and industry had long been embodied in heroic portraits of Benjamin Franklin and Thomas Jefferson, in glorious narratives about the conquest of the frontier, and in what David Nye has called the "American technological sublime."[14] What was new to the Cold War was the subsequent conclusion and its consequences: *because* democracy was the social expression of scientific and technical rationality, it was not only value neutral but *non-ideological*. Further, the alleged absence of ideology defined the fundamental difference between democracy, on the one hand, and fascism and communism on the other.

According to theories of totalitarianism developed by Hannah Arendt and other postwar intellectuals, ideology reigned in the Soviet Union and Nazi Germany because those nation-states based the legitimacy of their rule upon the presumption to know human nature. This judgment helps to situate the curious argument advanced by the philosopher Karl Popper, which differentiated between the domain of nature (knowable through rational scientific truth) and the domain of society and culture (in which any claims to rational knowledge led toward totalitarianism).[15] Like Merton, Popper believed that democracy could be nurtured by promotion of the mental habits of scientific thought. At the same time, he warned that truth claims regarding society and culture led to authoritarianism. Consequently, in order to preserve his faith in science, Popper had to shear science and technology away from the political, the social, and the cultural. Within the terms of this discourse, scientific and technological innovation became the standard of reason. But this was a peculiar definition of reason, one in which judgments about human nature were somehow purged and supposed to play no role. By abstaining from judgments about what human beings are and how they function socially, culturally, or politically, science and technology were supposed to provide (strangely enough) the form of thought appropriate for the one form of governance supposedly free of ideology: liberal democracy.

In the first 20 years of the Cold War, vilification of ideology spread far and wide—not just in American politics, but also in social-science scholarship. As Peter Novick has argued, "the disparagement of ideology and the concomitant celebration of American empiricism were among the forces which in the postwar years returned historiographic thought in the United States to older norms of objectivity."[16] Popper's argument resonated widely in the historical community, serving as an epistemological justification for the possibility of historical objectivity. In view of how central the separation of science and technology from society was to Popperian reasoning, it is not surprising that historians did not subject the *inside* workings of technology to scrutiny: its very epistemology supposedly precluded such scrutiny as a historical project.

The Cold War consensus did, however, necessitate a history of science and technology that demonstrated the eternal progress of rationality. Such a demonstration guided the tale of the British industrial revolution, which—together with its supposed link to the progressive liberalization of parliamentary government—was taken to be the "paradigm" of all modern industrialization. This conception of a predictable and necessary path of technological development had resonance far beyond the confines of academia (a resonance which in turn helped reinforce academic convictions). It lay at the heart of the Marshall Plan to reconstruct Europe, and it was fundamental to the actions of the World Bank, from its earliest European projects to its subsequent interventions throughout the world.[17] In the 1950s and the 1960s, Popperian positivism, coupled with modernization theory, suggested that the discipline of economics might yield durable laws akin to those of natural science. Bestowed with the legitimacy of scientific rationality, economics could also yield sound formulas for democracy—particularly when applied to "Third World" development.[18] Many hoped that economic "modernization" might prompt even "traditional" dictators to come round to democracy. Witness, for example, *A Proposal: Key to an Effective Foreign Policy*, put forward in 1957 by the American policy advisors W. W. Rostow and Max Millikan,[19] who were seeking to make "modern" societies of "developing countries" through the application of Western economic theory. Practically speaking, the primary vehicle of these policies would be technological and managerial know-how: the United States and the World Bank would export industrial rationality to underdeveloped areas. Ultimately, the aim of modernization was democratization.

But because both modernization and democracy—by virtue of their foundations in scientific and technical rationality—were construed as nonideological, aid to the "Third World" enacted through "technical assistance" was itself portrayed as "neutral with respect to the political issues which rouse men's passions."[20] "Modernizers" intended Western technology to produce economic growth in a manner that could soar above passion, fraction, and political dispute. In this way, economic and technical aid also served as a powerful weapon in the Cold War fight against communism.[21] Modernization theory justified interventions that might otherwise be construed as neocolonial, providing a means through which the United States could establish morally impeccable relationships with the "Third World."[22] In Europe too, "development" provided a rubric under which former colonial powers could transform the "civilizing mission" into programs that would prove more politically palatable in an era of decolonization.[23]

Thus, faith in the transformative, democratizing powers of technology not only provided the logic behind the arms race and the space race, it also undergirded Cold War geopolitics in its broadest forms. In this sense, technological determinism was a crucial aspect of the epistemology of US Cold War politics, which in turn rested upon a progressive view of technology's history that posited British and then US industrialization as the universal template. This model assumed that there existed only one true path for "development," one that culminated in large firms and mass production.[24] The geopolitical dominance of North America and Europe—symbolized by wealth and consumer goods—suggested that this was indeed the true path.

Perhaps it is no surprise that engineering schools were the first homes for the history of technology, or that the United States Armed Forces Institute sponsored the publication of *Technology in Western Civilization*.[25] But if the professionalization of the field benefited from the centrality of technology as both symbol and motor of the Cold War, the questions posed by historians of technology challenged that era's determinist epistemology.[26] In explorations of technological systems, the military-industrial complex, and industrialization, historians of technology demonstrated time and again that technology did not follow a single path, did not change society of its own accord, and did not inevitably promote democracy (even in democratic societies). Historians of technology also critiqued the radical technological determinism of cultural pessimists like Jacques Ellul. Technology did not inevitably lead to an Orwellian social order, any more than it did to a democratic utopia. Hence Melvin Kranzberg's formulation "Technology is neither good nor bad; nor is it neutral."

It is also unsurprising that one target audience of that first generation of historian of technology was engineers. The Cold War helped to shape the meaning and purpose of the institutional links that historians of technology forged to engineering education. In one of the ironies of "value-free" science and its supposed link to civic virtue, professional engineering societies began to bemoan the inadequate social and political skills of students trained entirely in the hard sciences and in technical courses. What better way to produce socially responsible engineers than to teach them technology's history?

In the 1980s, historians of technology pushed the edges of the discipline further by reaching toward sociologists of science and technology. Conversations between historians and sociologists resulted in a methodological paradigm that—although it had been emerging for a while—now came to be known as "social construction of technology," often abbreviated as SCOT. This acronym glossed over methodological differences among scholars, but it proved useful in articulating a common program to unpack the microrelations of technology. As several writers have observed, concluding that technology was socially constructed did not pack the same epistemological punch as similar conclusions about science.[27] But SCOT scholarship did pack political punch in the context of the Cold War. SCOT literature left readers with tantalizing suggestions that what had been done might also be undone. Nothing was inevitable about the arms race—or any other technological development.

Epistemologically, institutionally, and politically, the Cold War thus provided a compelling context for technology studies. On all these fronts, historians of technology had more than enough to keep them busy. Explicating the political and cultural construction of technological change proved an uphill battle, requiring careful attention to technical detail. Unfortunately, the considerable time that historians of technology spent on such detail probably contributed to the relative lack of interest exhibited by other historians. Indeed, historians outside the field did not seem to care much about matters technological. For one thing, they accepted (by and large) the Cold War premise that the power of technology was easily discernible and need not be subject to intensive political or cultural scrutiny. This was as true of economic historians (the only ones who, as a group, included technology in their analysis) as it was of others. Technical experts had power either because their knowledge was truly privileged or thanks to institutional structures (such as universities, professional societies, or government agencies) that produced and legitimated expertise and functioned as conduits for expert power. Either way, expert knowledge and the artifacts it produced appeared hermetic. This appearance was doubtless exacerbated by the division of the academy into "techies" and "fuzzies"—what C. P. Snow more formally called the "two cultures." Humanists allowed themselves to be intimidated by the disciplinary strategies of scientists and engineers. To most historians, the content of technology appeared culturally and politically inscrutable and therefore uninteresting. For another thing, "mainstream" historians were fighting their own battles, which were also epistemological and political in character. The middle and late years of the Cold War saw the rise of social history, a renewed questioning of historical objectivity, and a splintering of the field such that "every group [acquired] its own historian."[28] Questions about gender, race, and class dominated the cutting edge of historiography, mobilizing a new generation of historians against "the old guard." Historiographic battles reached a fever pitch with the emergence of the "new cultural history," which pitted cultural explanations of historical change against material ones. Disputes were especially intense among labor historians, who argued fiercely about the relationship between material life and the emergence of political consciousness and action. Did the emergence of industrial factory work lead to the creation and politicization of the working class, or did class consciousness emerge as the product of intersecting cultural discourses? Such questions posited the same divide between the material world and the cultural world that characterized the Cold War in general—a divide that polarized not just labor historians but many others. In this disciplinary climate, technology as an object of study was not merely inscrutable; except when viewed in symbolic terms, it was passé.

Fortunately, times are changing. Increasingly, historians are acknowledging the sterility of the material/cultural divide and recognizing the need to understand both how the material is cultural and how the cultural is material. Many still balk when the material side of the question involves analyzing what appears to be "merely" technical detail. But the growing interest in achieving a synthesis between material and cultural approaches leaves the door wide open for historians of technology to step in.[29]

Meanwhile, the field of technology studies as a whole is struggling through a comparable crisis of its own. The waning of Cold War concerns means that SCOT no longer packs even the political punch it once did.

Indeed, SCOT has recently come under attack for being "conservative," for "embracing everything while excluding nothing," and for being part of "pale-male constructions of the political."[30] SCOT has also come under epistemological fire for being socially determinist.[31] No matter how empirically rich, the refutation of technological determinism no longer provides a satisfying or sufficient conclusion.

These simultaneous disciplinary crises mean that the time is especially ripe for fruitful dialogue between historians of technology and the rest of the discipline. Of course, such disciplinary dialogue can revolve around many different themes, and we cannot pretend to address them all. The choice of topics here reflects the inspiration of our mentors: we thus focus primarily on technological systems and the actors who propel them. At the same time, all the essays treat a theme that has been an abiding subject of consideration in the historical profession at large: power, its practices, and its meanings. In one way or another, all the authors consider how a particular dimension of power relationships—the establishment and performance of authority—gets constituted by and performed through technological artifacts and knowledge.

Of course, "technology"—especially when used narrowly to refer to complex machines—is itself a power-laden term. Going back to the United States in the nineteenth and the early twentieth century, for example, we can see skills that white middle-class boys developed to design machinery were considered technogical by educators and the public, while skills developed by girls (such as sewing or cooking) were not. Similar conceptions endure today, when—in most contexts and for most people—"technology" denotes the latest machines and professionally vetted expert knowledge.[32] The Amish, for instance, are popularly portrayed as anti-technological, but we could more legitimately argue that they enthusiastically embrace Renaissance technologies. Perhaps more distressing is the way in which popular uses of "technology" today increasingly refer only to computer-related artifacts and networks. Such conceptions reveal that what counts as technology changes all the time, precisely because that designation continues to be an indicator of power and legitimacy.

As the Cold War continues to recede, we are left to reconsider some burning questions. What role do technological artifacts and knowledge play in reifying, reshaping, and performing power relationships? How do they become culturally identified as technological? How are they shaped by power relationships? What role do they play as agents of political or cultural change? Largely focusing on some dimension of the meanings of expertise and its relationship with authority, the essays here address these questions by examining the enactment of power and politics through specifically technological means—not just during the Cold War, but starting in the late nineteenth century.

One conclusion that emerges clearly from these essays is that the power of technical knowledge is neither purely cultural, nor purely institutional, nor purely technological. Rather, expert power is constituted heterogeneously and must be understood in hybrid terms. In order to express this hybridity, Gabrielle Hecht suggests that we think of the performance of power through technology as "technopolitics," which she defines as the strategic practice of designing or using technology to enact political goals. On the one hand, such practices are not simply politics by another name: their material, artifactual forms matter fundamentally to their success. On the other hand, calling these artifacts politically constructed technologies implies a static quality that does not adequately capture the dynamic ways in which they enact power. Other authors in this volume similarly seek to capture the hybrid and performative nature of technological power. For example, Amy Slaton and Janet Abbate show how technical standards perform labor relations in the nineteenth- and twentieth-century United States. And Hans Weinberger demonstrates how the acquisition of American military technology has belied Sweden's claims to neutrality, enacting a quite different set of national policies.

Of course, technology cannot embody politics in a conceptual or ideological vacuum. Edmund Todd's essay on electrification in Weimar Germany and Gabrielle Hecht's essay on nuclear power in postwar France both argue that the ways in which technologies perform expert power must be understood with reference to *beliefs* about how technological change operates in the world. This requires a shift in how historians approach technological determinism. Todd and Hecht argue that we should historicize it. Thus, instead of continuing to ask "Does technology drive history?" we should ask questions such as "When or why do historical actors believe or argue that technology drives history?" Addressing such questions leads us to view technological determinism—and other beliefs about the relationships between technology and social change—as political practices. To put it differently, Todd and Hecht argue that we need to move past evaluating the validity of ideas about technology and instead analyze these ideas as an integral part of the social and political life of technologies.

W. Bernard Carlson offers the earliest example in this volume of how ideas about the relationship between technology and politics shapes technological change, though additional and even earlier examples could certainly be found. His study of Gardiner Hubbard's role in developing the telephone identifies linkages between specific political ideologies and specific technological change. Carlson explores how Hubbard's vision of democracy shaped the technical choices he and Alexander Graham Bell made. Hubbard believed in the transformative power of technology. He thought that the telephone could be a direct instrument of democracy for the average middle-class American citizen. His vision of democracy shaped the kind of telephone system that he promoted. For him, telecommunications also required the active intervention of the state, and his conviction that technology would have a deterministic social and political effect was instrumental in driving this technological development.

Erik Rau's essay on the emergence of Operational Research as a systems science provides a more recent example of how ideas about the social and political relations of technology (and, in this case, science) shape its development and its political life. Rau demonstrates that the scientific origins of Operational Research lay in the Social Relations of Science movement in 1930s and 1940s Britain. Scientists in this movement argued that more active participation by the scientific elite in Britain's social and military policies would produce both better policies and better science. To legitimate their claims, they had to prove their worth by managing technological systems. Some of them found the opportunity during World War II, when they were called upon to analyze military tactics and strategies with a view to using British military forces more effectively. The reception of scientific recommendations varied a great deal according to local conditions, but in many cases this reception depended on how both scientists and the military commanders they dealt with conceptualized the appropriate role of scientific knowledge in making military decisions. Reading Rau's essay alongside Carlson's suggests ways in which the professionalization of expertise in the late nineteenth and early twentieth centuries reshaped conceptualizations of the relationships between technology and politics.

Beliefs about the social and political possibilities of technological change can be chilling, powerful forces in their own right, as Michael Thad Allen demonstrates in his analysis of Nazi modernity. Nazi leaders believed that their "one best way" of developing technological systems would inexorably lead to the best possible society. From among the many changes occasioned by their rapidly industrializing society, National Socialists sought to celebrate selective elements as part of a unified "German will." Others they branded "Jewish," "liberal," or "degenerate." An example is Bauhaus design, which the Nazis rejected. But that rejection did not make them any less modern, Allen argues. No less than the Bauhaus, Nazi designers privileged modern mass production, standardized norms, and machine culture. They merely differed fiercely over the content of that culture. Allen also argues that the Nazis' principles of modernity guided them toward the unique technological means of genocide—perhaps the most harrowing example of their effort to impose National Socialism as the culmination of German destiny and of all modern history.

Contests over aesthetics, technological change, and the modern social order have a long history. We can glimpse an earlier instance in Eric Schatzberg's analysis of debates over how to introduce streetcars into nineteenth-century American cities. Schatzberg shows that technological aesthetics provided a means for the public—that is, the supposed consumers of technological change—to shape the nature of that change. Tracing conflicts between inventors and ordinary citizens about overhead trolley wires, Schatzberg shows that urban residents and trolley designers framed their differences as competing symbols of progress. Each claimed to have a more compelling vision of modernity. At stake in these disputes was not just the aesthetic of urban areas, but control over the future of the city.

In all these cases, beliefs about the social or political dimensions of technological change shaped real social and political change. Statements about the nature of technology were thus themselves political or cultural strategies. Todd's essay on the Weimar Republic and Allen's on the Third Reich put in stark relief the fact that claims about the political (or apolitical) nature of technology cannot be taken as simple statements of truth. Modernity must be historicized. It must be viewed not as a unitary concept but rather as a set of multiple and conflicting debates. At their core, these contestations concern the connections between politics, culture, and technology. Very often, they center on the future of a nation, or even the planet. And the process of prescribing how technology, modernity, or national identity do or should relate to one another is itself political; that is, it involves negotiating and performing power relationships.

In part this is because modernity or claims advanced in the name of national identity always raise the same question: Who can best lead a society toward the future? This is an enduring question for technical experts, even as the role of experts relative to the state changes according to time and place. Carlson discusses the role of technical experts in nineteenth-century America, Todd in early-twentieth-century Germany, Rau in World War II Britain. Hecht's essay takes us to the postwar period, arguing that French technologists sought authority by claiming to provide the path to a modern national future. The need to reinvigorate or even reshape French national identity was widely agreed upon. By offering the mechanisms for this transformation, technologists secured positions of political power and cultural authority. In so doing, they sought to dissolve the boundaries between technology and politics as a way of claiming greater authority within the state. Their success in blurring these boundaries produced technopolitics; thus, expert authority within the French state must be understood both in terms of the technological systems built by those experts and in terms of the conscious association or dissociation of those systems with politics. In this case, technopolitics appears as a distinctive form of state power.

A version of this appears to be true for the Swedish Cold War state as well. Officially, Sweden's posture in the Cold War was one of neutrality: in principle, the government claimed, Sweden would respond symmetrically to any intrusion into its airspace, whether Western or Soviet. Simultaneously, Swedes' self-image as neutral mediators between East and West and as peaceful representatives of a viable Third Way lent their small nation a powerful sense of its role in world history. But Hans Weinberger shows that the real policy of the Swedish state cannot be understood in terms of diplomatic rhetoric. Instead, an examination of technological decisions shows that, in fact, Sweden's military infrastructure was designed to facilitate NATO's physical access to the Soviet Union. Sweden's military technological systems thus belied the official posture of neutrality. In the

Swedish case, as in the French case, technological systems formed *de facto* national policy. The parallels go further. In France, technology was explicitly conceived as a form of politics; this in turn shaped the technopolitical potential of the nuclear program. In Sweden, decision makers conceived of technology as neutral; this in turn made it possible for all technological decisions to appear as inherent vehicles of neutrality. In both cases, then, the way in which decision makers and the public conceptualized the valence of technology helped to shape the political life of specific technological systems. In both the French and the Swedish cases, we see a disjuncture between declared policy (policy as rhetoric) and enacted policy (policy as practice). In both cases, an understanding of actual technological practice is necessary in order to view the disjuncture clearly. One might say that technologies camouflaged actual political practices—except that the camouflage metaphor suggests a separation between technological and political practice that did not exist. A more precise formulation would be that these technological systems *performed* national politics.

Technopolitics during the Cold War may have had a distinctive flavor, but hiding political agendas and power relationships in technological artifacts, practices, or systems is nothing new. For example, Todd argues that German engineers in the early twentieth century constructed "shadow histories" with a technologically determinist thrust in order to legitimate their own choices and discredit those made by other engineers. Some engineers used shadow histories to justify centralized power grids; others used them to justify decentralized systems. Either way, the histories made the technologies seem nonpolitical when in fact centralization and decentralization would create basic differences in the political and social organization of regions and ultimately the nation. Todd maintains that these shadow histories did more than legitimate particular forms of electric power development: they also masked the political negotiations that engineers conducted with other industrialists and local officials in order to implement their technological plans. In one case, technological determinism obscured the convergence between National Socialism and the industrial development plans advocated by a prominent engineer.

Slaton and Abbate also probe the hidden political performances of technology. Their essay on technical standards compares the early twentieth century construction industry with the late-twentieth-century computer industry in the United States. What roles, they ask, do standards and spec-ifications play in encoding and performing hierarchies of technological knowledge? In both of the cases examined, they argue, standards encoded labor, produced or reproduced distinct relationships between expertise and authority, and redistributed work among different phases of production and consumption. Technical standards have a "hidden life," the perfor-mance of which has concealed effects. In the construction industry, these included the redistribution of expertise and authority among architects, materials suppliers, and tradespeople. In the computer industry, network standards displaced a great deal of labor from the designers of such net-works onto the users. Viewed on the surface, 1920s construction and 1980s information technology are worlds apart, but in both cases, Slaton and Abbate maintain, standards constituted a dynamic medium through which relationships between expertise and authority were renegotiated. Their analysis thus foregrounds another important theme: the contingent nature of the relationships between expertise and authority, and the per-formance of those relationships through technological practices. This point, we believe, is central to the contribution that the histories of tech-nology in this volume make to other historical fields. These relationships cannot be understood simply in institutional or discursive terms: they must be examined in terms of practice. As Rau also argues, such practices are always local. The nature and extent of power wielded by experts—within the state or anywhere else—is always contingent upon local conditions and negotiations, and must always be understood as the product of tensions between local and global forces and motivations. This is why we cannot fully understand the relationship between power and expertise in any given situation without understanding how both social and material practices constitute expertise. Many of the essays in this volume point to a central irony: that experts lay claim to power by blurring, overstepping, or redefin-ing boundaries drawn by other groups seeking authority. Time and again, we see that the practices and content of technological change are them-selves performances. They work out unequal social relations—that is to say, power dynamics.

In closing, we should reiterate that these contributions represent only a few of many possible ways in which the history of technology can contribute to scholarship in other historical fields. The essays all discuss artifacts, knowledge, and systems that are commonly recognized as "technological" precisely because they have served as instruments of expert power. We hope that conclusions drawn from this volume can illuminate the exploration of technology and authority in other areas as well. As scholars in our field have begun to demonstrate, for example, the technological performance of power relations involving gender, race, or colonialism is equally important to understanding these dynamics. The central challenge set by our mentors has always been that "technology is far too important to leave only to engineers; humanists, social scientists, historians, and citizens should also have their say."[33] By opening spaces (however small) in which historians of technology can contribute to other fields, we hope to meet that challenge. Technology is, indeed, far too important to be neglected by humanists.

Notes

1. Marc Bloch, *The Historian's Craft* (Vintage Books, 1953), p. 36.

2. Helmuth Plessner, *Die verspätete Nation* (W. Kohlhammer, 1935; reprinted 1959), esp. pp. 72–82. Compare similar arguments in Ernst Gellner's *Nations and Nationalism* (Cornell University Press, 1983).

3. Compare Jean-François Lyotard, *The Postmodern Condition* (University of Minnesota Press, 1979), esp. preface and p. 37.

4. We do not mean to imply that no one *tried* to reach mainstream historians, or that no one succeeded in doing so. See, for example, Judith McGaw, *Most Wonderful Machine* (Princeton University Press, 1987); Philip Scranton, *Endless Novelty* (Princeton University Press, 1997); Carroll Pursell, *The Machine in America* (Johns Hopkins University Press, 1995); David Nye, *Electrifying America* (MIT Press, 1990); Ruth Schwartz Cowan, *More Work for Mother* (Basic Books, 1983); Leo Marx, *The Machine in the Garden* (Oxford University Press, 1964).

5. Leo Marx credited Hughes as the only one to address this thorny issue, but he was being modest. Marx too has addressed this question—see his review of *In Context*, ed. S. Cutcliffe and R. Post (*Technology and Culture* 32, 1991: 394–396).

6. We are indebted to David Nye for suggesting this organization in his review of the volume.

7. Richard Overy, *Why the Allies Won* (Norton, 1996).

8. C. P. Snow, *Science and Government* (Harvard University Press, 1961).

9. Robert Merton, "Science and the Social Order," in *The Sociology of Science*, ed. N. Storer (University of Chicago Press, 1973), p. 257. See also Robert Merton, "The Normative Structure of Science" and "The Puritan Spur to Science," in the same volume. David Hollinger argues that Merton's celebration of the scientific ethos

holds a position in the history of ideas comparable to that held by Frederick Jackson Turner's "frontier thesis" in the history of the American West; see his "The Defense of Democracy and Robert K. Merton's Formulation of the Scientific Ethos," *Knowledge and Society* 4 (1983), p. 11.

10. For technological practice, modern "rational" engineering was distinguished from "craft" traditions, an equally heterogeneous category in historical fact. Whereas the mental state of rationality was supposed to tie all modern engineering together with science, the mental state of "irrationality" or "subjectivity" was supposed to link "traditional" practice to the habits of superstition. See David McGee, "From Craftsmanship to Draftsmanship: Naval Architecture and the Three Traditions of Early Modern Design," *Technology and Culture* 40 (1999): 209–236. On the dangers of stressing "mental states" as analytic categories in history, see Charles Tilly, *Big Structures, Large Processes, Huge Comparisons* (Russell Sage Foundation, 1984).

11. David Landes, *The Unbound Prometheus* (Cambridge University Press, 1969), p. 21.

12. For a discussion of historiography during the Cold War, see Peter Novick, *That Noble Dream* (Cambridge University Press, 1988).

13. See, for example, Robert N. Proctor, *Value-Free Science* (Harvard University Press, 1991). Democratic values were not associated with technology in other national cultures. See Plessner, *Die verspätete Nation*, or Joachim Radkau, *Technik in Deutschland* (Suhrkamp, 1989). Sometimes science was overtly associated with authoritarian regimes—for instance, in Peter the Great's Russia. See Alfred Rieber, "Politics and Technology in Eighteenth-Century Russia," *Science in Context* 8 (1995): 341–368.

14. See David Nye, *American Technological Sublime* (MIT Press, 1994); John Kasson, *Civilizing the Machine* (Grossman, 1976); Marx, *The Machine in the Garden*.

15. Karl Popper, *The Open Society and Its Enemies* (Harper & Row, 1962), pp. 57–59.

16. Novick, *That Noble Dream*, pp. 300.

17. Ashraf Ghani, "The World Bank," talk given at Ford Seminar, University of Michigan International Institute, February 3, 2000.

18. Popper, *The Open Society*, pp. 32–33, 67.

19. *A Proposal* appeared just a few years before Rostow popularized his famous modernization theory in *The Stages of Economic Growth: A Non-Communist Manifesto* (Cambridge University Press, 1960). Neither Rostow nor Millikan was a mere theorist; both had gotten their hands dirty in practical foreign policy. Millikan had worked in several subcommittees of the Marshall Plan; Rostow had worked as an advisor to the nascent institutions of what would become the European Community.

20. Rostow and Millikan, *A Proposal: Key to an Effective Foreign Policy* (Harper & Brothers, 1957), pp. 39–40. On technical assistance, see pp. 54 and 61–63. Under

technical assistance, Rostow and Millikan included managerial and administrative skills, themselves assumed to be transparently rational and thus "neutral."

21. This was the explicit message of Rostow's book *The Stages of Economic Growth*, for example.

22. For a discussion of the continuities between colonial ideologies and modernization theory, see the last chapter of Michael Adas, *Machines as the Measure of Men* (Cornell University Press, 1989). For more on the role of economics in these dynamics, see Arturo Escobar, *Encountering Development* (Princeton University Press, 1995).

23. Frederick Cooper, "Modernizing Bureaucrats, Backward Africans, and the Development Concept," in *International Development and the Social Sciences*, ed. F. Cooper and R. Packard (University of California Press, 1997).

24. This view of modern technological history was shared by socialists and communists. See Joseph Schumpeter, *Capitalism, Socialism and Democracy* (Harper, 1942) or Charles Maier, *Dissolution* (Princeton University Press, 1997).

25. Melvin Kranzberg and Carroll Pursell, eds., *Technology in Western Civilization* (Oxford University Press, 1967); Robert Multhauf, "Some Observations on the State of the History of Technology," *Technology and Culture* 15 (1974): 1–12; Eugene Ferguson, "Toward a Discipline of the History of Technology," *Technology and Culture* 15 (1974): 14; Carroll Pursell, correspondence with the authors, December 6 and 7, 1999. For a history of the institutionalization of the discipline, see Bruce Seely, "SHOT, the History of Technology, and Engineering Education," *Technology and Culture* 36 (1995), no. 4: 739–772. For an overview of its intellectual development, see John Staudenmaier, S.J., *Technology's Storytellers* (MIT Press, 1985).

26. Historians of technology were not the only ones to challenge the assumed links between technology and democracy. Cultural and political critics (among them Lewis Mumford, Jacques Ellul, and Jean Meynaud) also challenged these links, arguing that the juggernaut of technological change, far from providing the motor of democracy, threatened its very foundations. Yet such critiques remained, for the most part, within a technologically determinist framework.

27. See, e.g., the introduction to second edition of *The Social Shaping of Technology*, ed. D. MacKenzie and J. Wajcman (Open University Press, 1999). See also David Edgerton, "Tilting at Paper Tigers," *British Journal of the History of Science* 26 (1993): 64–75; "De l'innovation aux usages: Dix Thèses éclectiques sur l'histoire des techniques," *Annales* 45 (1998): 259–288.

28. For more on these developments, see chapter 14 of Novick, *That Noble Dream*.

29. For a few examples among cultural historians, see *Rethinking Labor History*, ed. L. Berlanstein (University of Illinois Press, 1993); Richard Biernacki, *The Fabrication of Labor* (University of California Press, 1995); Laura Lee Downs, *Manufacturing Inequality* (Cornell University Press, 1995).

30. See Langdon Winner, "Upon Opening the Black Box and Finding it Empty: Social Constructivism and the Philosophy of Technology," in *The Technology of*

Discovery and the Discovery of Technology, ed. J. Pitt and E. Lugo (Society for Philosophy and Technology, 1991); David Hounshell, "Hughesian History of Technology and Chandlerian Business History: Parallels, Departures, and Critics," *History and Technology* 12 (1995), p. 214. Hillary Rose, "My Enemy's Enemy Is—Only Perhaps—My Friend," in *The Science Wars*, ed. A. Ross (Duke University Press, 1996), p. 93.

31. Thomas Parke Hughes, "Technological Momentum," in *Does Technology Drive History?* ed. M. Smith and L. Marx (MIT Press, 1994), p. 104. Leo Marx, in his review of *In Context*, ed. S. Cutcliffe and R. Post (*Technology and Culture* 32 (1991): 394–396), asked whether any justification for the history of technology remains if all is constructed "in context" and technology is incapable of "determining" anything. Meanwhile, much of Bruno Latour's work has stressed that the "social" is a constructed analytic category, and thus cannot be used as an explanation for the "technological" or the "natural"—see *Science in Action* (Harvard University Press, 1987) and *We Have Never Been Modern* (Harvard University Press, 1993).

32. Judith McGaw, "No Passive Victims, No Separate Spheres," in *In Context*, ed. S. Cutcliffe and R. Post; Ruth Oldenziel, *Making Technology Masculine* (Amsterdam University Press, 1999); Nina Lerman, Children of Progress (manuscript, forthcoming); special "Gender and Technology" issue of *Technology and Culture* (1997), ed. N. Lerman, A. Mohun, and R. Olendziel.

33. Thomas Parke Hughes, "Shaped Technology: An Afterword," *Science in Context* 8 (1995), p. 455. This entire issue, edited by Hughes, addressed this very program.

The Telephone as Political Instrument: Gardiner Hubbard and the Formation of the Middle Class in America, 1875–1880

W. Bernard Carlson

"Politics." When I hear that word, I recall fondly one of my undergraduate engineering students, Ken, who used to drop by my office. More often than not, Ken would find me in deep conversation with a colleague about the latest departmental intrigue or what was going on in the discipline. After my colleague had left, Ken would come in, carefully sniff the air for the nonexistent cigar smoke, and say with a smile "Ah, politics."

With his appreciative sniff, my student demonstrated what many people know but what historians studying technology often downplay: that most human activities are essentially political. Whether one is talking about decisions affecting the future of nations, the strategy of giant corporations, the curriculum of university departments, or even what happens in a family, the dynamics are often political. We acknowledge that there is power and there are resources, and then we debate, plan, and scheme about how to shape the situation to advance our interests. Allies must be enrolled, bargains struck, campaigns mapped out. In the nostalgic past, these things were done in the smoke-filled back rooms of city halls. Hence Ken's sniffing for the nonexistent cigar smoke.

While the cigar smoke was nonexistent in my office, the air was heavy with ideology. What makes politics interesting is that it involves the clash of world views. On all levels of human interaction, individuals bring to a discussion differing notions of what to do and why. Whether we are debating the degradation of the global environment or who in a university department should teach which courses, the discourse is shaped by what we think constitutes the good society and how we think the good society will be achieved. The dynamic interplay of belief and action, I would explain to Ken, is what makes history—and people—so intriguing.

As I mentioned a moment ago, I'm not sure that historians studying technology understand as well as my student the role of politics in the world of technology. To be fair, historians of technology do look at politics in the sense of group dynamics. Using either the social-constructivist approach of Wiebe Bijker and Trevor Pinch or the actor-network framework of Bruno Latour, Michel Callon, and John Law, some technological historians have been busy identifying how different groups clash and shape technological designs in ways that advance their interests.[1] Equally, Thomas P. Hughes, in *Rescuing Prometheus*, introduced us to project managers who successfully built large-scale technological systems because they were willing to engage in the messy complexity of recruiting political allies.[2] As long as they are talking about politics on the micro level, inside the firm or system or marketplace, then historians of technology are comfortable.

But a discussion of micro-politics does not answer the question posed years ago by the political scientist Langdon Winner: "Do artifacts have politics?"[3] What Winner was asking was whether technological artifacts embody and carry forward the political ideology of different groups in society. Do political beliefs—as well as the profit motive—drive people to pursue one set of technological solutions over another? Bryan Pfaffenberger argued that the personal computer revolution was an extension of the political revolution of the 1960s, but what other technologies mirror political change?[4] I would argue that, for the most part, historians of technology have focused on economic and social explanations of technological change and have been reluctant to probe how political beliefs may have informed the actions of inventors, engineers, and designers. How many historians of technology are aware that Thomas Edison voted Republican? Even more to the point, most historians of technology would probably find this fact interesting but irrelevant.

It is unfortunate that many historians of technology have been reluctant to investigate the interplay between technological design and political ideology, because it isolates them from the issues and ideas that animate other branches of history. For many historians, the central task is to understand how different belief systems—arising from religious, ethnic, race, or class backgrounds—inform the actions of individuals and groups. In their caution not to invoke political ideology as a way of explaining technological choices, historians of technology are missing an opportunity to establish important connections with the rest of the historical profession.

It is important for historians of technology to make these connections because of the ways in which technology tends to get treated in political and social history today. In much of the literature of American history, technology is still treated as an autonomous force in the background. Periodically new technology appears, and various individuals and groups then have to alter the social and political order in order to cope with that new technology. Even though historians of technology believe they have effectively banished such a deterministic perspective from their narratives, technology is seldom interpreted in American historical narratives as being produced simultaneously with the social and political order.[5] Indeed, technological determinism is alive and well in many American history survey textbooks.

Consequently, my goals here are threefold. First, I want to tell a story in which ideology plays a major role in the development of a new technology. Second, I wish to show historians of technology the importance of bringing political beliefs into their narratives. Third, I want to challenge political and social historians to pay attention to technological stories. All too often, nontechnological historians concentrate on the discourse of ideas and fail to look at what happens when ideas are manifest in material objects. Equally, they assume that responses to industrialization and technological change take place outside the arena of business and technology, and I want to inquire here how reform could take place *inside* the world of business by using technology in new ways. What I want to investigate here is how attempts to use technology to realize political goals can lead historical actors in unexpected directions, directions that upset the nice, neat categories historians use to organize the past.

To bring together technology and political ideology, I will tell a story about the invention of the telephone in America in the 1870s. Most histories of the telephone treat its invention and development in purely economic terms: the telephone was invented because Alexander Graham Bell, Gardiner Hubbard, and Bell's other backers wanted to make money, and it was adopted by businesses because it increased efficiency. At the same time, most histories of America in the 1870s treat the political issues concerning the rise of big business and the appearance of new technologies such as the telephone, the phonograph, and the electric light as two separate phenomena; the product of heroic genius, these new technologies were unsullied by the political intrigue and corruption of the Gilded Age.[6]

Yet Bell and Hubbard did not develop the telephone in a political and social vacuum; indeed, their efforts were strongly shaped by the debates in the 1870s over the rise of big business, the appropriate role of the state in regulating business, and the formation of the middle class. I will discuss Gardiner Hubbard's role in promoting the telephone and how his efforts were based on his political ideology: that if democracy was to survive in America, then middle-class Americans had to have ready access to the information and technology needed in order to participate in both the political and economic arenas. In telling the story of Hubbard's ideology and the development of the telephone, I want to show that artifacts do have politics, and conversely that politics have artifacts.

Telegraphy and the Rise of Western Union

To appreciate Hubbard's political ideology and how it shaped his efforts to promote the telephone, it is necessary first to understand the telegraph industry in the United States in the middle of the nineteenth century. Here we will see how the rise of Western Union prompted Hubbard and others to think about the role of information in a democratic society.

Western Union had been founded by Hiram Sibley in 1851 as the New York and Mississippi Valley Printing Telegraph Company. During the 1850s, Sibley built up the firm by taking over smaller telegraph lines until he had gained control of the telegraph business in the Midwest (hence the name Western Union). In 1861, Sibley built the first transcontinental line to California and established Western Union as one of the major firms in the industry. In 1866, Western Union absorbed its two remaining rivals, US Telegraph and American Telegraph, achieving, more or less, monopoly control of the telegraph industry.[7]

Because of his success in reorganizing US Telegraph, William Orton (1826–1878) was named president of Western Union in 1867. At once, Orton began to convert Western Union from a confederation of independent companies into a single organization dedicated to a particular market strategy. Orton noticed that a significant number of the telegrams transmitted by the company were short business messages—market quotes, buy and sell orders to brokers, and brief instructions to salesmen in the field. For these messages, business customers chose the telegraph because it was quick and reliable, and they were not especially concerned about price.[8] Assuming that businessmen would send more of these messages as they pursued the new national market made available by the railroad, Orton decided to have Western Union concentrate on sending business messages between cities. In doing so, Orton made a distinct choice, since he could have pursued several other markets. In several European countries, for example, the bulk of the telegraph messages were either social messages or government reports. Likewise, Western Union could have also placed newspaper reports at the center of its strategy. Although he forged an alliance with the Associated Press, Orton never considered press dispatches as important as business messages.

To control and expand this business market, Orton took two steps. First, to secure information from capital and commodity markets for Western Union's business customers, Orton bought control of the Gold and Stock Telegraph Company. Gold and Stock had established itself by erecting local telegraph networks for transmitting prices from the floors of stock and commodity markets to the offices of brokers and investors. To make its service attractive, Gold and Stock encouraged Thomas Edison and Elisha Gray to develop printing telegraphs or stock tickers which brokers could easily read.[9] Because Orton saw Western Union in the long-distance, inter-city business and Gold and Stock operating in the local, intra-city business, he left Gold and Stock a separate firm. Control of Gold and Stock was valuable to Western Union because it gave Western Union access to the market information that business customers wanted. Second, in pursuing the short business message market, Orton invested selectively in new technology. Although several inventors (including Edison) had developed automatic telegraph systems that used punched tapes to send and receive messages more quickly, Orton refused to invest in them. From Orton's view, it was a waste of time and money to prepare a tape for a short message when an operator could just send it.[10] Instead, Orton encouraged the development of devices that could send multiple messages over a single wire. By being able to send two, four, or even more messages over a single wire, Western Union could increase the volume and speed of messages without having to make heavy investments in stringing new lines or hiring more operators. Consequently, Orton purchased Joseph Stearns's duplex (two message) patent in 1872 and supported Edison and Prescott's work on a quadruplex (four-message) system in 1874.[11]

Thanks in part to duplex and quadruplex, Western Union was able to maintain steady annual profits while increasing the number of messages and decreasing the cost per message. By 1873, Western Union was conducting 90 percent of the telegraph business in the United States.[12] However, this does not mean that it was all smooth sailing for Western Union in the early 1870s. Simultaneous with his efforts to establish Western Union's strategy and structure, Orton had to fight off the dangers of a rival network and a hostile takeover.

One threat came from Wall Street. Sibley and Orton had rapidly built up Western Union by erecting lines along railroads and placing telegraph offices in railway stations. However, this meant that, as new transcontinental railroads were built, railroad financiers could create their own telegraph networks and not ally themselves with Western Union. Jay Gould pursued this strategy twice, first in 1874–1878 and again in 1879–1881. In the first episode, Gould used his control of several railroads to help the Atlantic and Pacific Telegraph Company quickly build a rival telegraph system.[13] To steal business away from Western Union, the Atlantic and Pacific cut prices in January 1877 and forced Western Union to follow suit. Unfortunately, while the reduction in prices generated a large volume of business for Atlantic and Pacific, the firm lacked a sufficient number of lines to transmit the messages quickly and reliably. To cope with the message volume, Atlantic and Pacific tried using Georges D'Infreville's crude duplex and Edison's automatic system. Gould convinced Edison to join the firm briefly as its Chief Electrician. Despite these steps, Atlantic and Pacific performed poorly (it never paid a dividend), and Orton was able to force a merger in the spring of 1878.[14]

The other major challenge for Western Union came from Washington. No sooner had Western Union achieved national dominance in 1866 than individuals such as Gardiner Hubbard began attacking it as a threat to American democracy. For Hubbard and others, Western Union was the first national monopoly, and they did not believe that this corporate giant would exercise any restraint in raising prices or that it would serve the public interest. For these critics, the fact that Western Union was reluctant to cut prices and indifferent to some inventions (such as automatic telegraphy) suggested that the firm was pursuing private gain at public expense. Critics were especially concerned that Western Union had access to both market information and private business messages, and that the firm could use this information to manipulate markets in its favor and ruin individual businessmen. Finally, critics were worried that, by transmitting news for the Associated Press, Western Union could also interfere with freedom of the press.[15]

In response to these real and perceived problems, critics of Western Union attempted to persuade Congress to take action. Some thought that the Post Office should be permitted to erect its own telegraph lines; they cited the nationalization of telegraph lines in England, France, and Belgium as positive examples. In anticipation of possible nationalization, Congress passed the Telegraph Act of 1866, which gave the federal government the right to purchase the assets of all telegraph companies (at a mutually agreed upon price) in 5 years. Others, particularly Hubbard, disagreed with outright nationalization and instead thought the federal government should guarantee competition in the industry by underwriting the creation of a second telegraph network.[16] (I will say more about this proposed federally subsidized network in a moment when I discuss Hubbard's vision.)

Drawing on his experience in Washington as Commissioner of Internal Revenue during the Civil War and his ties to the Republican Party, Orton beat back these threats. To do so, Orton cultivated congressmen and regularly testified on Capitol Hill. In his testimony, Orton used a variety of arguments to defend Western Union, but he frequently returned to two themes. First, he argued that the need to earn a return on the huge capital investment made by stockholders drove Western Union to build and operate an efficient telegraph network. Second, he claimed that only a private organization could fully discipline its workforce and prevent the violation of private messages or the misuse of market information. In contrast, he suggested, a government telegraph system would have operators who were politically appointed, and this would inevitably lead to mischief.[17] Using arguments such as these, Orton was able to stymie attempts to either nationalize the telegraph or create a rival. In doing so, Orton helped establish the general pattern of telecommunications as a private industry that responds to federal regulation.

Orton's arguments can be seen as similar to the ideological position that Martin Sklar and other historians have come to call corporate liberalism.[18] In the years between the Civil War and World War I, rapid technological change unleashed a period of unprecedented economic growth. But with this growth came instability in the forms of unbridled competition, de-skilling and dislocation of workers, and a cyclical pattern of booms and recessions. To both businessmen and intellectuals, it seemed that market forces were exacerbating rather than relieving this instability. In response to the failure of markets, businessmen sought to eliminate competition, first through cartels and then by creating larger organizations. To explain their actions, businessmen and intellectuals drew on the intellectual traditions of laissez-faire economics and classical liberal political thought to frame a new ideology: corporate liberalism. If society wanted to reap the benefits of tech-nological change, businessmen and intellectuals came to argue, then soci-ety needed to trust business and permit it to create the organizations that could achieve the economies of scale possible with this new technology. Decisions about how to best use technology were to be made privately by owners of the means of production and not by the state, since the owners of the technology possessed both the motivation and expertise needed to deploy technology efficiently. The role of the state was to create, through laws, regulation, and tariffs, an environment that encouraged private enter-prise to use technology efficiently.

Along with his arguments advocating private control of technology, Orton employed a range of tactics—price competition, political lobbying, and hostile takeovers—to beat back the challenges posed by Gould and Hubbard. In this turbulent environment, a new tactic, technological inno-vation, came to play an important role. To maintain its dominant position, Western Union needed to adopt new inventions such as duplex and quadru-plex. At the same time, the challengers—Gould and Hubbard—also realized that innovations might be used to gain a foothold in the industry. An edi-tor of the *Telegrapher* observed in 1875:

. . . improved apparatus has become of vital importance, and, consequently, tele-graphic inventors who, for some years past, have been regarded as bores and nui-sances, suddenly find themselves in favor, and their claims to notice, recognition and acceptance, listened to with respectful attention. All parties are now desirous of securing the advantages which may be derived from a development of the greater capacity of telegraph lines and apparatus. The fact has become recognized that the party which shall avail itself to these most fully will possess a decided advantage over its competitor or competitors.

That this state of telegraphic affairs affords the opportunity for the inventive talent and genius of the country which has hitherto been wanting, is unquestionable.[19]

By the mid 1870s, the combination of Western Union's dominance and the possibility of launching a rival network created a unique and novel market for telegraph inventions. In fact, one could argue there was a demand for "blockbuster" inventions or patents that could be used by Western Union or its challengers. Thus, as Bell, Gray, and Edison were investigating devices that would become telephones, they were working not in a "normal" business environment but in a "hothouse" that favored a breakthrough.

The breakthrough sought by Orton, Hubbard, and others in the telegraph industry was the "next generation" of multiple-message systems. In the mid 1870s, several inventors in Europe and America began designing acoustic or harmonic telegraphs. Inspired partly by the work of the German physicist Hermann von Helmholtz, investigators thought it might be possible to send and receive several messages by assigning a separate acoustic tone to each message. Elisha Gray experimented with an acoustic system beginning in the winter of 1866–67 and publicly demonstrated a version in July 1874 in New York.[20] Because Gray was chief electrician for the Western Electric Manufacturing Company and Western Electric was one of Western Union's major suppliers, Orton quickly became aware of Gray's work.

In March 1875, Orton learned that another inventor, Alexander Graham Bell, was working on a similar harmonic scheme. At that moment, Orton was quite interested in new multiple-message systems since Edison had gone over to work for Atlantic and Pacific, and it was unclear who actually owned the quadruplex.[21] Orton had Bell demonstrate his apparatus, only to find Bell's arrangement inferior to Gray's design.[22] Nevertheless, Bell's work helped convince Orton that harmonic telegraphy might well be the next breakthrough. Consequently, when Edison sheepishly returned to work for Western Union in July 1875, Orton asked him to develop an acoustic telegraph.[23]

Although there were technical reasons why Orton turned down Bell in March 1875, strong personalities also played a part. Bell was promptly (but politely) escorted out of Orton's office when Orton learned that Bell was associated with his nemesis, Hubbard. To understand why the mere mention of Hubbard led Orton to refuse to work with Bell, let us examine Hubbard's opposition to Western Union.

Hubbard and the Postal Telegraph

Gardiner Greene Hubbard (1822–1897) was born in Cambridge, Massachusetts to an old New England family. He attended Phillips Academy (Andover), Dartmouth College, and Harvard Law School. After serving as a junior member of a law partnership with Charles P. and Benjamin R. Curtis in Boston, Hubbard went into practice for himself in 1848. His practice included some patent work, and he helped Gordon McKay secure patent coverage for his shoemaking machinery.[24]

Although Hubbard inherited a small fortune from his mother, he lost most of it speculating on a wheat deal in the late 1840s.[25] To overcome this financial loss, Hubbard supplemented his legal practice with real estate development in Cambridge. In 1849, Hubbard purchased about 45 acres south of Brattle Street, and in 1851 he built an imposing wood-framed Italianate villa. Although he initially planned this to be only his summer home, he and his family soon made the Brattle Street house their year-round residence. Hubbard landscaped the rest of his acreage, which he then sold as house lots. Hubbard was involved in developing subdivisions elsewhere in Cambridge, and in 1861 he joined other local businessmen in the East Cambridge Land Company, which struggled for years to fill in and develop the marshland along the Charles River as industrial property.[26]

Early on, Hubbard realized that his Cambridge real estate ventures would only succeed if he could help make Cambridge desirable to his fellow middle-class professionals. To accomplish this, Hubbard turned to technology. In 1849 he became the first president of the Harvard Branch Railroad, which provided steam-train service from downtown Boston to the Cambridge Common. Unfortunately, this commuter railroad was unable to compete with the horse-drawn omnibus service between Boston and Cambridge, which ran more frequently and charged lower fares. Ultimately, the Harvard Branch failed because there were too few passengers to pay off the substantial debts incurred laying the track. In 1853, Hubbard and his fellow Cambridge businessmen replaced the steam railroad with a horsecar line, and their new company, the Union Railway, proved to be a profitable venture.[27] In 1852, Hubbard also helped establish the Cambridge Water Works and the Cambridge Gas Light Company. Although none of Hubbard's transportation and utility ventures were the first of their kind in America, it is notable that in the 1850s such systems were generally found only in major cities, such as New York and Baltimore. It is remarkable that Hubbard established such services in a small city such as Cambridge, which had a population of only 15,215 in 1850.[28]

Undertaken in such a small city, Hubbard's transportation and utility ventures were serious economic risks. Just as the Harvard Branch Railroad failed, there was no guarantee that the populace would embrace the horse-car or gas lighting and generate sufficient income to pay off the costs of these systems. Indeed, there was always a gap between Hubbard's vision and business reality. Recalling Hubbard's contributions to Cambridge, one local historian wrote:

Mr. Hubbard was one of those forceful men who possessed a vision beyond the present. . . . Like many such men he made ventures without sufficient capital and met the difficulties such action is likely to cause. . . . They [i.e., his transportation and utility companies] all suffered for want of adequate capital in the early days, the public generally having little confidence in them as a business proposition.[29]

Yet, ever optimistic about the future, Hubbard promoted his ventures, confident that they would prove to be successful enterprises in the end.

In 1868, Hubbard moved from the local to the national stage. He turned his attention to the telegraph industry and to questioning Western Union's dominance. It is not exactly clear why Hubbard took up this cause, but he did share with Charles Francis Adams and other Bostonians a suspicion of large-scale organizations, especially those controlled by New York financiers.[30] As he did with other topics, Hubbard undertook a massive study of the telegraph business in Europe and America. Through this research, Hubbard concluded that Western Union was not serving the public interest: the company was pursuing only a business market, it was not using the most up-to-date technology (such as automatic systems), and it was not reducing its prices.[31]

Hubbard was concerned that Western Union was using its technology, size, and dominant market position to reap excess profits. Even more worrisome to Hubbard, Western Union was a threat to American democracy because of the firm's control over the dissemination of vital information. Because Western Union and Gold and Stock controlled the distribution of prices from stock and commodity markets, these firms exercised tremendous power over individual investors and small businessmen. Moreover, since Western Union transmitted news for the Associated Press, Hubbard and other reformers feared that Western Union could interfere with elections by tampering with the news. For Hubbard, Western Union was a threat to the power of individual Americans to pursue their own political and economic destiny.

For Hubbard, the solution to the problems created by Western Union's monopoly power involved two changes: first, one had to rethink the market for the telegraph; second, one needed to invent a new kind of organization which should serve that market. In Europe, telegraph networks handled a substantial volume of social messages and government reports; why not build a system to serve these markets as well as businessmen? To reach these additional groups, Hubbard proposed that telegraph offices be located not just in railway stations (Western Union's common practice) but also in post offices, which were much more convenient for the average citizen. Partly because the telegraph offices would be placed in post offices, Hubbard called his scheme the postal telegraph plan. Hubbard was confident that there was an enormous potential market for social and governmental messages and that this demand could be used to lower prices and expand telegraph service to every town and village in America. In doing so, Hubbard believed he was creating a telegraph system that would better serve the needs of the American public and advance democracy.[32]

However, for Hubbard it was not enough just to rethink the telegraph market. Reform would come only if a new type of organization were created to serve this market. To create an alternative organization, Hubbard did not hesitate to seek government support, as he had done with other causes. In the mid 1860s, concerned that deaf students such as his daughter Mabel were not receiving adequate public education, he convinced the Massachusetts legislature to support a school for the deaf. Hence, it is not surprising that Hubbard took his idea for a postal telegraph system to Washington and sought funds from Congress in 1868. What Hubbard asked Congress to provide was the capital for a private corporation which would build a new telegraph network and in turn enter into a contract with the Post Office.[33] This private corporation would be run by Hubbard and his associates.[34]

Viewed from Orton's perspective, Hubbard's postal telegraph scheme must have seemed puzzling, since the idea of a government-subsidized rival company denies the basic premise of corporate liberalism: that the private owners of technology know how to best deploy technology and reap the greatest good for society. However, for Hubbard it was not a foregone conclusion that private interests would necessarily serve the public good. Drawing on an older discourse that questioned the role of monopolies and special interests in a republic, Hubbard asked who should be controlling powerful corporations:

The time will come . . . when the people will rise in their might and crush these monopolies. It is necessary that the good men of both political parties . . . should take steps to correct the evil in these corporations, that they may be reformed rather than destroyed. Stockholders and officers . . . do not or will not understand their true relations to the public. They seem to think that their franchises were granted by the State solely for their own benefit. . . . They forget that these are not private but public corporations which have received certain privileges from the state, in consideration of their promise to perform certain duties to the public in return.[35]

Just as Orton's position can be related to corporate liberalism, so Hubbard's outlook corresponds to an ideological position that Mary O. Furner termed "democratic statist." Concerned about the problems posed by rapid industrialization and the rise of big business, many Americans believed that the solutions lay not in radically restructuring society and the economy but in experimenting with a variety of new organizations and institutions.[36] With great energy, Americans in the 1870s and the 1880s experimented with new organizations—government bureaus (for collecting information and enforcing regulation of the railroads and public health), labor unions (e.g., the Knights of Labor), movements for farmers (e.g., Farmers' Alliances), moral reform efforts (e.g., the Women's Christian Temperance Union), and educational institutions (e.g., the rise of research universities with professional schools).[37] Seen as alternatives to the existing political parties, these new organizations were the means by which "good men" such as Hubbard hoped to reform society.[38]

To ensure that these organizations had the resources and authority necessary to bring about change, individuals and groups did not hesitate to turn to the states or to the federal government for support. However, these reformers did not believe that the government should simply take over functions performed by the market; rather, they wanted the government to intervene and create a "space" for new organizations and institutions. For instance, Hubbard was opposed to the federal government's simply taking over the telegraph system, for he was convinced that nationalization would create a bureaucracy that would be just as unresponsive as Western Union. What was needed was a new kind of organization—a federally financed private corporation—that combined the efficiency of a private business organization with the authority of a government agency. Hence, I would argue that Hubbard's postal telegraph plan was part of a larger ideological movement—democratic statism—that sought to overcome the ills of industrialization through the creation of a variety of new organizations, some private and some state-supported.[39]

With great energy, Hubbard lobbied and persuaded congressmen to introduce bills for a postal telegraph scheme in sessions from 1868 to 1876.[40] Orton vigorously fought back, and the annual hearings for the postal telegraph bill came to be known as the "Wm. Orton and Gardiner Hubbard Debating Society."[41] In many ways, when Hubbard started his campaign for the postal telegraph, it was a reasonable proposition; after all, the federal government had just won the Civil War by actively creating all sorts of new practices and organizations. However, as the economy prospered and then collapsed in the mid 1870s (as a result of overexpansion in railroad construction), Orton and others came to argue effectively that the economy worked best when there was the least amount of government involvement. Although skillfully and repeatedly outflanked by Orton in the halls of Congress (the postal telegraph bill was never passed), Hubbard never gave up his belief in the need for an alternative telegraph system, and he instead looked for other ways to force Western Union to change. To Hubbard, there had to be other ways to utilize telecommunications to ensure American democracy.

Hubbard, Bell, and the Telephone

Hubbard found a new way to challenge Western Union in the creative ideas of Alexander Graham Bell, a young Scotsman who had emigrated with his parents to North America to help promote his father's system of visible speech. Visible speech was a system in which the deaf were taught to associate different sounds with symbols in order to learn to speak. Anxious to secure the best possible teaching techniques for his daughter and other deaf children, Hubbard invited Bell in 1872 to come Massachusetts to teach visible speech. Bell taught at the state-sponsored school for the deaf, then secured a professorship at Boston University.

Although a dutiful son, Bell did not wish to devote his life to advancing his father's system. In 1872, he turned to invention to make his own mark on the world. Bell may have read how Western Union had purchased Stearns's duplex, and this may have led him to decide to develop his own multiple-message telegraph.[42] Drawing on the extensive knowledge of acoustics he had acquired in teaching visible speech, Bell investigated a harmonic telegraph.

In the autumn of 1874, after he had begun courting Hubbard's daughter Mabel, Bell told Hubbard of his telegraph experiments. Hubbard took an immediate interest in Bell's efforts and encouraged him to perfect his harmonic scheme as quickly as possible. Hubbard knew that he could use the invention of a better multiple-message telegraph to bring about change in the telegraph industry.[43] With a strong patent for a multiple-message telegraph, Hubbard would have several options by which he could force change on Western Union. He could go back to Congress and argue for federal support of a new network built around this efficient invention, or he could imitate Gould and use the patent to found a new telegraph company.

Bell tried a dozen different arrangements but was never able to get his harmonic telegraph to work properly.[44] Bell realized that he lacked the manual skills to implement his ideas; however, rather than give up, he concentrated on being a "theoretical inventor" and perfecting the theory behind his multiple-message telegraph.[45] In the course of his theoretical musings, Bell came to be far more interested in the possibility of sending complex sounds (such as voices) over a wire than in sending individual tones. In February 1876, Hubbard filed Bell's patent application for a harmonic telegraph scheme, and this application included claims for a speaking telegraph.[46] Six weeks later, Bell succeeded in transmitting the voice and thus inventing the telephone.[47]

Although Hubbard was much more interested in getting Bell to complete his multiple-message telegraph, he followed Bell's telephone experiments, and he arranged for Bell to exhibit his inventions at the Philadelphia Centennial Exhibition in the summer of 1876. At the Centennial, scientists, inventors, and other dignitaries were fascinated more by Bell's telephone than his multiple-message telegraph, and Bell began to focus more attention on perfecting the telephone. Through the remainder of 1876, Bell tested his telephone on increasingly longer telegraph lines; he also began giving lectures at which he demonstrated the telephone.[48] For Bell and other scientifically minded people, the telephone was remarkable in that it illustrated the relationship between sound and electricity.

For almost a year (March 1876–February 1877), neither Hubbard nor Bell gave much thought about how to promote the telephone. For his part, Hubbard was quite busy heading up the Railway Mail Service, a branch of the Post Office, which had been established to speed up the delivery of mail. Under Hubbard's leadership, Theodore Vail and others developed special techniques for sorting the mail on board trains. As Richard John has pointed out, Hubbard's experience with the Railway Mail Service further convinced Hubbard of the importance of developing effective communications systems to foster democratic society.[49]

In view of his recent efforts with the postal telegraph, one might well have expected that Hubbard would have gone to Washington to seek some sort of government support for the telephone, but Hubbard did not do so. Most curiously, Hubbard at first thought it best to sell Bell's harmonic telegraph and telephone to Western Union. Perhaps Hubbard naively believed that Western Union lacked the technology needed to reform itself, and that if provided with breakthroughs such as the harmonic telegraph and telephone the company would change. Hubbard may also have wanted to see his future son-in-law established financially, and may have thought he could accomplish this by getting a good price for Bell's invention. Accordingly, in the autumn of 1876, Hubbard offered Bell's patents to Orton for $100,000.[50]

Much to the amazement of later historians, Orton turned down Hubbard's offer. Orton's decision makes sense in terms of his mindset and the resources he had at his disposal in the autumn of 1876. From Orton's viewpoint, the telephone was not suitable for Western Union's core business of sending short business messages between cities quickly and reliably. In 1876 Bell's telephone did not work reliably on circuits of more than 20 miles, and transmissions were somewhat muffled and indistinct.[51] Moreover, even though Bell invented it while pursuing a multiple-message telegraph, the telephone functioned in exactly the opposite way—rather than permitting several messages to be sent over a single wire, the telephone used an entire wire for one conversation. Hence, to Orton, the telephone would have reduced the throughput of his network rather than increasing it.

In the autumn of 1876, Bell's portfolio was also not very impressive: all that Hubbard could show Orton were a few crude instruments which were covered by three patents.[52] Of course, Orton had seen Bell's harmonic telegraph 18 months earlier and had concluded then that Bell's work was inferior to Gray's. Accustomed to the high-quality instruments developed by Edison, Gray, and other inventors, and knowing the importance of carefully phrased patents, Orton probably found it difficult to take Hubbard's offer seriously. Orton decided that it would not be difficult to have Edison continue his work on harmonic telegraphy, create a better device, and secure patents that could beat Bell in court. Consequently, at the end of 1876 Orton asked Edison to step up his investigation of harmonic telegraphy at Menlo Park. In March 1877, Edison signed a new contract with Western Union promising to develop new telephone patents.[53]

Rejected by Western Union, Hubbard next tried to convince several wealthy businessmen to form a company that would exploit Bell's patent.[54] In courting these investors, Hubbard argued that the telephone had the potential to remake the telegraph industry. Hubbard suggested that, by substituting the telephone for Morse keys and sounders, telegraph operators could send and receive messages much more quickly. "An operator by the Morse instrument ordinarily transmits about fifteen words a minute," he claimed, "while he can speak & therefore transmit by Telephone from one hundred & fifty to two hundred."[55] Likewise, Hubbard suggested that the telephone could be used on private lines to link two locations. Hubbard had already been contacted by several individuals who wanted to set up telephones between their homes and business establishments or between downtown offices and outlying factories. On these private lines, Hubbard believed that the telephone would be cheaper and quicker than printing telegraphs which often cost $250 each.

The first private-line telephone was purchased by Charles Williams, who connected his electrical workshop (where Bell had first experimented on his telephone) in downtown Boston to his home in the suburb of Somerville in May 1877. Within a few months, Hubbard was contacted by entrepreneurs who wanted licenses to install similar private lines. Several of these entrepreneurs wanted to imitate existing messenger, burglar, and fire alarm telegraph companies and connect all the telephones to a switchboard in a central office. The first of these exchanges was established in Boston in May 1877 by E. T. Holmes. Holmes had started a burglar alarm network, and he added telephones as a second service for his customers.[56] Holmes found that his telephone business grew quickly because a portion of his network served a neighborhood in which many "large grocers, confectioners, and cookery merchants" were located, and these businessmen found it convenient to use the telephone to conduct transactions among themselves.[57]

With great interest, Hubbard followed the efforts of Holmes and other local entrepreneurs as they installed telephones on private lines and set up exchanges. These developments convinced Hubbard in July 1877 to form the Bell Telephone Company, which was to hold the Bell patents and issue licenses to individuals who wanted to set up local telephone exchanges. Named as the company's trustee, Hubbard took it upon himself to promote the establishment of local telephone companies.[58] In the summer of 1877, Hubbard traveled across the United States as a member of the Special Commission on Railway Mail Transportation. He carried two telephones in his suitcase, enthusiastically demonstrating them to businessmen at every stop.

The Telephone as Middle-Class Invention

As Hubbard toured the country and demonstrated the telephone, one wonders how he reconciled this new campaign with his larger struggle against Western Union. How did his emerging market strategy mesh with his political ideology? Hubbard's efforts with the telephone make sense if one thinks of his opposition to Western Union because it mediated the access of individuals to information. For Hubbard, the success of American democracy hinged on the ability for individuals to secure news and market prices. The problem with Western Union was that it was so large and intrusive that it had the potential to prevent people from getting the information they needed. One way to minimize the intermediary was to place the technology squarely in the hands of the user and eliminate the telegraph operator, who controlled messages by encoding them. Since the first telephone was literally in the hands of the user, controlled and manipulated by him, Hubbard felt that the telephone eliminated the evils of the intermediary.[59] Moreover, Hubbard also believed that the telephone would be used for domestic and social purposes, and thus a telephone network would be more democratic than Western Union's business-oriented telegraph system. In his correspondence and his promotional efforts, Hubbard told of how middle-class and upper-middle-class people would use the telephone to coordinate servants, order groceries, and respond to social invitations. Although Hubbard recognized the business applications for the telephone, his imagination was fired by the ways in which the middle class could use the telephone.[60]

Hubbard saw the telephone as a device that would allow the middle class to create a space for itself in a chaotic world. That Hubbard saw the telephone in this way is not surprising when we recall that Hubbard used a variety of network technologies (gas, water, streetcars) to develop a middle-class community in Cambridge in 1850s. The telephone, Hubbard anticipated, would create not a physical community but, as we would say, a virtual one, since it was a means by which middle-class people could differentiate themselves from the poor and from the rich. To appreciate this, we need only review the typical uses of the telephone before 1880:

• Private lines between office and home connected the domestic (female) sphere and the (male) world of work while also maintaining protection.
• Private lines between downtown offices and outlying factories maintained the distinction between the middle-class owners and the workers.
• In the home, the telephone was used to direct servants and to order goods from shopkeepers, hence maintaining social distinctions between the middle class and the working class.
• Telephone exchanges connected middle-class peers across a community in a technological network that manifested the common set of interests that middle-class people were trying to define.

One might well ask why Hubbard and the middle class chose to deploy the telephone in these ways. I would argue that they did so to shore up their position as American culture underwent a series of traumatic changes in the 1870s.[61] During that decade, a new rich business elite appeared who had made their fortunes on Wall Street by financing the railroads. For Hubbard and other Boston Brahmins, this new financial elite was especially troubling because it was based in New York. At the same time that they were concerned with the appearance of a powerful upper class, middle-class people were also frightened by the lower orders. Not only was immigration swelling the ranks of the poor and working classes, but the decade was marred by labor unrest. Most notably, 1877 was marked by the first nationwide railroad strike—a strike that was broken only by calling out the Army. As the United States celebrated its centennial in 1876, its political institutions were not working very well; the presidential election of that year, between Rutherford B. Hayes and Samuel J. Tilden, was so close that it was contested by the Republicans and settled only by a special electoral commission. The country was in the grip of an economic recession caused by the Panic of 1873, federal troops were still occupying the South in the aftermath of the Civil War (the federal occupation ended only in 1877), and there was the unsettling experience of thousands of people moving to the Western frontier.

Collectively, these developments may have signaled to middle-class Americans that progress was not guaranteed. The basic elements of American society—its political institutions, economy, and social structure— seemed to them to be either changing or in disarray. It was in this sense that I would suggest that middle-class Americans were experiencing cultural trauma—their world was being radically altered.

As they were experiencing this trauma in the 1870s, middle-class Americans may have shifted their hopes for progress from social and political institutions to invention and science. News of Bell's invention of the telephone, seen as a product of science and genius, may have been reassuring to Americans, suggesting that the mind of man could bring order to the world. Writing in 1878 about another blockbuster invention, the phonograph, a reporter in *Scribner's Monthly* expressed this view:

The invention has a moral side, a stirring optimistic inspiration. "If this can be done," we ask, "what is there that cannot be?" We feel that there may, after all, be a relief for all human ills in the great storehouse of nature. . . . There is an especial appropriateness, perhaps in its occurring in a time of more than usual discontent. It is a long step in a series of modern events which give us justly, in the domain of science, wholesale credulity.[62]

Anxious to find some signs of progress, Americans were quick to seize on several inventions of the late 1870s—the telephone, the phonograph, the electric light—and to make them into cultural icons.

What is notable about this tendency to celebrate blockbuster inventions is that Americans did this several years *before* these inventions had widespread economic impact. In part, this may have been due to the fact that in early demonstrations the telephone and the phonograph did not always work, and, instead of reporting such results, reporters wound up speculating on the invention's potential impact. For example, in the spring of 1877, during a major lecture in New York's Steinway Hall, Bell was unable to get his telephone to operate; the reporters described how unruly the crowd became and then proceeded to speculate on the telephone's wonderful social potential. Practical applications did not exist, so cultural meanings took shape in this way.

This cultural framing was central to the development of the telephone because it was through this process that it acquired a set of positive attributes which made it attractive to businessmen and investors. Because the telephone promised to permit the middle class to create a social space for itself, it was an appealing investment. Consequently, the telephone "took off" as a commercial technology. By early 1880 there were approximately 300 telephone exchanges. By then, only nine American cities with populations over 10,000 did not have telephone exchanges.[63] Of course, as Claude Fischer argued in *America Calling*, the telephone only became a widespread consumer or domestic technology in the twentieth century.[64] My point here, however, is that the telephone acquired its basic political and social meanings before 1880, and that these meanings informed the commercial development of the telephone through the remainder of the nineteenth century and into the twentieth.

Conclusion

The development of the telephone was not shaped exclusively by technical and economic considerations; there were also political factors. In the formative years, before 1880, the Bell telephone acquired its particular characteristics because its principal promoter, Gardiner Hubbard, was deeply involved in efforts to reform the telegraph industry. Hubbard, moreover, believed in using technology to create a space for the middle class.

Generally given short shrift in popular and scholarly accounts of the development of telecommunications, Hubbard played a pioneering role in defining the telephone. It was Hubbard who closely monitored Bell's early experiments and pushed him to patent his inventions. It was Hubbard who sorted through the marketing possibilities, including selling the patents to Western Union, building private lines, selling versus leasing telephones, and establishing exchanges. Hubbard was a classic entrepreneur who coordinated the resources necessary for promoting telephone exchanges. He recruited local agents, convinced Charles Williams to manufacture telephones, raised capital from other New England businessmen, and helped found the Bell Telephone Company. Hubbard was endlessly optimistic about the telephone, and his enthusiasm infected local entrepreneurs and Boston investors and prepared the way for the Bell group to begin installing telephones in 1877.

But aside from his organizational contributions, Hubbard shaped the telephone by linking it to a political and social vision. Unlike Western Union, which viewed telegraphy exclusively as a technology to be used by private interests to serve the businessman, Hubbard saw telegraphy as a technology that should serve the average citizen. He opposed Western Union because he saw the company coming between citizens and the news and information they needed. Initially, Hubbard tried to redress the problems with telegraphy by lobbying Congress to create an alternative telegraph system, only to be stopped by Orton. Consequently, Hubbard turned to a technological solution: the telephone invented by Bell. In the telephone, Hubbard saw a telecommunications system that would overcome the defects he saw in Western Union. Literally in the hands of the user, the telephone seemed to eliminate Western Union and its operators as a meddlesome intermediary. Installed in homes, shops, and offices, the telephone could be used by individuals for both social and business messages. Finally, Hubbard encouraged entrepreneurs to set up local exchanges. In doing so, he may have felt he was aiding democracy by taking a "grass roots" approach and letting individuals deploy the telephone in ways suitable to their communities.

Understanding Hubbard's political vision is essential for understanding the choices that he and Bell made in developing and promoting the telephone. The point for technological historians is that ideology does matter. Technological artifacts are not "neutral" tools whose design and utilization are determined by economic necessity or efficiency. Indeed, artifacts do have politics, and the telephone came to embody meanings that were central to the middle class. Anxious to ensure that the businessmen and entrepreneurs who constituted the emerging middle class had access to the information they needed, Hubbard both sought to reform Western Union and promoted the telephone. For Hubbard and others in the middle class, the telephone was a means for reifying—for manifesting in a concrete way—their position in society.

Hubbard's experience with the telegraph and the telephone enriches our understanding of the rise of the middle class in nineteenth-century America. Several historians, particularly Stuart Blumin and Olivier Zunz, have traced how individuals formed this class, utilizing work, consumption, residential location, associations, and family structure to create a place for themselves in society.[65] Although Zunz touches on how middle-class executives shaped the architecture of the office and the skyscraper to manifest their values, social historians have generally not considered how specific inventions (such as the telephone) were utilized by the middle class to define who they were or to manifest their values. Yet, as we have seen here, the telephone was influenced by Hubbard's vision of the middle class. A similar argument, I would suggest, might well be made for the rise of other major technologies of the late nineteenth century, including the phonograph, electric lighting, and the automobile. The development and diffusion of these technologies might well be interpreted in terms of their appropriation by the middle class. The current explosion of popular interest in the Internet and the World Wide Web may well be due to the fact that these technological developments are now being used by middle-class Americans to define who they are and what they believe.

Investigating how the middle class was (and continues to be) concerned with technology also raises the possibility of furthering our understanding of the political behavior of this class. For the most part, self-conscious political action by the middle class seems strangely absent in American history.

"As ethnocultural political historians point out," Stuart Blumin observed, "political movements based explicitly on the grievances or aspirations of intermediate social classes are indeed rare in American society."[66] The lack of political action by the middle class is often explained in terms of the paradox this class embodies: How can a group that celebrates individualism exercise collective action? Rather than this standard explanation, however, perhaps the reason why historians have found so few episodes of middle-class political action has more to do with where they were looking. Though the middle class does not necessarily express its aspirations in normal political channels, it may well do so in other areas such as technology. The story of Hubbard and the telephone hints that the way to see the middle class in political action is to pay attention to the ways in which they try to direct technology. For the middle class, political action through technology may be more palatable in the sense that many of the arguments turn on empowering the individual and hence do not appear to be based on narrow class interest. Hubbard certainly made this sort of empowerment argument for the telephone, yet at the same time he certainly thought that the telephone should be deployed in ways that maintained social boundaries between the middle and lower classes. Glancing again at the contemporary situation, one wonders how much of the Internet "revolution" can be interpreted as political action on the part of the middle class.

Hubbard's experience with the telephone raises another point for political historians. All too often, scholars concerned with the responses to industrialization in late-nineteenth-century America trace the evolution of ideas in the writings of politicians and reformers in their efforts to shape legislation and policy, and in the institutions they help create.[67] The assumption is that responses to industrialization must have taken place outside of the realm of business and that the responses certainly were nontechnological. Hence, the typical political narrative of Hubbard might focus on his postal telegraph articles and his lobbying efforts in Congress, and it would treat his efforts in promoting the telephone as something entirely different and even irrelevant.[68] Surely, according to typical thinking, political reform and business entrepreneurship are two separate activities, the first concerned with the good of society and the second with the good of the individual. But, as we have seen here, Hubbard did not separate his political and technological efforts but rather saw both as integral to his efforts to reform democratic society. Thus, it is important not to limit our investigation of individuals such as Hubbard to an analysis of either their political or their technological efforts; rather, we need to follow their actions wherever they lead and seek explanations that connect them. Indeed, Hubbard challenges the basic assumption that Americans coped with industrialization exclusively by changing the government (through regulation and new agencies) or by creating non-business institutions (such as schools, hospitals, foundations); perhaps we should enlarge our understanding of late-nineteenth-century America by looking at how reformers such as Hubbard participated in the business world.

A final lesson that may be drawn from Hubbard's efforts concerns the pervasive role of the state. Throughout his struggle with Western Union and his efforts to introduce the telephone, Hubbard employed different elements of the state. He lobbied Congress for his postal telegraph scheme, promoted the telephone while traveling on government business, and, of course, resorted to the courts to protect Bell's patent monopoly. Significantly, Hubbard saw the government not as an externality with which he had to cope but rather as a set of resources which he could partly control and direct toward his goals. Although it is tempting to see nineteenth-century America as an unregulated, free-market paradise in which heroic individuals built great technological systems and business empires without interference from the government, the early development of the telephone reveals that major technological changes were intimately bound up with the evolving political and legal environment.[69] Consequently, if we are to understand how inventors and entrepreneurs created the technology of the second industrial revolution, we must pay more attention to how entrepreneurs interacted with the state.

For social and political historians, then, the message is that technology matters. If we wish to fully appreciate the political beliefs of different groups in society, then we need to inquire about how material objects illustrate and confirm these beliefs. Political ideology should not be explored merely in the realm of ideas.

As the twenty-first century dawns, Americans often wish their society were as creative as it was at the end of the nineteenth century. In doing so, Americans fantasize that individuals driven by personal ambition should somehow come up with new technologies that will automatically make

American business more efficient and competitive in the global economy. However, I would caution Americans against assuming that personal ambition and greed are the only factors driving technological change. We need to recognize that the stories of technology in America are genuine political stories, stories in which we see how different groups and individuals used technology to shape their identities. Only when we are able to fully grasp the ways in which groups used technology to manifest their political ideologies will we complete the task posed to us by Thomas Hughes in *American Genesis:* to understand how America went from being nature's nation to being technology's.[70]

Acknowledgments

The research used for this essay was supported by fellowships at the Dibner Library at the Smithsonian Institution, the Dibner Institute for the History of Science and Technology in Cambridge, Massachusetts, and the Commonwealth Center for Cultural and Literary Change at the University of Virginia. Previous versions were presented at the Business History Conference, at the annual meeting of the Organization of American Historians, and to the History of Science Unit at the University of Manchester, and I am grateful to the audiences on those occasions for their comments and suggestions. I wish to thank Michael Allen, Brian Balogh, Jane Fewster, Gabrielle Hecht, and Richard John for advice on how to strengthen the argument.

Notes

1. Wiebe Bijker and Trevor Pinch, "The Social Construction of Facts and Artifacts: Or How the Sociology of Science and the Sociology of Technology Might Benefit Each Other," in *The Social Construction of Technological Systems*, ed. W. Bijker et al. (MIT Press, 1987); Bruno Latour, *The Pasteurization of France* (Harvard University Press, 1988); Michel Callon, "Some Elements of a Sociology of Translation: Domestication of the Scallops and Fishermen of the St. Brieuc Bay," in *Power, Action, and Belief*, ed. J. Law (Routledge & Kegan Paul, 1986); John Law and Michel Callon, "The Life and Death of an Aircraft: A Network Analysis of Technical Change," in *Shaping Technology/Building Society*, ed. W. Bijker and J. Law (MIT Press, 1992).

2. Thomas P. Hughes, *Rescuing Prometheus* (Pantheon, 1998).

3. Langdon Winner, "Do Artifacts Have Politics?" in Winner, *The Whale and the Reactor* (University of Chicago Press, 1986).

4. Bryan Pfaffenberger, "The Social Meaning of the Personal Computer: Or, Why the PC Revolution Was No Revolution," *Anthropological Quarterly* 61(1988): 39–47.

5. Merritt Roe Smith and Leo Marx, eds., *Does Technology Drive History?* (MIT Press, 1994).

6. For an example of how politics and technological developments can be treated as separate phenomena in American history, see Page Smith, *The Rise of Industrial America* (McGraw-Hill, 1984).

7. For the early history of Western Union, see Robert Luther Thompson, *Wiring a Continent* (Princeton University Press, 1947). One has to be careful in claiming Western Union had complete monopoly control of the telegraph industry. First, like dominant firms in other industries, Western Union found it useful to permit a few small firms to exist (for example the Franklin Telegraph Company). Second, as we shall see, it was possible for new rivals to spring up by building new lines along major railroads.

8. Paul Israel, *From Machine Shop to Industrial Laboratory* (Johns Hopkins University Press, 1992), p. 129.

9. Ibid., pp. 125–127.

10. Edison's efforts to develop an automatic telegraphic are documented in volume 1 of *The Papers of Thomas Edison*, ed. R. Jenkins et al. (Johns Hopkins University Press, 1989). On Orton's opposition to automatic telegraphy, see Israel, *From Machine Shop to Industrial Laboratory*, pp. 132–134.

11. Israel, *From Machine Shop to Industrial Laboratory*, pp. 135–140.

12. Michael F. Wolff, "The Marriage That Almost Was," *IEEE Spectrum* 13 (February 1976), p. 41.

13. Gould was apparently especially motivated in attacking Western Union because it was controlled by his rival William H. Vanderbilt. See Matthew Josephson, *The Robber Barons* (Harcourt, Brace, 1934), pp. 205–206; Maury Klein, *The Life and Legend of Jay Gould* (Johns Hopkins University Press, 1986), pp. 197–205 and 277–282.

14. James D. Reid, *The Telegraph in America* (New York, 1879), pp. 586–587; Israel, *From Machine Shop to Industrial Laboratory*, pp. 146–147.

15. Charles A. Sumner, *The Postal Telegraph* (San Francisco, 1879).

16. Lester G. Lindley, *The Constitution Faces Technology* (Arno, 1975).

17. David A. Wells, *The Relation of the Government to the Telegraph* (New York, 1873); [William Orton], *Argument of William Orton on the Postal Telegraph Bill* (New York, 1874), Box 3, Western Union Telegraph Company Collection, Archives Center, National Museum of American History, Washington.

18. Martin J. Sklar, *The Corporate Reconstruction of American Capitalism, 1890–1916* (Cambridge University Press, 1988).

19. "The Progress of the Telegraphic Contest," *Telegrapher* 11 (30 January 1875), p. 28.

20. For a detailed discussion of Gray's work on the harmonic telegraph, see Michael E. Gorman, M. E. Mehalik, W. B. Carlson, and M. Oblon, "Alexander Graham Bell, Elisha Gray, and the Speaking Telegraph," *History of Technology* 15 (1993): 1–56. See also David A. Hounshell, "Bell and Gray: Contrasts in Style, Politics and Etiquette," *Proceedings of the IEEE* 64 (1976), 1305–1314; "Elisha Gray and the Telephone: On the Disadvantages of Being an Expert," *Technology and Culture* 16 (1975): 133–161; "Two Paths to the Telephone," *Scientific American* 244 (January 1981): 156–163.

21. Edison's work on the quadruplex and his complex relationships with Western Union and Atlantic and Pacific are documented in volume 2 of *The Papers of Thomas A. Edison*, ed. R. Rosenberg et al. (Johns Hopkins University Press, 1991).

22. Bell to Papa and Mama, 5 March and 22 March 1875, Box 5, Bell Family Papers, Library of Congress, Washington.

23. Edison, "Reis Telephone Drawings" [July 1875], in *The Papers of Thomas A. Edison*, volume 2, pp. 524–526.

24. [Biographical Sketch of Gardiner Hubbard], n.d., Box 16, Hubbard Family Papers, Library of Congress, Washington.

25. [Mabel Bell], undated reminiscence in files of Cambridge Historical Commission.

26. Samuel Atkins Eliot, *A History of Cambridge, Massachusetts (1630–1913)* (Cambridge Tribune, 1913), pp. 117–119; Bainbridge Bunting, "Brattle Street: A Resume of American Residential Architecture," *Cambridge Historical Society Proceedings* 43 (1973–1975), pp. 45–46; Lois Lilley Howe, "Dr. Estes Howe: A Citizen of Cambridge," *Cambridge Historical Society Proceedings* 25 (1938–39), pp. 139–140.

27. It actually took Hubbard and his Cambridge associates two attempts to establish a successful horsecar line. Their first company, the Cambridge Railroad, was formed in 1853 and was only able to raise enough capital to lay track between Cambridge and Boston. Their second company, the Union Railway, was organized in 1855, and it leased the Cambridge Railroad's track.

28. Robert W. Lovett, "The Harvard Branch Railroad, 1849–1855," *Cambridge Historical Society Proceedings* 38 (1959–60): 23–50; *The Cambridge of Eighteen Hundred and Ninety-Six*, ed. A. Gilman (Riverside, 1896), pp. 380, 396–399; Foster M. Palmer, "Horsecar, Trolley, and Subway," *Cambridge Historical Society Proceedings* 39 (1961–1963): 78–107; Harding U. Greene, "The History of the Utilities in Cambridge," *Cambridge Historical Society Proceedings* 42 (1970–1972): 7–13. Partly as a consequence of these improvements in transportation and utilities, the population of Cambridge nearly doubled in the 1850s; in 1860 the city boasted 26,060 inhabitants.

29. George Greer Wright, "Gleanings from Early Cambridge Directories," *Cambridge Historical Society Proceedings* 15 (1920–21), p. 39. Hubbard recalled

local opposition to his utility and transportation ventures in "A Comprehensive Plan. . . . An Interesting Interview with Mr. Gardner [*sic*] G. Hubbard" (*Cambridge Tribune*, 14 August 1886, p. 1; in files of Cambridge Historical Commission).

30. Hubbard linked his concerns about Western Union with Adams's efforts to reform the railroad industry on p. 84 of "The Proposed Changes in the Telegraphic System," *North American Review* 117 (July 1873), 180–107. On Charles Francis Adams's concerns about railroads and large-scale organizations, see Thomas K. McCraw, *Prophets of Regulation* (Harvard University Press 1984), pp. 1–56. For a discussion of the concerns of other Bostonians in the post-Civil War era, see Peter Dobkin Hall, The Organization of American Culture, 1700–1900 (New York University Press, 1984), pp. 261–270.

31. Gardiner G. Hubbard, Letter to the Postmaster General on the European and American Systems of Telegraph, with Remedy for the Present High Rates (Boston, 1868).

32. Gardiner G. Hubbard, Postal Telegraph. An Address Delivered by the Hon. Gardiner G. Hubbard, before the Chamber of Commerce of the State of New-York, April 3, 1890 (Box 11, Hubbard Papers).

33. Just as the Post Office contracted with private railroads, reasoned Hubbard, so the Post Office could contract with his new company to transmit telegrams.

34. In proposing that the federal government provide the capital for his company, Hubbard comes across as a genuinely puzzling character—as both a grasping entrepreneur and high-minded reformer. Somehow he was perfectly comfortable combining a crusade for the public good (fighting the Western Union monopoly) with pursuing private gain (establishing a private for-profit company). It is as if one combined Ralph Nader with Lee Iaccoca. Elsewhere, Hubbard has been seen solely as a grasping entrepreneur and his reform efforts as insincere; see Lindley, *The Constitution Faces Technology*.

35. Hubbard, "Proposed Changes in the Telegraphic System," p. 82.

36. Mary O. Furner "The Republican Tradition and the New Liberalism: Social Investigation, State Building, and Social Learning in the Gilded Age," in *The State and Social Investigation in Britain and the United States*, ed. M. Lacey and M. Furner (Woodrow Wilson Center Press and Cambridge University Press, 1993). See also Louis Galambos, "Theodore N. Vail and the Role of Innovation in the Modern Bell System," *Business History Review* 66 (1992): 95–126.

37. Richard L. McCormick, "Public Life in Industrial America, 1877–1917," in *The New American History*, ed. E. Foner (Temple University Press, 1990); William R. Brock, *Investigation and Responsibility* (Cambridge University Press, 1984); Laurence R. Veysey, *The Emergence of the American University* (University of Chicago Press, 1965).

38. John G. Sproat, *The Best Men* (Oxford University Press, 1968).

39. Indeed, what is interesting and new about locating Hubbard within this ideological movement is that Furner and others tend to assume that democratic statism took shape in the 1880s; what Hubbard reveals is that this response to industrialization may have started 10 years earlier.

40. Lindley, *The Constitution Faces Technology.*

41. "Congress and the Telegraph," *Telegrapher* 10 (6 June 1875), p. 135.

42. Robert Bruce, *Bell: Alexander Graham Bell and the Conquest of Solitude* (Little, Brown, 1973), p. 93.

43. Ibid., pp. 125–127.

44. Gorman et al., "Bell, Gray, and the Speaking Telegraph," pp. 5–14.

45. Alexander Graham Bell, *The Multiple Telegraph* (Boston, 1876), Box 274, Bell Family Papers, p. 8.

46. Alexander Graham Bell, "Improvement in Telegraphy," US Patent 174,465 (filed 14 February 1876, granted 7 March 1876).

47. Alexander Graham Bell, Entry for 8 March 1876, Notebook, "Experiments made by A. Graham Bell (Vol. I)," Box 258, Bell Family Papers.

48. Bruce, *Bell*, pp. 188–214.

49. Richard John, "Theodore N. Vail and the Civic Origins of Universal Service," presented at Business History Conference, Chapel Hill, March 1999.

50. Hubbard's efforts to interest Western Union in Bell's inventions are mentioned in two letters to his wife, Gertrude, 16 Oct. and 16 Dec. 1876, Hubbard Papers. In these letters, Hubbard spoke of taking Bell's "inventions" to Western Union, meaning presumably both his harmonic telegraph and his telephone. These letters do not mention specifically that Hubbard offered Bell's patent for $100,000; this figure comes from Thomas A. Watson's recollections; see *Exploring Life: The Autobiography of Thomas A. Watson* (New York, 1926), p. 107. See also Bruce, *Bell*, p. 229.

51. Hubbard described using Bell's telephone as follows: "Conversations can be easily carried on after slight practice and with the occasional repetition of a word or sentence. On first listening to the Telephone, though the sound is perfectly audible, the articulation seems to be indistinct; but after a few trials the ear becomes accustomed to the peculiar sounds and finds little difficulty in understanding the words." (Source: "The Telephone," printed handbill, May 1877, Box 1097, AT&T Historical Collections, AT&T Archives, Warren, New Jersey).

52. The crude nature of Bell's early telephones in comparison with Edison's telegraph devices was made apparent to Mike Gorman and me in the course of studying artifacts from both inventors in the AT&T Historical Collections. In the autumn of 1876, Hubbard would have had three patents to offer Orton: Bell's US patent for harmonic telegraph and telephone (174,465); "Improvements in Transmitters and Receivers for Electric Telegraphs," US Patent 161,739 (filed 6 March 1875, granted 6 April 1875); and "Telephonic Telegraph Receivers," US Patent 178,399 (filed 8 April 1876, granted 6 June 1876).

53. Israel, *From Machine Shop to Industrial Laboratory*, p. 141.

54. Bruce, *Bell*, p. 29.

55. Hubbard to John Ponton, 21 February 1877, Ponton Collection, AT&T Archives, Warren, NJ. See also Hubbard to Bell, 22 February 1877, Hubbard Family Papers.

56. Frederick Leland Rhodes, *Beginnings of Telephony* (Harper, 1929), pp. 147–148.

57. Gertrude M. Hubbard to Mabel [Bell], 19 Oct. 1877, Hubbard Family Papers.

58. American Telephone and Telegraph Co., "The Early Corporate Development of the Telephone," printed pamphlet, [dated after 1935], Box 71, Western Union Telegraph Company Collection, pp. 8–12.

59. Clearly as telephone exchanges grew in terms of the number of subscribers in the early 1880s, it was necessary to hire telephone operators. However, in this early period (1876–1879), it was generally assumed that the user would be directly connected by a single wire to another location and that there was no operator involved.

60. For an early discussion of the domestic and social uses of the telephone, see Kate Field, *The History of Bell's Telephone* (London, 1878).

61. Robert Bruce, *1877: Year of Violence* (Bobbs-Merrill, 1959; Ivan R. Dee, 1989); Alan Trachtenberg, *The Incorporation of America: Culture and Society in the Gilded Age* (Hill & Wang, 1982).

62. "A Night with Edison," *Scribner's Monthly* 17 (1878), November, p. 88.

63. To be sure, the development of telephone exchanges in the late 1870s and the early 1880s was influenced not only by this cultural framing but also by the fact that Bell Telephone won a significant legal victory over Western Union in 1879 in the Dowd case. For a discussion of this landmark decision and its impact on the development of the telephone industry, see W. Bernard Carlson, "Entrepreneurship in the Early Development of the Telephone: How did William Orton and Gardiner Hubbard Conceptualize this New Technology?" *Business and Economic History* 23 (1994): 161–192.

64. Claude Fischer, *America Calling* (University of California Press, 1992).

65. Stuart Blumin, *The Emergence of the Middle Class* (Cambridge University Press, 1989); Olivier Zunz, *Making America Corporate, 1870–1920* (University of Chicago Press, 1990).

66. Blumin, *The Emergence of the Middle Class*, p. 9.

67. Morton Keller, *Affairs of State* (Harvard University Press, 1977); Samuel P. Hays, *The Response to Industrialism, 1885–1914* (University of Chicago Press, 1957); *The Gilded Age*, ed. H. Morgan (Syracuse University Press, 1970).

68. Lindley, *The Constitution Faces Technology*.

69. Richard John, *Spreading the News* (Harvard University Press, 1995); "Governmental Institutions as Agents of Change: Rethinking American Political Development in the Early Republic, 1787–1835," *Studies in American Political Development* 11 (1997): 347–380.

70. Thomas P. Hughes, *American Genesis* (Knopf, 1989).

Culture and Technology in the City: Opposition to Mechanized Street Transportation in Late-Nineteenth-Century America

Eric Schatzberg

Between 1888 and 1898, Americans mechanized urban street transportation, converting thousands of miles of horse-drawn street railroads to the new electric trolleys. Perhaps no other modern innovation has diffused as rapidly as the electric streetcar. Electric lighting, in contrast, faced determined competition from gas and reached only a minority of urban homes well into the twentieth century.[1] More than any other product of the early electric power industry, the trolley demonstrated the practical impact of electricity on daily life.

Yet this rapid diffusion was not accomplished without opposition. In America's largest cities, urban residents waged a brief but vigorous campaign against the most common form of the electric streetcar, the trolley powered by overhead electric wires. Opponents objected to the environmental consequences of the new streetcars, especially the aesthetics of overhead wires. This opposition encompassed broad segments of an increasingly diverse urban community, bringing together civic reformers, small shopkeepers, and residents of the affected streets.

Opposition to overhead trolley wires shows that the American public did not always acquiesce to the onslaught of technological progress. But most history of technology is written from the victors' viewpoint. The widespread opposition to overhead wires was quickly forgotten and has received little attention from historians, at least with regard to the United States.[2] Indeed, this opposition had little long-term impact; ultimately, it failed to halt the adoption of overhead trolley wires except in Manhattan and in central Washington DC.

The history of opposition to the overhead trolley has a significance beyond its direct influence on a particular transportation technology. This controversy illustrates how the methodological insights developed by historians of technology can enrich other areas of history, especially American urban history. Many of the pivotal events of American history have technological issues at their core, especially in the late nineteenth and the early twentieth century, when Americans experienced the rise of corporate capitalism, the resulting bitter labor struggles, the Populist revolt, the urban reform movements of the Progressive Era, and the emergence of American imperial entanglements.

Historians of technology inspired by the path-breaking work of Thomas P. Hughes bring a fundamental insight to these events: refusing to take technology as a given, they insist on viewing it as a malleable product of human history, no more and no less given than culture or politics. Although few historians would quibble with this view of technology, historians have been slow to change their practices, continuing the tendency to treat technology as an external factor that sets the stage for the more proper subjects of scholarship. Historians of technology, in contrast, provide concrete examples of how to treat technological change as a consequence as well as a cause of historical conflicts. When historians are sensitized to the historical contingency of technology, numerous political and social conflicts reveal themselves to involve struggles over technological choices. From this perspective, Progressivism is to be seen not as a "response to industrialism" but rather as a struggle over who would bear the costs and reap the benefits of technological change.[3] By their very nature, struggles over costs and benefits involved questions of technological choice, directly shaping the development and diffusion of specific technologies. The controversy over trolley wires was precisely this sort of struggle to control a new technology.

Historians of technology have more to offer than an exhortation to treat technology as a historical product, however. In the past 20 years, historians and sociologists of technology have elucidated the complex web of social factors shaping technological change, building on Hughes's comparative study of electric power systems.[4] More recently, historians of technology have brought culture into the causal equation, demonstrating how cultural values and symbolic meanings influence technological choice.[5] Yet historians of technology have not been content to expose technology as a social and cultural construct. To make technology an integral part of mainstream history, historians must also acknowledge technology's power to transform social relationships and culture. Much recent work has done just that, especially in the cultural history of technology, where historians have demonstrated the complex and contested ways in which new technologies gain cultural significance.[6]

Full integration of technology into mainstream history requires a synthesis of both these insights: the idea of technology as a contingent historical product and as a powerful factor shaping history. Historians can achieve such a synthesis by acknowledging that all technological change involves two distinct but simultaneous processes, the mutual production of material artifacts and cultural meanings.

My analysis of trolley wires provides one example of this type of synthesis. Most fundamentally, the struggle over trolley wires demonstrates the simultaneous shaping of technology and culture through the mediation of politics, a process that transformed both the material technology and its cultural meanings. Opposition to overhead wires shaped streetcar technology by encouraging inventors to develop underground-conduit systems for electric streetcars, systems that were adopted in Washington, New York, and a good number of European cities. At the same time, the debates over trolley wires illustrate the malleability of culture. Both proponents and opponents of overhead wires struggled to possess the symbolism of progress, each side seeking to identify its solutions as the most modern. The entire battle was fought on the terrain of urban politics, a terrain that favored the trolley companies despite strong stirrings of early Progressive Era opposition to the unchecked power of publicly sanctioned monopolies.

History of Technology and Urban History

The struggle over trolley wires provides methodological lessons for both urban historians and historians of technology. For urban historians, this story suggests a way to integrate the cultural history of cities with the history of urban infrastructure. For historians of technology, this story demonstrates how culture can shape the development of a new technology through explicitly political processes.

With the possible exception of business history, no historical field has been more receptive to history of technology in the past 20 years than American urban history. Lewis Mumford stressed the essential link between technology, culture, and urban form in the 1930s, but Mumford's urban writings suffered much the same fate as his earlier *Technics and Civilization*, provoking much comment but few scholarly imitators. In the 1960s, urban historians embraced the new social history exemplified by the work of Sam Bass Warner, perhaps the most influential postwar American author in the field of urban history. Although technology was not central to most of these works, the new social history did clarify the centrality of urban infrastructure to everyday life in the city. Warner in particular provided clear quantitative evidence of the relationship between patterns of urban settlement and the diffusion of new transportation technologies in the nineteenth century, along with a critical assessment of the resulting urban pathologies. But Warner paid little attention to the technology itself, treating it like a machina ex deus that almost magically remade the structure of the city.[7]

In the 1970s, a number of historians began to delve deeper into the significance of technology for urban history, quite self-consciously forging links between urban history and history of technology. These urban historians moved beyond "impact" studies like Warner's to adopt the contextual approach advocated by historians of technology, demonstrating how social interests and political processes have molded urban transit, sewer systems, housing patterns, and other aspects of urban life.[8] For historians of urban infrastructure, the work of Thomas Hughes has been particularly influential, especially his comparative study of electrification in Berlin, Chicago, and London.[9] By the 1980s, no historian of urban technology could safely neglect the centrality of technological systems.[10]

When it comes to integrating the history of urban technology and urban culture, however, urban historians have made little progress. In most studies of urban infrastructure, culture plays no role as an explanatory variable, even in works that feature "culture" in their titles.[11] Moreover, when urban historians do invoke culture to explain the development of urban form, culture is viewed as an autonomous agent that shapes technology and urban form from the outside. This tendency is particularly strong in works that seek to explain the distinctive structure of American cities, with their sprawling suburbs and their dependence on the automobile. This urban pattern is often explained by reference to some innate American cultural preference for individual over collective action.[12] In this type of argument, autonomous culture replaces autonomous technology as the main explanatory factor. As Gabrielle Hecht has recently argued, neither culture nor technology can function as autonomous historical agents; both are in fact mutually produced in the process of technological change.[13]

While urban historians can benefit from the history of technology, historians of technology can learn much from urban history. The city provides historians of technology with an ideal site for examining the interaction of technology, culture, and politics. Cities have long nurtured technical innovation. Aspiring young inventors in nineteenth-century America flocked to the cities, although a majority of Americans continued to live in rural areas.[14] Even when Edison removed his inventive activities from the city, he retained close ties with the metropolis, situating his Menlo Park and West Orange laboratories only a short railroad trip from Manhattan. But even more than a source for invention, cities in the late nineteenth and the early twentieth century were prime sites for the diffusion of new technologies. The technologies of the second industrial revolution—especially electrical technologies—were largely urban. Whereas the railroad and the telegraph facilitated intercourse between cities, the telephone and electric power operated primarily within cities. "It is largely due to the development of the modern city," a prominent electrical engineer commented in 1894, "that the electrical inventions of the present century have been so successful."[15]

For historians of technology, cities are ideal research sites not just for the technologies they encompass but also for what they reveal about the intersection of politics, culture, and technological change. Until recently, historians of technology paid little explicit attention to politics.[16] For historians of urban infrastructure, however, politics is almost impossible to ignore. In the rapidly expanding cities of the late nineteenth century, urban technological choices almost invariably became matters of public policy. Much Progressive Era reform had its origins in the struggles of urban residents to regulate the new technologies controlled by private utility companies, none more so than street railroads.[17] New urban infrastructures almost invariably required the use of public spaces, especially the streets, which made them subject to the legislative authority of state and municipal governments. Public debates over the implementation of these new technologies exposed them to a level of scrutiny rarely faced by other owners of capital, even when companies bypassed public opposition through bribery of elected officials. These debates made explicit the often hidden cultural values shaping technological choices, especially with regard to how the benefits and costs of new technologies should be distributed. At the same time, these cultural values themselves became contested. In political struggles over new infrastructures, both technology and culture were up for grabs; that is, both the physical technology itself and the cultural meanings attached to it could be changed in the process. The struggle over trolley wires produced just such changes, simultaneously shaping both technology and culture.

New Technologies and Urban Streets before the Trolley

Opposition to overhead wires was part of a long struggle over control of city streets, a struggle that involved urban residents, manufacturers, small and large retailers, teamsters, and transportation companies. These groups entered into a variety of coalitions that fought over everything from methods of street paving to locations of railroad stations.[18] The struggles often took the form of opposition to mechanized street transportation, an opposition that emerged with the first intercity steam railroads. In both the United States and Europe, almost all large cities banned the new steam locomotives from urban streets. Americans enthusiastically embraced the steam railroad for intercity travel but would not countenance it within the city. The three largest American cities in the 1830s, New York, Philadelphia, and Boston, all prohibited steam locomotives from the existing urban areas. New York at first permitted steam locomotives to operate north of 14th Street, but as the city grew it expanded the area in which they were prohibited, banning them south of 32nd Street in 1844 and south of 42nd Street in 1854. Because of the compactness of pre-industrial cities, the distance from the railroad terminus to the city center was generally short. In Philadelphia and New York, railroad lines continued into the center of the city but used horses instead of locomotives.[19] Urban residents sometimes objected to these lines as well, arguing that they were merely a prelude to the introduction of steam locomotives. The courts generally sided with abutting property owners, ruling that intercity rail travel was an impermissible use of city streets.[20]

In retrospect, banning steam locomotives from city streets hardly seems to require much historical explanation. Steam locomotives were noisy, emitted large volumes of steam and smoke, and posed dangers of fire and explosion. They were, in fact, about as objectionable as an urban freeway. Yet even when companies like the Baldwin Locomotive Works developed environmentally benign steam "dummies" for use on street railroads, these vehicles remained confined to suburban routes. Furthermore, nineteenth-century cities were not hygienic paradises, especially with regard to animal wastes, and conditions worsened as the century progressed. Nevertheless, both Americans and Europeans remained unwilling to trade the excreta of horses for the excreta of steam engines.[21] This refusal was a direct consequence of an urban culture that did not blindly embrace the machine.

The technology of the steam railroad did transform urban transportation after 1850, but through the use of iron rails rather than steam power. Although cities prohibited steam traction, horse-drawn vehicles could dramatically increase their efficiency by using iron rails and flanged wheels. The standard raised T-rail used by mainline railroads, however, created a significant obstruction in city streets, interfering with pedestrians, horses, and wheeled vehicles. In the 1850s, inventors developed a variety of grooved rails that lay flush with the street, making street railroads less objectionable to urban residents. By 1860, most major American cities had acquired networks of horse railroads, which quickly displaced the horse-drawn omnibuses that had provided most public transportation. The horse railroads significantly increased the supply and quality of urban passenger transportation, which helped promote the growth of the urban area.[22]

Thus, in the third quarter of the nineteenth century one finds an interesting irony. Rapid urbanization, which was largely a product of the intercity steam railroad, led to a huge increase in the use of animal power for travel within cities. Yet urban steam locomotives remained anathema, despite the widespread use of stationary steam engines within cities.

Beginning in the early 1860s, inventors developed steam streetcar engines that produced little smoke, steam, or noise. Although these "dummy" engines found a significant niche on suburban streets in both Europe and the United States, they rarely penetrated the urban core. In England, regulations issued by the Board of Trade in 1875 imposed a rigid set of requirements on steam streetcars, permitting no visible steam or smoke—quite a stringent regulation, in view of the relative lack of interest in controlling other sources of air pollution. Similar regulations were imposed in Continental Europe and in the United States.[23] Meeting these regulations raised costs so much that steamcars lost any economic advantage they may have had over horsecars. In John McKay's analysis, the steamcar proved unable to meet simultaneously the economic and the environmental-aesthetic conditions required to replace animal traction.[24] Except for the New York elevated, cities were willing to tolerate steam-powered transit only if the steam engine remained stationary, as it did in the cable-car systems adopted by most large American cities in the 1880s. From a present-day perspective, it seems that cities imposed much more stringent environmental requirements on the steam dummy than on the horse. The electric streetcar, in contrast, would manage to avoid such strict environmental regulation—but not without a struggle.

Before electric streetcars became practical, though, cities became involved in another fight to regulate electrical technologies in the streets: the fight against overhead telegraph, telephone, and electric-lighting wires. By the early 1880s, a haphazard web of overhead wires cluttered the downtown streets of every large American city. These wires not only disfigured the street but also interfered with fire departments and building trades. Overloaded poles sometimes collapsed under the weight of hundreds of wires. With the spread of high-voltage electric arc lighting in the 1870s, overhead wires became not only unsightly but deadly. Cities had almost no regulations to ensure the safety of overhead wires, and broken telegraph wires often fell across poorly insulated arc-lighting wires, sometimes with fatal consequences to people in the street below. Beginning in the late 1870s, city governments began demanding that the telegraph and telephone companies put their wires underground. The companies refused, insisting the underground wiring was technically unfeasible.[25]

Concern over the proliferation of overhead wires was widespread on both sides of the Atlantic. In 1883 the trade journal *American Architect* approvingly reprinted, under the title "A Nuisance and a Danger," an English article denouncing the spread of overhead wires. The author condemned "the telegraph engineer" as "the spider of modern civilization . . . ceaselessly weaving his metallic web above our busiest thoroughfares [and] day by day . . . adding fresh wires to those which already intersect each other at all imaginable angles, until it promises to be difficult for a Londoner to catch a glimpse of sky except through the messes of wire." The article noted efforts in New York and Continental cities to place all wires underground, and pronounced overhead wires "doomed."[26]

Opposition to overhead wires was especially strong in New York, which probably had the largest concentration of such wires in the world. By 1880, this opposition was sufficient to convince Thomas Edison to use underground wiring for his first central electric-lighting system at Pearl Street in lower Manhattan. The primitive state of electrical insulation gave Edison considerable trouble, but by 1882 he had demonstrated that electrical power could be reliably transmitted through underground cables.[27]

Throughout the 1880s, urban reformers continued to demand underground wires in urban areas, with increasing success. Chicago began requiring some companies to use underground cables by 1882, and in 1885 New York established an Electrical Subways Commission with the power to compel companies to put overhead wires in city-owned conduits. Nevertheless, technical problems and political corruption slowed the removal of overhead wires in New York and other cities. In March 1888, a snowstorm in New York brought down so many wires that "the metropolis of this country was entirely cut off from all communication with the rest of the country"[28] for several days, prompting renewed demands for underground wires. After the gruesome electrocution of a lineman the following year (caused by faulty overhead wires), New York authorities moved decisively, obtaining legal permission to chop down hundreds of unsafe poles. In 1892, a presidential commission recommended that all electrical wires in Washington be placed underground in city-owned conduits. By the early 1890s, most large American cities had plans to eliminate overhead wires from their urban centers.[29] But these plans were threatened by demands for overhead wires to power the new electric trolley.

Origins of the Electric Streetcar

The development of electric traction goes back to experiments with battery-powered vehicles in the 1830s. A major conceptual advance occurred in the 1840s, when inventors realized that they could keep the power source stationary and transmit current to the cars through wires.[30] The main difficulty then became the design of a reliable system for maintaining electrical contact between a moving streetcar and a stationary wire. Beginning in the early 1880s, inventors worked hard to solve this problem. They developed three main types of systems: overhead, third-rail, and underground. Other inventors continued to experiment with battery-powered streetcars.

Of the three systems using wires, the third-rail system was suited only to segregated rights-of-way, because the exposed ground-level conductor posed an obvious hazard to people and animals. For street railroads, the bare conductor had to be out of reach, whatever one's opinions on the danger of electricity. The earliest systems used overhead conductors connected to the car through a pole or a trailing wire. By the mid 1880s, there were a number of overhead systems, some using a single wire, some with two wires, some with sliding contacts, others with rolling contacts, some with contacts riding on top of the wire, and others with contacts pressing up on the underside of the wire.[31] None of these systems proved particularly reliable until 1888, when Frank Sprague completed what was then the world's largest electric streetcar system, in Richmond, Virginia. Sprague's system used an underrunning trolley, a small grooved wheel that rode along the underside of the overhead wire while attached to a pole on top of the streetcar. A strong spring on the trolley pole kept the wheel firmly in contact with the overhead wire. The apparent success of the Richmond installation encouraged many street railroads in small cities and suburban areas to adopt the overhead trolley. By 1890, electric street railways, almost all using overhead conductors, accounted for 16 percent of street-railway mileage in the United States[32]

Despite the promise of overhead-conductor systems, many inventors continued to develop streetcars that drew power from underground conduits or batteries. These inventors were motivated in part by technical considerations but mainly by the widespread opposition to overhead wires, which convinced many of them that cities would never permit overhead conductors for streetcars. In 1882, an editorial in *American Architect* noted the problem posed by overhead conductors while commenting on the trial of a prototype electric streetcar in Pittsburgh: "The main objection to this, as to all street railways operated by electricity, would naturally be the obstruction caused by the line of posts which carry the main conducting wire; and in most cases such an objection would prevent even the consideration of the plan."[33] Even Sprague initially believed that few large cities in the United States would permit the overhead system, and that this system was simply out of the question in cities where opposition to overhead wires had already developed.[34]

Because of the intense public objections to overhead wires, underground-conduit systems remained popular among inventors, accounting for a large percentage of electric streetcar patents well into the 1890s.[35] The first commercially operated electric street railroad in the United States, installed in 1884 by the Bently-Knight Railway Company in Cleveland, used an underground conduit.[36] In a typical conduit system, a small device called a plow hung from the bottom of the car into the conduit through a narrow slot in the street. The conduit contained two bare wires mounted on insulators. The plow had contacts that slid along the wires, thus completing the circuit from the streetcar motor to the dynamo. As early as 1888, there was considerable technical evidence that underground-conduit systems would be feasible, although they would be considerably more expensive than overhead systems because of the need to excavate the street for the conduit. In 1890, the city of Budapest opened the first successful conduit system in Europe, using a system developed by Siemens and Halske. By 1892 the Budapest system had proved itself profitable, and American inventors soon developed similar systems. In the early 1890s, opponents of the trolley frequently pointed to the Budapest system as a viable alternative to overhead wires.[37]

Battery systems provided another alternative to the overhead trolley. In these systems, each car contained its own batteries, which were typically recharged at a central power station. The battery system gave each car independent operation, making system-wide failures unlikely. Battery systems could often use existing horsecar tracks without the need for an expensive network of trolley wires or conduits. Alexander Julien, a Belgian engineer, installed a battery-powered streetcar line in Brussels in the mid 1880s. In 1886 he visited the United States to promote his system, and he convinced a number of street-railway companies to equip routes with battery-powered cars.[38]

Despite these attempts to develop alternatives to the trolley, by about 1890 American streetcar companies had come to view battery and conduit systems with disfavor. In their view, the trolley had definite technical advantages over the alternatives. In battery systems, each car typically carried over 3000 pounds of batteries, reducing acceleration and ability to climb hills. The batteries themselves proved expensive and short-lived, requiring replacement after 2 years or less.[39] Conduit systems suffered from a different set of costly problems. Conduit systems cost much more to build than overhead systems because they required major excavation of the street. Overhead wires also provided much easier access for maintenance in case of broken wires. The conduit systems had to have excellent drainage to keep water off the bare wires, and the conduits also tended to accumulate dirt and mud, which interfered with the operation of the plow.[40] But the principal objection to the conduit system was clearly its higher first cost.

Debating the Trolley

Beginning in 1888, overhead trolleys spread rapidly in smaller American cities. But except for Boston, large cities were reluctant to become early adopters of the new technology. By 1890, the electric trolley appeared to have demonstrated substantial savings over horse traction in terms of operating costs, at least for routes in smaller cities and suburbs.[41] Street-railroad companies in large cities salivated over the potential profits to be gained from electrification, which included new opportunities for stock watering and insider construction contracts. Urban residents also demanded improved transportation, as continued urban growth overburdened the existing horse railroads.

Under such conditions, one would think that electrification would serve both public need and private greed. Street railroads, however, were natural monopolies, and market mechanisms could not ensure the congruence of need and greed. Because the street railroads operated on public property with governmental consent, their technical choices were mediated by politics as well as by the market. When the companies entered the political arena to obtain approval for electrification, they found that a substantial portion of the public did not feel that the potential benefits of the new technology outweighed its objectionable aspects.[42]

Whether street railroads needed governmental permission to convert to electricity depended on the language of their charters. In most large cities, companies required the approval of the municipal government before they could install the trolley system, although ultimate authority lay with the state legislature.[43] In nearly every large city in the country, the proposal to install the trolley system sparked vigorous debate, including angry editorials in reform-minded newspapers, public hearings, and petition drives. In the early 1890s, New York, Chicago, Philadelphia, Brooklyn, and Washington all experienced major campaigns against the introduction of the trolley.

The opposition to overhead wires arose principally from residents of the urban core, who were motivated by aesthetics, concerns over safety and noise, and general distrust of the petty robber barons of the street railways. Aesthetic arguments provided the strongest objections to the overhead trolley. Opponents rarely articulated their objections in clear terms, in part because almost everyone agreed that the wires were unattractive. Shopkeepers, homeowners, and US senators insisted that the poles and wires were unsightly. Even engineers accepted the aesthetic objection to overhead wires. In Richmond, for example, the city engineer had nothing but praise for Sprague's system, "with the exception of the unsightly appearance of poles and wires." The electrical engineer M. B. Leonard, also from Richmond, denounced Sprague's overhead trolley as a "grievous eyesore," although Leonard approved of Sprague's technical accomplishments in general. In Richmond and other small cities, the street railways commonly installed inexpensive wooden poles that began to fail under the weight of the wires. Even the staunchly pro-industry *Street Railway Journal* recognized this problem, noting that "a glance along a street where wooden poles have been in use any length of time often reveals them pointing in all manner of angles, like a file of tipsy soldiers."[44] This opposition appears to have been based on the same aesthetic values as opposition to earlier overhead wires: dislike of the visual clutter introduced by poles and wires. Because these values were widely shared, they were rarely made explicit. In the early 1890s, dislike of visual clutter in cities was growing with early stirrings of the City Beautiful movement, which was in part inspired by Baron Hausmann's transformation of Paris. Overhead wires had no place in the grand avenues of Paris or their American counterparts.[45]

Manufacturers of trolley systems took a number of steps to deflect aesthetic objections. Sprague and other manufacturers developed a number of designs for ornamental poles made of cast iron rather than wood. In narrow streets, trolley companies often attached cross-wires directly to the sides of buildings, avoiding the use of poles altogether.[46] Yet photographic evidence does show that trolley wires, even when well designed, did contribute to visual clutter. New York's Third Avenue line, for example, used an underground-conduit system in Manhattan below 125th Street but switched to an overhead trolley in the Bronx. Photographs of the streetcars below 125th Street reveal broad boulevards unobstructed by poles and wires; photographs of the Bronx clearly show a cluttered maze of overhead trolley wires.[47]

The second major objection concerned safety, both with regard to the overhead electric wire and the mechanized streetcar itself. The urban public naturally feared the dangers of a bare trolley wire carrying 500 volts above their heads. Electricity was still relatively unfamiliar, and its dangers not well known. In addition, the first execution using the electric chair had occurred in August 1890; it was the culmination of a campaign by associates of Thomas Edison to discredit the higher voltages used by the Westinghouse lighting system.[48] Newspapers in Chicago, New York, and Philadelphia denounced the dangers of overhead wires.[49] The US Senate debated the safety of the system for Washington, and one senator described the wires as "demoniac" in the "constant menace that they present to life and limb."[50]

Electrical "experts" often belittled the public's fear of electricity, labeling it irrational and superstitious. Many historians have uncritically echoed this viewpoint.[51] Yet the public's fear concerning the safety of the overhead wire was quite legitimate. A number of horses were killed by downed wires from early trolley systems. Telegraph and telephone wires would often fall over the trolley wires and then become charged with 500 volts. Fallen wires frequently got hot enough to cause fires. The overhead wires also interfered with the ability of firemen to fight fires in tall buildings. Manufacturers endeavored to improve the safety of their systems by adding guard wires above the trolley wire to protect it from falling telegraph or telephone wires, implicitly acknowledging the danger.[52]

In the 1890s, very little was known about the physiological effects of electricity. Despite this ignorance, prominent electrical experts did not hesitate to pronounce the 500-volt trolley wires safe. Through the early 1890s, they insisted that there was no documented case of a fatality caused by a trolley wire. Many leading engineers claimed to have taken the full 500 volts numerous times with no ill effects, although none would demonstrate the experience in public. According to modern research, direct current is, as Edison claimed, significantly safer than 60-Hz alternating current, requiring approximately three times more current to induce fatality. But a bare 500-volt direct-current conductor is still very dangerous and can certainly deliver a fatal current.[53]

Another major concern of trolley opponents was the safety of pedestrians. Even the *Street Railway Journal* condemned the epidemic of injuries to people (both pedestrians and passengers) who fell under the wheels of streetcars. These fears applied to all types of streetcars, but the greater speed of the trolley cars increased the danger. The *Street Railway Journal* urged the companies to develop safety devices to reduce the carnage, such as improved fenders, but the street railroads resisted such measures for decades.[54]

A third objection involved the noise of the trolley cars. Although modern electric streetcars are fairly quiet, the early trolley cars were quite noisy. In 1891, Elihu Thomson of Thomson-Houston, one of the largest suppliers of trolley systems, complained that the noise of the Thomson-Houston trolley cars was "simply appalling." Early trolleys emitted a loud, high-pitched whine from the double-reduction gearing used almost universally on trolley cars before 1892. Not until these gears were sealed in oil-filled cases and replaced with single-reduction gearing was the noise pollution significantly reduced.[55]

Another strong source of anti-trolley sentiment was the hostility of reform-minded groups to the rapacity and corruption of the street-railway companies. These urban reformers often did not object to the trolley per se, but merely to the granting of a highly remunerative use of municipal property (that is, the city streets), while the city obtained nothing in return. Electrification promised major cost savings to the owners of street railroads, but the companies refused to pass these savings on in the form of lower fares or increased payments to the city. When public anger threatened to block electrification, the street railroads frequently resorted to bribery of elected officials. In effect, many opponents of the trolley were fighting to control how mechanization's benefits were to be distributed, as well as to minimize its negative effects.[56]

The anger of these groups was augmented by the very real arrogance of the street-railroad companies, which remained unmoved by public protest. Spokesmen for the companies insisted that street railroads had the right to define what technology was best. For example, an editorial in the *Street Railway Journal* denounced opposition to overhead wires as unreasonable, claiming that the public demanded perfection but was unwilling to pay for it. In fact, the journal insisted, the street-railroad companies shared the interests of the community; therefore, "the public should let them alone, and leave them free to follow their own best judgment." The journal's editors denied that the public had any role in the choice of transportation technology: "So we say hands off; let a company put in any system they like, provided they are willing to bear the expense."[57]

Spokesmen for the street-railroad industry insisted that cities would gain moral benefits from electrification, which supposedly would give the working classes access to open spaces in the suburbs.[58] But in fact this reformist rhetoric concealed a strategy to expand investment opportunities in the suburbs while refusing to pass on lower costs to residents of the urban center. In 1890 the electrical engineer T. C. Martin made this strategy explicit. According to Martin, the trolley created a great investment opportunity by reducing costs as much as 50 percent. But this saving was not to be passed on in the form of lower fares: "The public is intelligent enough to know that other things are more necessary." Instead of lowering fares, Martin argued for building "hundreds of new roads" into suburban districts. Such a strategy, Martin argued, would blunt radicalism among the working class while aiding capitalists by forestalling the tendency of the rate of profit to fall by opening up new land for profitable investments.[59] In fact, this strategy disproportionately benefited the middle-class families who could afford to move to the suburbs and who paid the same 5-cent fare for their long commute as did urban workers for short trips within the city.[60]

Supporters of the streetcar companies were not just cynically manipulating reformist rhetoric to hide their rapacity; they embraced electrification in part because it symbolized technological progress. David Nye has ably documented the powerful enthusiasm and utopian expectations that accompanied electrification.[61] But this fascination was perhaps greatest among those who expected to profit from its use. One would think that street-railway men of the 1880s, comfortable with the world of horses, iron rails, and cobblestone paving, would have approached electricity with considerable skepticism, demanding proof of its economy and reliability before risking an investment in the new technology. Some such skepticism existed, especially before 1888, but it was overshadowed by palpable excitement over the potential of electric traction and by widespread belief in the certainty of its application to street railroads. In 1883, the president of the American Street Railway Association (ASRA) insisted that the steam engine and the telephone had "ceased to be classed as wonders," and that "the last and greatest discovery of the century" would be "electricity as a motive power."[62] The following year, Calvin A. Richards, manager of the Metropolitan Railroad of Boston, likened electric traction to an infant, "which the Creator desires to confer as a new blessing on the world."[63] In the next decade, the street-railway press continued to insist that electric traction was more progressive than steam: "Electricity is constantly knocking at the door of the domain so long ruled with undisputed sway by the steam engine, and while the locomotive still snorts in derision, electricity is hopeful and progressive."[64] Such faith gave street-railway promoters the moral certainty they needed to defy public opposition to the trolley.

This identification of electrification with progress did much to structure the debate over the trolley's introduction. Proponents of the trolley drew on the symbolism of progress to cast the struggle as one between tradition and modernity, castigating opponents as technically ignorant, narrow-minded traditionalists bent on protecting narrow interests. As early as 1887, when no electric streetcar system had yet demonstrated reliable operation, proponents of electric traction denounced a "spirit of conservatism" for slowing its adoption.[65] Other supporters of the trolley ridiculed the "many and odd" objections to the electric streetcar as based on "ignorance or prejudice."[66]

But opponents of the trolley in Philadelphia, New York, and elsewhere did not simply cede the symbols of progress to the street railways. Instead, these opponents directly attacked the association of overhead wires with progress. They insisted that other paths to mechanized urban transit, such as underground-conduit systems, subways, or battery cars, were more "progressive" than the trolley. Nowhere was this struggle over the meanings of progress more evident than in Philadelphia.

Philadelphia versus the Trolley

All the elements in the trolley debate were present in Philadelphia in the spring of 1892. The city had one of the largest streetcar systems in the United States, most of it controlled by the politically powerful Philadelphia Traction Company headed by the Peter Widener and William Elkins.[67] Under the prodding of the PTC, the city councils began holding hearings in March 1892 on legislation to permit installation of the trolley system.

Opposition to overhead wires in Philadelphia was well established by 1892. The city had banned overhead electric wires in the early 1880s, but the ordinance contained many loopholes and the wires proliferated anyway. Philadelphians also had little love for the city's street railroads, which fixed prices, distributed stock to politicians, and fought bitterly in the courts against the limited obligations required of them, such as street paving.[68]

When the city councils began active consideration of the trolley ordinances, a number of groups immediately mobilized and formed a "Union Committee" to coordinate opposition. The Union Committee brought together of a wide variety of groups, including urban reform organizations like the Municipal League, small-business trade associations such as the Master Builders, the Lumbermen, the Druggists, and the Grocers, and the Board of Trade, predecessor to the Chamber of Commerce. Another founding member was the Wheelmen's Street Improvement Association, an organization of bicycle enthusiasts.[69] The Union Committee thus united patrician reformers and organizations of the old middle class, whose members both lived and worked within the urban core. From this base, the Union Committee collected petitions and organized meetings against the trolley ordinances, and also presented testimony before the city councils. The committee collected signatures from more than 7000 people living along proposed trolley routes, including almost all the property holders on these streets.

Over several days of hearings, which began on March 14, the city councils heard arguments for and against the trolley. The well-organized opponents of the trolley marshalled all available arguments against the new technology, starting with the trolley's objectionable environmental consequences. A group of the city's artists presented a petition "against the disfiguration of the streets." The spokesman for the Union Committee noted that the trolley would undermine the growing movement for urban beautification in Philadelphia. The Union Committee also obtained letters from prominent citizens in other cities where the trolley was already in operation about the dangers of the system and its negative effect on property values. A lawyer from Pittsburgh declared the noise there "unendurable," describing it as "a hissing sound that is especially racking on the nerves." Two more correspondents from Pittsburgh had similar objections, reporting the noise loud enough to prevent conversation while the streetcar passed.[70]

The Union Committee also presented evidence of the dangers of the trolley. It introduced a newspaper report on a recent fall of some trolley wires in Boston that had shut down the trolleys, endangered pedestrians, and caused a small fire. Correspondents from Pittsburgh reported that the trolley wires there often broke and fell to the ground. The chief engineer of Philadelphia's fire bureau, after examining the trolley system in Pittsburgh, declared the trolley wires a menace to firemen. The Philadelphia Fire Underwriters' Association also adopted a resolution against the overhead trolley on grounds of fire safety. Finally, the opponents of the trolley noted the increase in accidents that accompanied its introduction. For example, the Massachusetts Board of Railroad Commissioners reported that electric streetcars in Boston had twice as many accidents per vehicle-mile than horsecars.[71]

Perhaps the greatest opposition, however, derived from the fact that the trolley ordinances placed almost no conditions on the street-railway companies. In the view of the Union Committee, the city of Philadelphia was granting the companies a highly profitable use of public property while demanding no compensation, despite the objectionable aspects of the new technology. Although electrification was supposed to reduce costs up to one half, there were no provisions for reducing fares or compensating the city for this valuable use of the streets. The ordinances did not even require the companies to pave the streets, an obligation that the city insisted was already required under existing franchises. The only restriction on the companies was a vague clause giving the Department of Public Safety a role in ensuring the safety of the system. The city did not even retain a right to order the poles to be removed in case the city later decided on an alternative system.[72]

The opponents of the trolley did not limit themselves to decrying the innovation, however. They also endorsed alternative systems of electric traction, in particular battery systems and the underground conduits. The opponents admitted that the alternatives were more expensive than the trolley, but insisted that the absence of overhead wires was worth the higher cost. They frequently mentioned the conduit system in Budapest, which had been operating profitably for several years. William D. Marks, the head of the Edison Electric Light Company of Philadelphia, provided expert testimony in favor of an underground-conduit system, citing the success of the Budapest system. Other opponents argued for battery systems after visiting an apparently successful line in operation in Washington.[73]

The Traction Company, represented by its New York lawyers, mobilized technical arguments in response to the trolley's opponents. The company presented an impressive list of expert witnesses and testimonials to the benefits of the trolley. These witnesses included George Westinghouse, whose firm had recently entered the electric-streetcar business, and Oliver T. Crosby, a prominent electrical engineer and the head of the streetcar department at Thomson-Houston, one of the largest manufacturers of electric streetcars. Crosby echoed the industry line on safety, denying that the overhead wires posed any threat to human life. He also claimed, quite implausibly, that the trolley car made less noise than the horse-drawn streetcar, and insisted that the trolley posed no fire danger. Westinghouse seconded Crosby on the safety of the trolley. Crosby and Westinghouse also sought to discredit the alternatives, arguing that batteries remained costly and unreliable and that the success of the Budapest conduit systems was anomalous.[74]

But the defenders of the trolley did not limit themselves to technical arguments; they also attacked their opponents as enemies of progress, "ignorant of the features of the [trolley] system . . . and opposed to any change."[75] In his testimony, George Westinghouse expressed amazement that the public was opposing the trolley, rather than seeking to compel the companies to introduce it.[76] As part of this progressive image, the trolley's supporters cast themselves as allies of the working man, a strategy that suggests that the Union Committee had significant working-class support. For example, the trolley magnate Peter Widener, in a Saint Patrick's Day dinner speech, attacked "obstructionists" who wanted to "block progress." Widener's obstructionists included the well-heeled residents "between Chestnut and

Pine streets, [who] think themselves better than the rest of the world," as well as small shopkeepers "who have gotten a little wealth and protest against everything." Widener denounced the municipal reform groups, which "seem to resolve against everything."[77] In a hearing at the mayor's office, Rufus E. Shapley, a lawyer for Union Traction, echoed Widener's remarks, branding trolley proponents elitists who would deny improved transit to workers. "The streets are public highways, and the men who live in the back alleys have as much right on them as the aristocrats," Shapley proclaimed. Of course, Shapley was not offering to reduce fares, which remained too high for daily commuting by the average worker. In the same hearing, A. K. McClure, editor of the Philadelphia *Times*, dismissed opposition to the trolley as typical of "objections usually made against innovations"; similar objection, he claimed, had been raised earlier against gas lighting and horsecars.[78]

Opponents of the trolley did not acquiesce to its symbolic association with progress; rather, they sought to undermine its progressive image. Leading opponents like Frederick Fraley, president of the Board of Trade, insisted that they were not opposed to electric traction itself, but rather to the specific type of system proposed. Opponents argued that the trolley was an imperfect, transitional system that would soon be replaced by alternatives already in development. In their vision of technological progress, batteries or the conduit system represented the future, not the trolley. The reform-minded *Philadelphia Inquirer*, a strong opponent of the trolley, made this argument repeatedly in its editorials. The trolley system, claimed the *Inquirer*, was merely cheap, not progressive. It was not even a recent invention but rather "an already out-of-date system that will very shortly be swept out of existence in the progressive cities of the land."[79] To install the trolley system would be "a step backwards—a step into the dark ages of old Philadelphia and not a progressive movement."[80] The *Inquirer's* editors even argued that the streets could never provide real rapid transit, implying that elevated or underground trains were preferable.

For the trolley's opponents, its initial success in smaller cities provided another set of symbolic associations that contradicted its progressive image. Opponents viewed the trolley as a provincial technology unworthy of use in a great metropolis. The *Inquirer* referred to the trolley as a "country system" suited only to "country towns."[81] When a supporter of the trolley claimed that 300 cities had already adopted the system, George Mercer, a representative of the Union Committee, responded that the United States did not have 300 places "which deserve the name of cities." The trolley was suitable only for small towns with wide streets, said Mercer, not for a great metropolis like New York or a capital city like Washington. "The trolley system is provincial," concluded Mercer, and "neither the city of Washington nor any one of the beautiful cities of Europe would ever think of adopting such a system."[82]

Indeed, neither New York nor Washington ever accepted the trolley in its most developed urban areas. But Philadelphians did not succeed in stopping the trolley. Although public sentiment seemed overwhelmingly against the trolley, the city councils voted in favor of the ordinances by a wide margin. The *Philadelphia Inquirer* then led an editorial campaign to get Mayor Edwin S. Stuart to veto the measures, urging its readers to sign a petition against the trolley, which the newspaper printed on its front page. Other city newspapers joined the *Inquirer* in urging the veto, and thousands of outraged citizens attended meetings against the trolley ordinances. The trolley interests responded to the opposition by offering to pave and light 40 miles of city streets if the mayor signed the bills, but this offer did not placate opponents.[83]

Mayor Stuart, a reformer, was sympathetic to the trolley's opponents but had kept silent throughout the controversy. On the last day of March, Stuart vetoed the trolley bills. In his veto message, the mayor insisted on his support for improved transit but claimed that he had not been convinced that the trolley offered the best alternative. Stuart also condemned the lack of municipal control provided by the trolley ordinances, suggesting that he would have vetoed the legislation even if he had supported the trolley system. The city councils then passed the bills over the Mayor's veto with almost no debate. The *Inquirer* hinted at bribery of the council members by the street-railway companies. Although direct evidence of corruption was not forthcoming in this case, bribery was a standard practice in obtaining municipal franchises in the Gilded Age. Furthermore, bribery would explain the lack of debate by the city councils and the reluctance of council members to explain their votes to a hostile public. The disgusted editors of the *Inquirer* condemned the "trolley infamy" and called on the voters to defeat the "traitors" in the city councils who had supported the trolley.[84]

In the end, the trolley's opponents did receive some slight compensation. The Traction Company stood by its agreement to pave 40 miles of street adjacent to its tracks, and other companies did the same when they electrified. This street paving transformed Philadelphia from one of the worst-paved to one of the best-paved American cities.[85] Philadelphians also benefited from improved service and from the extension of lines into outlying areas. However, fares remained at 5 cents (owing to the lack of competition), and this permitted the companies to retain much of the money that the new technology had saved them.

Successful Alternatives in New York, Washington, and Europe

Was the trolley really necessary for electrified transit in large American cities? Proponents of the trolley, including most electrical engineers, insisted that the overhead-wire system was the only practical solution. Yet two cities in the United States, and many more abroad, succeeded in electrifying transit lines without using overhead wires in central urban districts. These cities used the underground-conduit system, typically in combination with the trolley for suburban routes.

The two American cities that partially banned overhead trolley wires were New York and Washington.[86] In both cities, special conditions helped the opponents to victory. In the early 1890s, when New York began considering electric streetcars, the city had finally achieved some success in placing wires underground after 10 years of intense struggle against telegraph, telephone, lighting and alarm companies. Because of the intense opposition to overhead wires in Manhattan, New York's first trolley system was installed in the Bronx (then called the "Annexed District") in 1892, but even there it generated considerable opposition. A few lines were installed in upper Manhattan despite protests, but after 1894 the trolley system was banned in all of Manhattan. Thus, in the early 1890s, when most cities were installing electric streetcars, Manhattan street-railroad companies continued to operate and even expand cable railroads.[87]

A similar situation unfolded in Washington, but with an important difference. In the 1890s, the District of Columbia was governed by the US Congress. This situation, paradoxically, made Congress more receptive to popular protest, since congressmen did not depend on the city's local businessmen and political bosses for electoral support. In addition, many congressmen took seriously Washington's role as a capital city, and overhead wires had no place in their aesthetic vision. Congress has requested information regarding electric traction for the District of Columbia as early as 1888, and it received a detailed report from Captain Eugene Griffin, an electrical engineer with the Army Corps of Engineers. Griffin provided a ringing endorsement of electric traction in general, claiming that "horsecars belong to a past era." Yet even Griffin recommended against overhead wires in the urban part of the District of Columbia.[88]

Since the District's street-railroad charters were acts of Congress, changes in motive power became occasions for congressional debate. The first debates occurred in 1888 when the District's Commissioners approved a new streetcar line to be powered by an overhead trolley, defying the spirit if not the letter of a congressional resolution directing the commissioners to prohibit further overhead wires. This action prompted a lengthy debate in the Senate that mirrored later debates in larger cities over the safety and aesthetics of the overhead-wire system. The *Washington Star* played the role of the *Philadelphia Inquirer*, giving voice to the residents of Washington who opposed the trolley. The issue came up again in 1890, and after more debate the Congress voted to prohibit overhead trolley wires within the city limits of Washington (which at that time included the area of the District of Columbia south of Boundary Avenue, the present-day Florida Avenue). Congress was willing to permit electric streetcars, but only if these used underground wires.[89]

It became increasingly clear in the early 1890s that neither New York nor Washington would relent on overhead wires in order to reap the benefits of electrification. In both cities, cable railways continued to operate on major streets, despite the high fixed cost of these systems and frequent service interruptions due to problems with the cables. Yet even though underground conduits promised cheaper operation and lower capital costs than cable systems, no major American electrical manufacturer tried to develop a conduit system. No doubt the manufacturers had little desire to offer cities an alternative that might have slowed the profitable business of electrifying street railroads. Not until 1894 did the Metropolitan Street Railroad in New York begin working closely with the General Electric Company to develop an underground-conduit system. The Metropolitan Street Railroad acquired the services of Fred Stark Pearson, an accomplished electrical engineer who had supervised the construction of Boston's first large-scale trolley system. Pearson traveled to Budapest, where he examined the Siemens-Halske conduit system. He then returned to the United States, where he developed a somewhat more robust system suited to New York's streets. General Electric installed the system on Lenox Avenue in 1895. In Washington, meanwhile, a number of companies had experimented with various systems developed by underfinanced inventors, but none had proved entirely successful. With prodding from Congress, Washington's Metropolitan Railroad began installing the General Electric conduit system shortly before the New York line went into operation. Both lines began running in 1895, and both proved successful. By the end of the century, most of the horsecar and cable-car operations in Washington and in Manhattan had been converted to the underground-conduit system.[90]

The underground-conduit system proved to be a viable alternative to overhead systems. Although underground conduits cost approximately 4 times as much to install as overhead wires, neither Washington nor New York appeared to have suffered terribly as a result. Street railroads in both cities operated convertible streetcars that switched from underground to overhead systems when passing into suburban districts. The historian Charles Cheape claims that the ban on the trolley in New York restricted transit, especially on the less-traveled crosstown routes, because the street railroads found these too costly to electrify with the conduit system. Instead, the companies used horses long after they had been abandoned in other large cities, and later they used unreliable storage-battery cars.[91] Yet there is little evidence that New Yorkers or Washingtonians clamored for the overhead system so that they could enjoy the supposedly ample transit available in other cities; both cities stuck with conduits until they replaced their streetcars with buses after World War II, along with most other American cities.

Conduit systems were also favored by many European cities, whose residents also opposed the overhead conductors, as John P. McKay has so ably documented in his multinational study of European mass transportation. McKay's analysis of Europe in many ways mirrors the approach I have used for the United States. McKay clearly demonstrates the cultural shaping of streetcar technology, both in the adoption of the conduit system by many

European cities and in the attempts to minimize the aesthetic impact of poles and wires. Yet McKay errs when he suggests that greater European opposition to overhead wires explains both the wider adoption of conduit systems there and the slower pace of electrification in general.[92] The arguments used against the trolley in Europe were almost identical to those used in large American cities.[93] In most European states, alternatives to the overhead trolley were seriously considered only in the capital and a few other large cities.[94] The same situation existed in the United States, where the trolley was banned in both Washington, the capital, and New York, the largest city. Berlin prohibited the trolley only in a few historic areas, proving more amenable to it than either New York or Washington. Although European cities used the conduit system more often than American cities, conduits were generally required only for short distances in the most crowded or historic parts of the city.[95]

The greater use of the conduit system in Europe may simply reflect the fact that in the late 1890s, when most European street railroads electrified, that system had already demonstrated its practicality. Also important was the fact that Europeans cities in the 1890s were much less compliant to the wishes of private corporations than American cities. But McKay's explanation has another problem. In effect, he essentializes the national differences between European and American aesthetic perceptions of the trolley. Differences did exist, but they were not fixed. Rather, they emerged through a struggle in the public sphere to shape the cultural meanings of the trolley. Participants in this struggle drew on shared set of symbolic resources, as demonstrated by the common rhetoric on both sides of the Atlantic. Europeans may well have had stronger aesthetic objections to the overhead-wire system, but this cultural difference was as much a contingent result of technological change as were specific design differences between European and American streetcars.[96]

Conclusion

The successful banning of the trolley in Washington, New York, and some European cities drew little attention from urban Americans, whose cities had almost all adopted overhead conductors by the early 1890s. In most cities, the initial opposition was quickly forgotten after the grand civic celebrations that marked the opening of new trolley lines. Ridership on the new streetcars increased by the millions, and urban residents grew accustomed to the unsightly wires. Massive investments in new electric lines increased track mileage roughly five-fold from 1890 to 1903.[97] These new lines gave middle-class urbanites easy access to bedroom communities that emerged beyond the edge of the older urban core, and they did much to create the class-segregated neighborhoods that characterized American cities well before the arrival of the automobile.[98] Urban residents still had many grievances against the trolley companies, but the complaints were mainly about crowded cars, infrequent service, and high fares, not the aesthetics of overhead wires.[99] Even after New York and Washington proved that the underground-conduit system was practical, other cities did not demand similar installations. New York and Washington seemed anomalous, New York because of its unique urban density and Washington because of its status as the capital and the willingness of its congressional overseers to impose their aesthetic judgments on the city. From this perspective, opposition to the overhead-wire system seems merely an epiphenomenon in the history of technology, a curiosity of interest only to antiquarians.

But such an interpretation would be seriously misguided. In opposing the trolley, urban residents were not blindly fighting technological change, but rather demanding that such change be guided by values other than the maximization of corporate profits. The fact that opponents succeeded in excluding the trolley from the United States' capital and its most populous city was in itself a significant achievement. Even where opponents failed to restrict the trolley, their efforts forced street railroads to spend more money on building aesthetically pleasing and safe overhead systems. This opposition also shaped research and invention in streetcar technology, encouraging the development of underground-conduit systems and quieter gearing. Finally, opponents were often able to extract significant concessions from the streetcar companies (for example, street paving in Philadelphia). Although these changes were not dramatic, they clearly illustrate the shaping of technology by cultural values.

Still, the question remains as to why the public objected so strongly to a technology that we now regard as environmentally desirable, especially relative to the automobile. In the late nineteenth century, most American cities had inadequate sewage systems, offered spotty refuse disposal and street cleaning, permitted appalling air pollution from coal-burning factories, and left many streets practically unpaved.[100] Why urban residents should have objected so strongly to a technology that promised to improve urban transportation while eliminating tons of horse droppings remains puzzling, at least from a present-day perspective.

In large part, the answer lies in the culture of the street. For the mass of urban residents, mechanized street transportation threatened to destroy the traditional function of the streets as spaces for social interaction. In most nineteenth-century cities, the houses fronted directly onto the streets. Back yards were often rendered useless by refuse and by leaking privy vaults. Whenever they could, urban residents used the streets of their neighborhoods for socializing and recreation. The streets were especially important as play areas for children. Horse-drawn vehicles interfered somewhat with the social function of the street, but their low speed kept the interference to tolerable levels. Mechanized street transportation, in contrast, threatened these social functions. Urban residents spent a great deal of time in the streets, and thus they had good reason to fear the noise, danger, and aesthetic disruptions of electric traction.[101]

The triumph of the trolley represented a transformation of urban culture as well as of urban technology—a change in the meaning that city residents associated with the street. The ascendancy of the trolley was a victory for the instrumental approach to urban streets: streets came to be viewed primarily as transportation arteries rather than centers of social life. After electrification, reformers demanded that parents keep their children off the streets to shield them from the lethal dangers of the new technology. Instead of making the streets safer, reformers now sought to segregate children in playgrounds, where they would be protected from dismemberment by streetcars.[102] The triumph of this instrumental view of urban streets transformed urban culture, literally paving the way for urban America's capitulation to the automobile. As an agent of cultural as well as technological change, the electric streetcar in a sense represents the slippery slope that led to urban freeways.[103]

The struggle against overhead trolley wires thus illustrates in microcosm the joint construction of artifact and culture, of the physical technology of urban transportation and the symbolic meanings that gave it significance. Even after the triumph of the trolley, both the technology and its meanings remained contested within the public sphere of urban politics. The progressive aura of the trolley quickly faded in the early twentieth century as public opinion turned even more sharply against the arrogant management of the trolley companies. Between the world wars, the trolley faced both a cultural and a technological threat from the automobile and the motor bus. On the cultural plane, the trolley's supporters sought, with little success, to counteract the perception of the motor bus and the automobile as "modern" alternatives to the trolley. On a technological plane, trolley manufacturers countered the motor bus with improved trolleys—most important, the PCC car, which entered service in 1936. Ultimately, the trolley's supporters were fighting a losing battle, and after World War II most major American cities abandoned their trolley systems in favor of motor buses.[104] Yet as the example of trolley wires implies, the abandonment of the trolley cannot be understood primarily as the triumph of a superior technology, or as an expression of some uniquely American set of cultural values, or as an exercise of political power by particular social groups. Both the trolley's triumph and its demise must be understood as contingent outcomes of political and social struggles that simultaneously shaped both physical artifacts and cultural meanings.

Acknowledgments

My thanks to the editors of this volume for their helpful comments, and also to Jennifer Bannister for letting me use her excellent undergraduate research paper on opposition to overhead wires in New York and Washington. Portions of this research were conducted while I was a postdoctoral fellow at the Center for the History of Electrical Engineering. This material is based in part upon work supported by the National Science Foundation under grants 9311124 and 9601371.

Notes

1. David E. Nye, *Electrifying America* (MIT Press, 1990), p. 239; *Historical Statistics of the United States* (US Government Printing Office, 1960), p. 510. On the competition between gas and electric lighting, see Wolfgang Schivelbusch, *Disenchanted Night* (University of California Press, 1988), pp. 48–50.

2. In a neglected classic in the history of urban technology, *Tramways and Trolleys* (Princeton University Press, 1976), John P. McKay provides an excellent analysis of this opposition in Europe, although he slights its significance in the United States. It was McKay's work that drew my attention to the similar circumstances in the United States. Charles W. Cheape mentions this opposition on pp. 62, 119–120, 172–173 of *Moving the Masses* (Harvard University Press, 1980), but his Chandlerian framework prevents him from recognizing its significance.

3. For a balanced overview of Progressive historiography, see Arthur S. Link and Richard L. McCormick, *Progressivism* (Harlan Davidson, 1983). The quote is a reference to Samuel P. Hays, *The Response to Industrialism* (University of Chicago Press, 1957). Applying insights from the history of technology could cast new light on the history of Progressivism, especially by undermining the technological determinism implicit in both the organizational-synthesis thesis and the corporate-liberalism thesis. For the classic statement of the organizational synthesis, see Louis Galambos, "The Emerging Organizational Synthesis in Modern American History," *Business History Review* 44 (1970): 279–290.

4. For a collection of essays marking the arrival of this new historiography, see *The Social Construction of Technological Systems*, ed. W. Bijker et al. (MIT Press, 1987).

5. For a discussion, see my *Wings of Wood, Wings of Metal* (Princeton University Press, 1999), pp. 5–21.

6. In the 1990s, three exemplary works in this genre received SHOT's Dexter Prize for the best book in the history of technology: Donald Reid's *Paris Sewers and Sewermen* (Harvard University Press, 1991), Claude S. Fischer's *America Calling* (University of California Press, 1992), and David Nye's *Electrifying America* (MIT Press, 1990).

7. Lewis Mumford, *The Culture of Cities* (Harcourt Brace, 1938); Mumford, *The City in History* (Harcourt Brace Jovanovich, 1961); Sam Bass Warner, *Streetcar Suburbs* (Harvard University Press, 1962). On Warner's influence, see Carl Abbott, "Reading Urban History: Influential Books and Historians," *Journal of Urban History* 21 (1994): 33–34.

8. Josef W. Konvits, Mark H. Rose, and Joel A. Tarr, "Technology and the City," *Technology and Culture* 31 (1990): 284–294.

9. Thomas P. Hughes, *Networks of Power* (Johns Hopkins University Press, 1983), pp. 175–261.

10. Two representative works are *Technology and the Rise of the Networked City in Europe and America*, ed. J. Tarr and G. Dupuy (Temple University Press, 1988) and Mark H. Rose's *Cities of Light and Heat* (Pennsylvania State University Press, 1995).

11. See, for example, an otherwise excellent study by Stanley K. Schultz, *Constructing Urban Culture* (Temple University Press, 1989). For critical reviews that emphasize the importance of integrating urban technology and urban culture, see Bill Luckin, "Sites, Cities, and Technologies," *Journal of Urban History* 17 (1991): 426–433; Julie Johnson-McGrath, "Who Built the Built Environment:

Artifacts, Politics, and Urban Technology," *Technology and Culture* 38 (1997): 690–696.

12. This is the essence of Scott Bottles's argument in *Los Angeles and the Automobile* (University of California Press, 1987). Ruth Schwartz Cowan makes a similar argument when she attributes the failure of housework alternatives to a general preference "for privacy and autonomy over technical efficiency and community interest" (*More Work For Mother,* Basic Books, 1984, p. 150). Clay McShane, in a much more nuanced account of urban automobility (*Down the Asphalt Path,* Columbia University Press, 1994), argues that urban culture both shaped and was shaped by changes in transportation technology. For a general critique of simplistic cultural explanations of the differences between American and European cities, see Moshe Adler, "Ideology and the Structure of American and European Cities," *Journal of Urban History* 21 (1995): 691–715.

13. See Gabrielle Hecht, *The Radiance of France* (MIT Press, 1998), esp. pp. 8–11. Charles Tilly has criticized urban historians for their reliance on such "unhistoricist" explanations as "prevailing national attitudes" ("What Good Is Urban History," *Journal of Urban History* 22 (1996), p. 709). On the mutual construction of technology and culture from the perspective of cultural studies, see Paul du Gay et al., *Doing Cultural Studies* (Sage, 1997).

14. For evidence of the urban orientation of inventors in this period, see Hughes, *American Genesis* (Viking Penguin, 1989), pp. 24–47. Although Hughes suggests that inventors sought to avoid entanglements in the day-to-day business of the large companies who funded them, sometimes by moving to rural areas, the vast majority continued to work in close proximity to major urban centers.

15. Thomas Commerford Martin, "Electricity in the Modern City," *Journal of the Franklin Institute* 138 (1894), September, p. 199.

16. Mark H. Rose, "Machine Politics: The Historiography of Technology and Public Policy," *The Public Historian* 10 (1988), spring: 27–47. Rose cites Hughes's analysis of the diffusion of electric power systems in *Networks of Power* as an exception to this neglect of politics.

17. See e.g. Paul Barrett, *The Automobile and Urban Transit* (Temple University Press, 1983); Clay McShane, *Technology and Reform* (State Historical Society of Wisconsin, 1974), chapter 7.

18. My analysis of the role of streets relies heavily on Clay McShane's "Transforming the Use of Urban Space: A Look at the Revolution in Street Pavements, 1880–1924," *Journal of Urban History* 5 (1979): 279–307.

19. George Rogers Taylor, "The Beginnings of Mass Transportation in America, Part I," *Smithsonian Journal of History* 1 (1966), autumn, pp. 34, 38; Harry J. Carman, *The Street Surface Railway Franchises of New York City* (Columbia University, 1919), pp. 19, 25–27. New York did grant exceptions to its ban on steam below 42nd Street.

20. Philadelphia Committee of Residents and Property Holders, *Address and Remonstrances Against Fifth and Sixth Street Railways* (Philadelphia: Crissy &

Markley, 1857); Philadelphia Citizens, *Memorial to the Legislature of the State of Pennsylvania, Against the Navy Yard, Broad Street and Fairmount Rail Road Company, Against a Freight, Baggage and Passenger Rail Road on Broad Street* (Philadelphia: King & Baird, 1863); Edward Q. Keasbey, "Poles and Wires in the Streets for the Electric Railway," *Harvard Law Review* 4 (1891): 245–270.

21. Historians writing on the condition of urban streets often apply twentieth-century standards of cleanliness, which fail to explain why urban residents would oppose mechanical forms of transportation that promised to eliminate large volumes of animal wastes from city streets. See Stanley K. Schultz, *Constructing Urban Culture* (Temple University Press, 1989), 117–118; Joel A. Tarr, "The Horse—Polluter of the City," in *The Search for the Ultimate Sink* (University of Akron Press, 1996), pp. 323–333; Jon C. Teaford, *The Unheralded Triumph* (Johns Hopkins University Press, 1984), pp. 231–232. On steam dummies, see below.

22. Taylor, "The Beginnings of Mass Transportation in America, Part II," *Smithsonian Journal of History* 1 (1966), fall: 39–52; Warner, *Streetcar Suburbs*.

23. John H. White, "Grice and Long: Steam-Car Builders," *Prospects* 2 (1976): 25–39; McKay, *Tramways and Trolleys*, pp. 27–30; Richard Clere Parsons, "The Working of Tramways by Steam," *Minutes of Proceedings of the Institution of Civil Engineers* 29 (1885): 99–103; Massachusetts Street Railway Commission, *Evidence before the Street Railway Commissioners* (Boston: Wright & Potter, 1864), pp. 89–126; Daniel Kinnear Clark, *Tramways, Their Construction and Working* (Lockwood, 1878), pp. 312–414.

24. McKay, *Tramways and Trolleys*, pp. 30–32, 34.

25. Teaford, *The Unheralded Triumph*, p. 231; Joseph P. Sullivan, "Fearing Electricity: Overhead Wire Panic in New York City," *IEEE Technology and Society Magazine* 14 (1995), fall: 8–16; William Orton, *A Review of the 'Opinions of Experts' as to the Necessity for the Poles Now Erected in Tenth Street, in the City of Philadelphia, by the Western Union Telegraph Company* (Russell Brothers, 1876).

26. "A Nuisance and a Danger," *American Architect and Building News* 11 (1882), 17 June, p. 284.

27. Robert Friedel and Paul Israel with Bernard Finn, *Edison's Electric Light* (Rutgers University Press, 1986), pp. 179, 196–198.

28. *Engineering News* 19 (1888), March 24, p. 238.

29. Teaford, *The Unheralded Triumph*, p. 231; Sullivan, "Fearing Electricity," pp. 11–13; George F. Swain, "Underground Telegraph-Wires," *American Architect and Building News* 16 (1884), December 13: 285–286; "The Electrical Subways," *Engineering News* 16 (1886), September 11: 173–174. For a detailed survey of efforts to require underground wires, see *Report of the Electrical Commission Appointed to Consider the Location, Arrangement, and Operation of Electric Wires in the District of Columbia*, US Congress, House Exec. Doc. no. 15, Serial Set, vol. 2949 (US Government Printing Office, 1892).

30. Robert C. Post, "Some Traction Prehistory: Robert Davidson and Charles Grafton Page," in *Pioneers of Electric Railroading*, ed. J. Stevens (Electric Railroaders' Association, 1991), pp. 1, 3.

31. William D. Middleton, *The Time of The Trolley* (Golden West Books, 1987), pp. 52–65.

32. McKay, *Tramways and Trolleys*, pp. 50–51.

33. *American Architect and Building News* 12 (1882), 26 August, p. 94.

34. "Remarks of Mr. F. J. Sprague on Electricity as a Motive Power," *Proceedings of the American Street-Railway Association* 6 (1887), p. 65.

35. Robert David Weber, Rationalizers and Reformers: Chicago Local Transportation in the Nineteenth Century, Ph.D. dissertation, University of Wisconsin, 1971, p. 07.

36. McKay, *Tramways and Trolleys*, pp. 44–45.

37. Carl Hering, *Recent Progress in Electric Railways* (W. J. Johnson Co., 1892), pp. 167–188; H. W. Bartol, "The Electric Railway at Buda Pesth," *Journal of the Franklin Institute* 133 (1892), February: 125–130; "The Electric Street Railways of Budapest: An Object Lesson for American Cities," *Review of Reviews* 11 (1895), March: 287–289. For a survey of electric streetcar operations up to 1888, including conduit systems, see *Letter of the Board of Commissioners of the District of Columbia . . . on the Subject of Electricity as a Motive Power for Street Cars*, US Congress, Senate Misc. Doc. no. 84, 50th Cong., 1st sess., Serial Set, vol. 2516 (20 March 1888).

38. Herring, *Recent Progress in Electric Railways*, 189–206; Alexander Julien, discussion comment to T. C. Martin, "Electric Street Cars," *Transactions of the American Institute of Electrical Engineers* 3 (1886): 40–43.

39. McKay, *Tramways and Trolleys*, p. 97; Hering, *Recent Progress in Electric Railways*, pp. 192, 196.

40. Hering, *Recent Progress in Electric Railways*, pp. 168–170; "Safety of Overhead Electric Wires," *Street Railway Journal* 5 (1889), May, p. 126.

41. R. J. McCarty, "Special Report Concerning the Relative Cost of Motive Power for Street-Railways," *Proceedings of the American Street-Railway Association* 9 (1890): 70–74.

42. On the split between pubic and private interests as a political issue in this period, see Richard L. McCormick, "The Discovery That Business Corrupts Politics: A Reappraisal of the Origins of Progressivism," *American Historical Review* 86 (1981), p. 257. McCormick sees this issue emerging with the 1893 depression, but the trolley controversy suggests that the conflict was an important aspect of local urban debates before the onset of the depression.

43. Edward Quinton Keasbey, *The Law of Electric Wires in Streets and Highways* (Callaghan & Co., 1892), chapter 10.

44. Remarks by Senator Manderson, *Congressional Record*, 50th Cong., 1st Sess., vol. 19, pt. 8 (17 August 1888): 7603 (first and third quotes); M. B. Leonard, "Some

Objections to the Overhead Conductor for Electric Railways," *AIEE Transactions* 5 (1888), September, p. 404 (second quote); "Metallic Poles for Electric Railways," *Street Railway Journal* 5 (September 1889): 254 (fourth quote).

45. Charles Milford Robinson, *The Improvement of Towns and Cities, or The Practical Basis of Civic Aesthetics* (Putnam, 1901), pp. 55–57; Jon A. Peterson, "The City Beautiful Movement: Forgotten Origins and Lost Meanings," *Journal of Urban History* 2 (1976): 415–434.

46. "Some New Styles in Poles," *Street Railway Journal* 6 (1890), May, pp. 217–218; McKay, *Tramways and Trolleys*, pp. 100–106. McKay contrasts European attention to aesthetic concerns with American indifference, but American street railroads also sought to minimize the aesthetic impact of overhead wires.

47. Compare photographs on pp. 17 and 25 of Frederick A. Kramer's *Across New York by Trolley* (Quadrant, 1975). In many of the black-and-white photographs, however, the overhead wires are barely noticeable.

48. Hughes, *Networks of Power*, pp. 107–109.

49. For Chicago see Weber, "Rationalizers and Reformers," pp. 103–105. For New York see Jennifer B. Bannister, "Opposition to the Overhead Trolley System in New York City and Washington, D.C.," paper for History Seminar, Rutgers University, May 5, 1992. For Philadelphia, see below.

50. Remarks by Senator Teller, *Congressional Record*, 50th Cong., 1st Sess., vol. 19, pt. 8 (17 Aug. 1888): 7654.

51. For example, Cheape, *Moving the Masses*, p. 119. Cheape states that in Boston "opposition [to the trolley] stemmed from fear and ignorance," as well as sentiment against the traction monopolies. For an egregious echoing of the industry's arguments on safety, see Weber, "Rationalizers and Reformers," pp. 105–106, 109. Weber denounces "the irrationality of popular opinion" and "the unfounded fear of the 'deadly trolley'" for delaying the trolley's introduction in Chicago (p. 109).

52. John Trowbridge, "Dangers from Electricity," *Atlantic Monthly* 65 (1890), March: 413–418; "Safety of Overhead Electric Wires," pp. 125–126; "That Electricity Does Not Kill," *Street Railway Journal* 5 (1889), September, p. 266; "Safety Guard Wire for Overhead Electric Railways," *Street Railway Journal* 5 (1889), November, p. 367; "The Danger Elements in Electric Traction," *Street Railway Journal* 6 (1890), May, p. 232; Henry Morton, "The Dangers of Electricity," *Engineering News* 24 (1890), 6 September, pp. 206–207.

53. For a good example of expert testimony insisting on the safety of 500 volt trolley lines, see *Evidence before the Massachusetts Committee on Street Railways as to the Safety of Overhead Electric Wires* (Boston: R. H. Blodgett, 1889). For evidence of the limited understanding of electrical injuries into the early twentieth century, see Charles Alpheus Lauffer, *Electrical Injuries* (Wiley, 1912). For a present-day view of electrical injuries, see *Electrical Trauma*, ed. R. Lee et al. (Cambridge University Press, 1992).

54. "The Frequency of Accidents," *Street Railway Journal* 5 (1889), September, p. 266; Tom Sitton, "The Los Angeles Fender Fight in the Early 1900s," *Southern California Quarterly* 72 (1990): 139–156.

55. Elihu Thomson to E. W. Rice, 20 July 1891, Elihu Thomson Papers, American Philosophical Society, Philadelphia.

56. For examples from Philadelphia, see below. For a discussion of Progressive Era reform of street railroads in Milwaukee, see McShane, *Technology and Reform*, chapters 7 and 8.

57. "Opposition to the Introduction of Mechanical Traction," *Street Railway Journal* 6 (1890), March, p. 121.

58. See Joel A. Tarr, "From City to Suburb: The Moral Influence of Transportation Technology," in *The Search for the Ultimate Sink*.

59. T. C. Martin, "The Social Side of the Electric Railway," *Street Railway Journal* 6 (1890), April, p. 203.

60. Warner, *The Private City*, pp. 169–172.

61. Nye, *Electrifying America*.

62. H. H. Littell, "Address of the President," *ASRA Proceedings* 2 (1883), pp. 8–9.

63. Remarks of C. A. Richards, *ASRA Proceedings* 3 (1884), pp. 138–140.

64. "Electricity vs. Steam, " *Street Railway Review*, April 15, 1894, p. 234.

65. "Electricity for Street Railroads," *Railroad and Engineering Journal* 56 (1887), July, pp. 309–310.

66. "Objections to Electricity," *Street Railway Journal* 5 (1889), September, p. 266.

67. Cheape, *Moving the Masses*, pp. 162–165; Harold E. Cox and John F. Meyers, "The Philadelphia Traction Monopoly and the Pennsylvania Constitution of 1874: The Prostitution of an Ideal," *Pennsylvania History* 35 (1968): 411–414.

68. Cheape, *Moving the Masses*, p. 172; Edmund Stirling, "Competition Soon Gave Way to Mergers in Car Lines," *Public Ledger* (Philadelphia), 17 February 1930; "Address by Mr. John C. Bullitt," in *The Overhead Electric Trolley Ordinances* (Philadelphia, March 1892), pp. 7–8; "Traction's Methods Unveiled," *Philadelphia Inquirer*, 30 March 1892, p. 1.

69. Cheape, *Moving the Masses*, pp. 172–173; "Business Men Fight Traction," *Philadelphia Inquirer*, 10 March 1892, p. 4; "Fighting Trolley Wires," *Philadelphia Inquirer*, 11 March 1892, p. 5.

70. *Philadelphia Inquirer*, 15 March 1892, p. 2; 17 March 1892, p. 3; *The Overhead Electric Trolley Ordinances*, pp. 12, 27, 55–57, 58.

71. *The Overhead Electric Trolley Ordinances*, pp. 52, 54, 57, 58, 61.

72. See esp. "Address by Mr. John C. Bullitt," "Address of Mr. Finley Acker," and "Resolutions Adopted by the Citizens' Municipal Association, of Philadelphia," all in *The Overhead Electric Trolley Ordinances*.

73. "Address by Professor William D. Marks," ibid., pp. 35–42; "Use the Storage Battery System," *Philadelphia Inquirer*, 9 March 1892, p. 4.

74. "Trolley Battle Fiercely Waged," *Philadelphia Inquirer*, 15 March 1892, p. 2.

75. "Electric Railway Construction in Philadelphia," *Street Railway Journal* 10 (1894), January, p. 1.

76. "Trolley Battle Fiercely Waged," *Philadelphia Inquirer*, 15 March 1892.

77. "Magnate Widener Roasts the '400': The Traction King Creates a Flurry at the Cloverites' Hibernian Dinner," *Philadelphia Inquirer*, 18 March 1892.

78. "Urging the Veto," *Public Ledger* (Philadelphia), 29 March 1892.

79. "Citizens, Can You Stand This?" *Philadelphia Inquirer*, 25 March 1892 (quote). See also "The Trolley System Out of Date," ibid., 29 March 1892; "Use the Storage Battery System," ibid., 9 March 1892; "Opposition to the Trolley," ibid., 11 March 1892; "Marching to the Crack of the Traction Whip," ibid., 18 March 1892.

80. "The Protest Against the Trolleys," *Philadelphia Inquirer*, 24 March 1892.

81. "Opposition to the Trolley," *Philadelphia Inquirer*, 11 March 1892.

82. "Citizens Ask the Mayor for His Veto," *Philadelphia Inquirer*, 29 March 1892

83. "An Outrage and an Insult," *Philadelphia Inquirer*, 23 March 1892; "An Outraged Constituency Speaks Out," ibid., 24 March 1892; "Citizens Ask Mayor Stuart for a Veto," ibid., 25 March 1892; "Traction's Methods Unveiled," 30 March 1892; "Urging the Veto," *Public Ledger* (Philadelphia), 29 March 1892; "The Short Road to Rapid Transit," ibid.

84. "Passed Over the Veto," *Public Ledger* (Philadelphia), 1 April 1892; "The Trolley Infamy," *Philadelphia Inquirer*, 2 April 1892. On the general corruption in late-nineteenth-century cities, see Ernest S. Griffith, *A History of American City Government* (Praeger, 1974), chapter 8.

85. Edmund Stirling, "New Controversy Attended the Switch to Electricity," *Public Ledger* (Philadelphia), 20 February 1930.

86. My analysis of New York and Washington draws heavily on Bannister, "Opposition to the Overhead Trolley System in New York City and Washington, D.C."

87. Stearns Morse, "Slots in the Streets," *New England Quarterly* 24 (1951), March: 9–11; Cheape, *Moving the Masses*, pp. 61–62; Bannister, "Opposition to the Overhead Trolley System in New York City and Washington, D.C.," pp. 12–14.

88. *Letter of the Board of Commissioners*, p. 18.

89. *Congressional Record*, 50th Cong., 1st Sess., vol. 19, pt. 8 (16, 17 Aug. 1888): 7601-5, 7650–7660; Bannister, "Opposition to the Overhead Trolley System in New York City and Washington, D.C.," pp. 18–24; LeRoy O. King, *100 Years of Capital Traction* (Taylor, 1972), pp. 17, 31. According to Bannister (p. 24), there was no single act of Congress forbidding wires below Boundary Avenue; rather, this provision was placed in the charters of the individual streetcar companies.

90. Morse, "Slots in the Streets," pp. 8–11; King, *100 Years of Capital Traction*, pp. 31–32. For a balanced contemporary discussion, see Dwight Whitney Bowles, "The Deadly Trolley," *Harper's Weekly* 36 (1892), November 12, p.1091.

91. Cheape, *Moving the Masses*, p. 68.

92. McKay does argue that differing aesthetic values are only part of the explanation of the lag in European electrification. He also singles out "differing institutional arrangements [as the] decisive factor," claiming that "most companies in the United States were able to electrify their lines . . . without even having to modify existing franchises," whereas Europeans companies were subject to strict regulation ("Comparative Perspectives on Transit," p. 13). On the contrary, when American horse railroads converted to electricity they almost invariably had to obtain permission from the city councils, and the courts insisted that this municipal action be sanctioned under state law, which often explicitly granted street railroads the right to use electric motors with the approval of the municipality. (Edward Q. Keasbey, *The Law of Electric Wires in Streets and Highways*, Callaghan, 1892, pp. 16–23). McKay is nevertheless quite correct in arguing that American street railroads enjoyed much more freedom in the choice of motive power because compliant legislatures and city councils rarely opposed the trolley.

93. For McKay's summary of European objections, see *Tramways and Trolleys*, pp. 84–89.

94. See McKay's examples from Germany, France, and Great Britain (*Tramways and Trolleys*, pp. 95–191).

95. McKay, *Tramways and Trolleys*, pp. 75–76, 99–100, 139–140.

96. Here I am drawing on a concept of culture as a collection of resources (or "tool kit") that social actors use selectively in concrete situations. See Ann Swidler, "Culture in Action: Symbols and Strategies," *American Sociological Review* 51 (1986): 273–286.

97. McKay, *Tramways and Trolleys*, pp. 50–51.

98. This process is described nicely in Warner, *Private City*.

99. Nye, *Electrifying America*, pp. 97–102; Scott Bottles, *Los Angeles and the Automobile*; Barrett, *The Automobile and Urban Transit*.

100. For an overview see Tarr, *The Search for the Ultimate Sink*.

101. On the social uses of the street, see McShane, "Transforming the Use of Urban Space," pp. 279–307; François Bedarida and Anthony Sutcliffe, "The Street in the Structure of the City: Reflections on Nineteenth-Century London and Paris," *Journal of Urban History* 6 (1980): 379–396; Penelope J. Corfield, "Walking the City Streets: The Urban Odyssey in Eighteenth-Century England," *Journal of Urban History* 16 (1990): 132–174.

102. See Viviana A. Zelizer, *Pricing the Priceless Child* (Basic Books, 1985).

103. This analysis draws in part on Clay McShane's *Down the Asphalt Path* and "Transforming the Use of Urban Space."

104. The best account of the struggle between streetcar and bus remains David J. St. Clair, *The Motorization of American Cities* (Praeger, 1986). On New York see Zachary M. Schrag, "'The Bus is Young and Honest': Transportation Politics, Technical Choice, and the Motorization of Manhattan Surface Transit, 1919–1936," *Technology and Culture* 41 (2000): 51–79. Schrag's fine account provides plenty of evidence for the role of symbolic meanings in the choice of bus over streetcar, but he fails to use these meanings as an explanatory resource, thus illustrating the need for historians of urban technology to pay more attention to the causal role of cultural meanings in technological change.

The Hidden Lives of Standards: Technical Prescriptions and the Transformation of Work in America

Amy Slaton and Janet Abbate

Standards, specifications, and similar technical protocols have pervaded scientific and industrial operations for more than a century, but scholars have only recently noted their tremendous transformative power for American industry. These bodies of knowledge and practice have largely maintained their character as "stealth technologies," pervasive but well below the radar of historical inquiry. In some ways this invisibility is not difficult to understand. After all, their proponents have generally defined standards as instruments of reduction: reducing complexity and variety in products and processes, reducing costs, reducing the time and effort required for efficient industrial operation. Accordingly, economic historians have identified standards primarily as by-products of twentieth-century capitalists' search for economic control—as incidental though sometimes potent advances in the process of industrial rationalization. Yet, although standards have certainly been used to streamline the operations of modern business, there are also many cases in which they *add* complexity and even conflict to the workplaces in which they are used. As instruments that encode knowledge and order labor relations, standards deserve the attention of historians of American work. A few historians of labor and technology have begun the task of locating the myriad implications of standards for modern management and the experiences of workers, but a great deal of uncharted territory awaits investigation.[1]

If we begin simply with the idea that standards bring economic control to their users, many possibilities for historical inquiry emerge. Control of revenues and profits suggests control of opportunities and economic advantages; control of the conditions of production also implies the deliberate design of work processes. Standards emerge from, and direct in powerful

ways, divisions of manual and intellectual labor. In the process, they bring
and eliminate risk, credit, and blame for different participants in the modern productive sector. They may well be mundane in the sense of dictating
routine procedures, but the very importance of routine in twentieth-century production (which many historians have remarked) should be an
immediate clue that standards can offer powerful insights into the nature
of work in modern industry. In addition, standards themselves are a means
of capturing labor—whether physical techniques or "knowledge" work—
and they provide a medium for redistributing the responsibility for this
work among groups of workers, between industry sectors, or between producers and consumers. If we look at the larger system of specifying, producing, marketing, and using goods and services, we can see that the
adoption of standards may simplify some aspects of the system while creating a demand for more skilled labor elsewhere. One aim of this article is
to explore how and why such tradeoffs are made.

The overarching value of labor history is that it enables us to understand
the power relations implicit in and supported by systems of production.
Cultivated among social history subdisciplines in the 1960s and the 1970s,
this focus on social conflict and consensus subsequently attracted historians of technology. For several decades scholars have been striving to depict
the varying impacts that mechanization, mass production, and automation have had on those who institute technological change in industry and
those who are directed to work with new technologies. The sites where
"machines and people meet" have been subject to rich interpretation by
both historians of labor and historians of technology.[2] It seems to us that
some techniques and emphases developed in the latter field might be
brought to bear on projects of the former. Of course, the division of these
works into separate disciplines is to a large degree an artifact of university,
funding, and publication structures, and the exercise of "lending" an
approach from one historical field to another involves some spurious clas-
sifications. However, we can recognize certain directions in studies of work
undertaken by self-described historians of each type. Traditionally, labor
history has defined as its subject activities occurring within certain locations within the capitalist economy, most often the literal sites of material
production such as factories, railroads, and mines. In part through the
growth of history of technology, many other sites in which physical labor

is undertaken, including households, the military, and even agricultural settings, have recently garnered scholars' attention. Historians of technology identify "skill" in extremely broad terms, unrestricted by categories of economic organization, and through these studies previously unnoticed social negotiations have been newly explicated. The study of standards can expand the universe of labor studies further, drawing attention to knowledge systems, commercial instruments, and even consumption patterns that also profoundly affect the lives of persons involved in production. We want to suggest here some of the richness of these subjects for the study of social authority and mobility, which have long been major concerns of labor history.

We have paired case studies of industrial standards and post-industrial "information age" standards to suggest the range of issues involved and their persistence over time. Examining early-twentieth-century materials standards and specifications, we have found that these written regulations not only ensured efficient technical operations but also instantiated a hierarchy of technical expertise in production contexts. As written codes of "best practice," standards and specifications conveyed a particular distribution of labor from a centralized source (a product supplier, government body, or academic laboratory) to the dispersed sites of industrial production. The process of fabricating an elevator or erecting a building occurred with tasks of design, assembly, and supervision firmly attached to different levels of personnel. A contrasting set of issues are revealed by a study of data communications systems of the late twentieth century. In this very different work environment characterized by skilled labor, the focus has been on ensuring a standardized product rather than on rationalizing the process of producing computer software. If process standards do not necessarily make production more "efficient" in the sense of eliminating labor, this is even more true of product standards. Since they add new constraints on the product, standards for quality, safety, or interchangeability can add significantly to the labor required to create it.

Product standards create both a need for more careful production and a need for evaluation of the finished product, changes that may disrupt existing work practices. Even more significant, product standards can shift responsibilities between the creators of the technology and its users, who must often contribute their own unpaid expertise and effort in order to make new technologies fit their needs. This aspect of standardization has received little attention from historians, because standards have been the domain of economic rather than labor historians but also because the activities of end users tend to be classed as "consumption" rather than "work." In most analyses of the effects of technology on labor conditions, the labor of users is invisible. Ruth Schwartz Cowan and Lizabeth Cohen have focused much-needed attention on the work involved in consuming certain technologies, but labor tradeoffs between users and producers have yet to be explored.[3] Users of computer networking products, though they might characterize their general activities as "research" or "education," were often also expected to provide the labor needed to install and operate the network systems they used. Under these circumstances, such expert users hoped that adopting standardized networking products would minimize the effort required of themselves. Standards can thus become a means of negotiating the very nature of the product: What features does the producer guarantee, and what must the user bring to the technology?

Throughout our case studies, we emphasize that standards and specifications have multiple performative and legislative functions of interest to historians. In cases of both blue-collar and white-collar work environments we find that technical knowledge and authority in commercial exchanges are inextricably linked. Whether these exchanges are among competing business concerns, employers and employees, or producers and consumers, standards and specifications instantiate power as they define and reproduce practices. In particular, we draw attention to efforts to control the distribution of technical responsibility among persons of similar standing in the work environment: technical professionals, business owners, managers. We argue that these "lateral" negotiations for commercial control, which have usually been left to business historians to scrutinize as features of commercial competition, are as relevant to labor historians as the more traditional vertical interactions between management and labor. Professional workers' use of standards to claim authority and establish expertise brings social advantage and contributes to the larger system of power relations in the workplace—issues beyond the scope of most business historians but of central concern to historians of labor.[4]

Our project of finding historical meaning in industrial standards proceeds along an interdisciplinary path. In part, we follow a challenge, posed by Phil Scranton in 1988, that is still enticing: to integrate fully the study of labor and workplace technology.[5] While applauding the democratizing impulses of recent labor history, with its focus on the experiences of workers and the politics of machine use, Scranton notes that the field shows only a limited understanding of issues of technology diffusion, market conflict, and standardization. For Scranton, the latter topics subject themselves nicely to the techniques of history of technology, with its emphasis on how technical problems are chosen and solved (or not solved), rather than on reductive economic explanations. Ideally, each discipline's strengths will inform the other. We also make use of the conception (developed by John Law, Tom Hughes, and others) of technology as a heterogeneous intellectual undertaking in which a technical practitioner pursues not merely material but also economic and cultural goals, crafting bodies of technical knowledge to conform to conditions well beyond the laboratory or the shop floor. For Hughes most famously, the inventor is as much entrepreneur or reformer as engineer, and to succeed the inventor must approach technological enterprises as systemic undertakings. We combine these different strands of inquiry to look for consequences of such "heterogeneous engineering" for workers well below the level of trained technicians, as well as for the people who use the end products of industry.[6]

Standards present a remarkably fruitful focus for the project of integrating the history of science, technology, and labor.[7] Since standards are widely used, they attach to broad social patterns. From a methodological standpoint, because standards are written and published, historical materials are easily obtained, which helps the historian of industry avoid reliance on the censored or selective corporate records about which David Nye warns.[8] Finally, standards are vivid historical markers of social authority being established. At the moment a standard or specification is adopted, some piece of technical knowledge is granted legitimacy and some person is granted authority in the workplace or the marketplace, while other actors find themselves freed—or robbed—of their former responsibilities. The end is not always predictable: some standards build in flexibility to accommodate local conditions; others fail when confronted with resistant users. In the case studies that follow, we offer examples of intertwined systems of knowledge, labor, and technique within the hidden lives of standards.[9]

Specifications and the Organization of Modern Construction

Standards and specifications became essential tools of American commerce with the growth of large-scale industry in the late nineteenth century. As they grew, manufacturing, food processing, construction, and most other types of productive enterprises found themselves faced with unprecedented logistical challenges. Standards and specifications first emerged within companies seeking to establish uniform procedures and products and soon became the purview of trade associations and government bodies wishing to coordinate operations in an ever-larger sphere of technical activity. From 1900 on, the American Society for Testing and Materials, the American Society of Civil Engineers, the National Bureau of Standards, and individual city permitting agencies published collections of recommendations intended to regularize commercial relations. "Standards" generally addressed the physical characteristics of raw materials or finished goods, often incorporating scientific data; "specifications" addressed precise requirements for individual jobs or products or listed testing and inspection procedures to be used in the execution of a commercial contract. (To a degree the terms are interchangeable, since each type of document might incorporate the other.[10])

On one level, standards and specifications controlled physical aspects of production, ensuring the quality of raw materials, the interchangeability of manufactured parts, or the performance of an engine or a machine. This kind of control had implications for the ethical status and profitability of businesses: worker safety and efficiency (on the one hand) and consumer safety and satisfaction (on the other) improved when written regulations for industrial operations came into existence. Standards and specifications also regularized economic relations at the point where the supplier and the buyer of a service or a product met. They outlined expectations, obligations, and legal recourses available to parties in commercial exchanges, and they remain a pervasive feature of commercial exchange today. In many senses, standards and specifications are texts that communicate and help to enforce particular behaviors in the marketplace.

As did the systems of reports, memoranda, and statistical analysis adopted by ambitious turn-of-the-century businesses, standards and specifications offered means of managing information in an economic environment of increasing complexity. The logistical and economic aims of these written instruments include significant controls not just on the material and fiscal features of commerce, however, but also on organizations of productive labor and its associated structures of reward.[11] Historians have described how, as mechanization and routine became by-words of modern business practice, the range of employment options shrank in many technical occupations.[12] Industrial employers used standards and specifications to define work processes and divisions of labor: exactly who did what on the shop floor or construction site. While ordering a huge and almost infinitely complex array of technical projects for industry, these protocols also dictated employment opportunities and occupational advantage. We have written elsewhere on materials standards of this period, such as the immensely influential body of regulations produced by the ASTM and the social utility of those standards to the scientists who helped write them. For many materials experts, standards offered a means of disseminating technical expertise to a broad clientele without diluting the authority of their own scientific specialties.[13]

We want to consider here the role played by such written regulations in a different set of competitive concerns: those that occupied so-called white-collar technical trades (such as manufacturers, architects, and engineers), and the skilled occupations (such as machinists, plumbers, electricians) that worked for them early in the twentieth century. While high-level scientific knowledge brought basic research to bear on routine matters of manufacturing or construction, a vast body of less esoteric knowledge kept factories, mines, processing plants, and engineering firms running smoothly. If science-based standards constituted a sort of idealized code of best practice for industry, specifications mustered specific, practical applications of standards or other technical knowledge. From the thickness of plywood to the fineness of flour to maintenance procedures for electrical turbines, the material details of industry found systematic expression in specifications that formed the basis of contracts, plans, and orders. These guidelines for raw materials and finished goods included dimensions, composition, and design features desired by purchasers or—depending on who was writing the contract—offered to purchasers by suppliers. The specifications ranged from brief, almost crude invoices casually exchanged between buyer and seller to elaborate documents intended to provide full financial and legal coverage for both parties. In all cases they represent delineations of expected technical practice.

The true social power of these commercial instruments is well illustrated by examples drawn from the construction industry, an ambitiously modernizing American venue in the period 1900–1920. All but the smallest commercial building concerns in this period showed a growing commitment to state-of-the-art production and management techniques. The construction site became a place of streamlined work, often planned and overseen by a new type of engineering or architectural firm that exercised close control over every feature of construction work. In their search for efficiency and economy, these companies exploited the shift in many building materials from individually fabricated to standardized products manufactured off-site by specialized companies. By the 1880s, architectural features ranging from window frames to roofing tiles had ceased to be products only of conventionally skilled workmen operating on a "batch" basis and had become products of factories in which mass-production techniques were used to achieve high output. By 1910, staircases, chimneys, and many other larger architectural elements had joined this list. By the 1920s, entire prefabricated sheds and warehouses had come on the market, implying an almost complete displacement of traditionally skilled construction workers for such utility buildings.[14] The availability of these reliable and relatively inexpensive products meant that carpenters, masons, roofers, painters, and many other long-established trades, steeped in traditional systems of apprenticeship and broad-based knowledge of building methods, found themselves first in positions of lowered pay and less autonomy than they had previously known and then, by the 1920s or the 1930s, excluded from some building sites altogether. The strategic large-scale engineering, architecture, or construction firm of the twentieth century had little commitment to the highly trained, relatively highly paid (and often unionized) construction worker, preferring to hire whenever possible, low-paid workers with either a very narrow set of technical skills or no building experience at all.[15]

As Gwendolyn Wright and a few other historians have shown, this modernization of the building industry brought with it a new set of business relationships. First, as rationalization and economies of scale pervaded the building industry, suppliers of materials, contractors, engineers, and architects needed to find ways to garner opportunities and control risks in a highly competitive market. Potential profits for a large-scale construction enterprise were steadily growing, but so were risks, and a heavily capitalized company hoping to thrive beyond a local market had to fight for a secure client base. Commonly, occupations with related fields of expertise, such as architects and designing engineers or electrical engineers and electricians, competed for the same customers. At the same time, members of the traditionally skilled, less elite building trades, not surprisingly, adopted an adversarial stance toward the radical economies sought by their employers. Deep rifts appeared between associations of construction workers and those of the building industry. Wright has articulated the complex shifts of occupational identity and allegiance that pervaded the construction industry after 1900.[16]

Specifications, used in business exchanges among all these groups, shed light on both the origins and the consequences of those shifts. Here we will briefly discuss three cases, each of which suggests a possible function for technical specifications in the consolidation of occupational advantage in modern industry. The first case characterizes the competitive relationships among producers of building supplies and apparatus; the second addresses relationships among the skilled occupations vying for work in commercial building; the third illustrates the general devaluation of craft labor by both materials suppliers and building designers. In all instances, the formulaic nature of specifications promised to carry particular visions of the construction process into common practice, reproducing the arrangements of skill and opportunity desired by their proponents. These written instruments represent not simply the achievement of material and commercial regularity, but the instantiation of social stratifications.

Specifications for Elevators and Dumbwaiters: Controlling Inter-Firm Competition

In the first two decades of the twentieth century, elevators and dumbwaiters passed from the status of intriguing but relatively unusual architectural features to common use in America's proliferating office buildings, apartment buildings, hotels, and hospitals. Their origins lay in the simple horse-powered or hand-powered lifts used in factories and warehouses in the 1830s. By the 1850s, a number of entrepreneurial inventors had seen the commercial potential of safe and efficient elevators and investigated increasingly sophisticated design possibilities, including steam-driven and hydraulically operated passenger elevators. Elisha Gray Otis patented his pioneering safety brake in 1851, founding a business continued by his heirs into the profitable era of tall (12–15-story) office buildings late in the century. By 1900, engineers had adapted electrical technologies employed in street railways to elevators, and Otis's company and its competitors were exploiting and furthering the craze for still taller buildings (which maximized the status and the rental income of their developers). In less architecturally dramatic structures, such as hotels, hospitals, and large apartment buildings, elevators and dumbwaiters fulfilled important logistical functions. As the layout of factories and office buildings of the new century reflected the carefully planned, labor-saving agendas of modern organization, so these other large institutions sought new ways of moving people and supplies. Elevators and their small-scale counterparts, dumbwaiters, provided a crucial infrastructural innovation that brought efficiency, and at times great luxury, to service-centered buildings.[17]

If the person commissioning a new building in 1900 or 1910, or the designer of that building, wished to include an elevator or a dumbwaiter, he (or occasionally she) turned to a company that specialized in producing and installing such devices. Otis was one of several large firms with branches nationwide; many smaller firms worked locally. All modern hoisting equipment was technologically complex and involved the use of electrical, mechanical, and structural knowledge. Thus, a firm's experience and reputation were of great concern to those commissioning elevator installations. However, establishing a commercial reputation is a historically variable task. Credibility must be constructed from evidence that is meaningful to potential clienteles, and the nature of such evidence changes over time.[18]

Let us look in detail at the actual means by which a successful elevator company secured customers and at the role of specifications in this process. First, a word about general arrangements commonly made for building design and construction in the first quarter of the twentieth century. For most buildings of significant size, an architect or an engineer designed a structure to meet the owner's general requirements. The designer would produce plans and specifications to be given to contractors, who performed the actual work of erecting a structure. The designer might issue specifications to a number of specialized firms (which would provide foundations, window frames, water towers, elevators, etc.) or to a "general contractor" (who would coordinate the specialized suppliers). In either case a "supervising engineer" in the employ of either the designer or general contractor usually oversaw all activity on the work site, implementing the owner's or the architect's instructions and using any specific supply firms that the designer had requested. Increasingly after 1900, larger commercial buildings were erected by firms that employed their own architects, engineers, and a construction staff, but even these companies brought in outside suppliers for materials and mechanical systems. Only the simplest vernacular structures (such as small homes or rural buildings) entirely evaded this connection with the multi-layered world of commercial building expertise.

Specifications were a channel by which a designer of buildings communicated with a contractor about which materials, or which subcontractor, to use in the erection of a building.[19] The supervising engineer was entrusted with carrying out the specifications to the letter. Thus, any supplier of building materials would essentially be selling to the building's designer in an effort to be included in that designer's general specifications, since that is how particular products and providers found their way to actual application. Catalogs, trade expositions, and appearances in architectural or engineering trade publications (in editorial or advertising matter) all bolstered the supplier's presence in the market, but a building supply company could also go further by including its own specifications in such promotional materials. In so doing, a supplier secured its control of the field conditions under which its products were disseminated. Beyond simply achieving a sale and offering convenience to the architects and engineers buying its products, a supplier could, through specifications, preempt appropriation of its technical authority.

In a move typical of building supply companies of the period, elevator and dumbwaiter companies solidified their hold on the market by including technical specifications in their catalogs and in advertising copy. These so-called facsimile specifications were intended for reproduction, "boilerplate" style, in the general specifications created by architects or designing engineers for conveyance to contractors. They listed required physical conditions for elevator installation and equipment, and they included both operational and decorative features. Blanks were left for the architect's insertion of the exact model number desired or for related points such as dimensions, speed, and power (figure 1). Not surprisingly, the facsimile

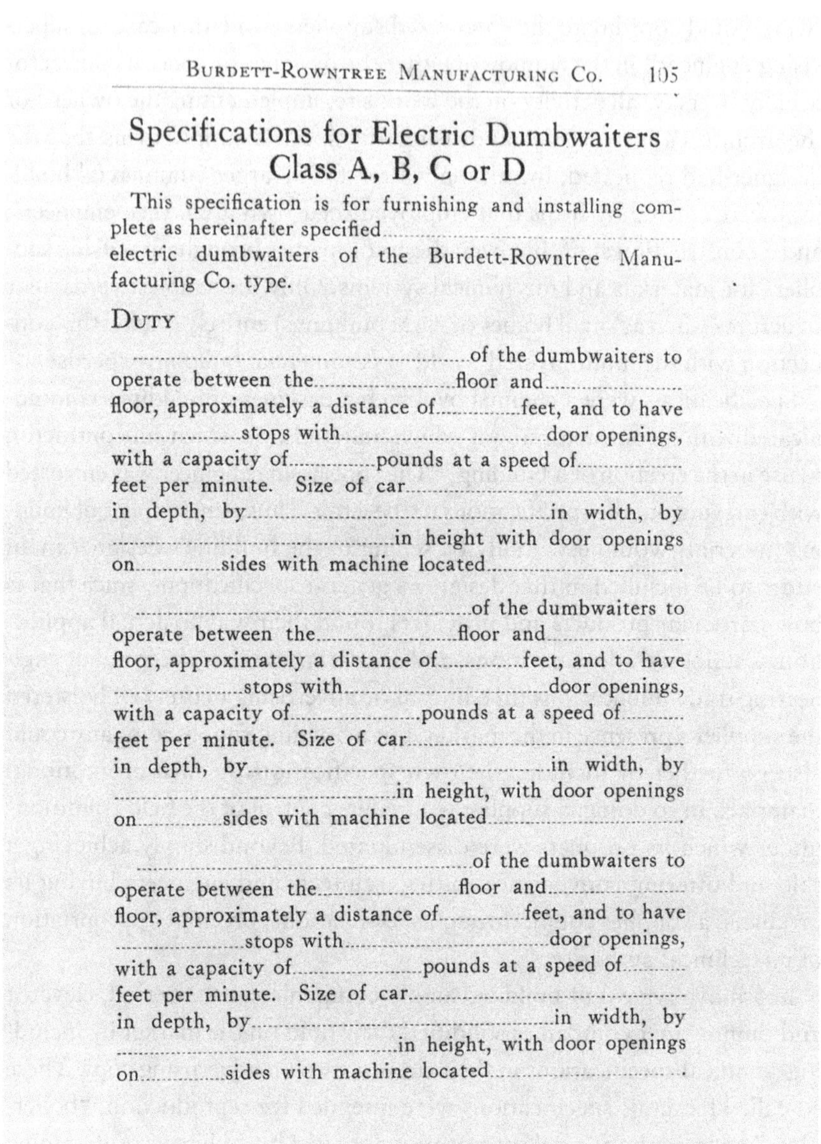
BURDETT-ROWNTREE MANUFACTURING CO. 105

Specifications for Electric Dumbwaiters Class A, B, C or D

This specification is for furnishing and installing complete as hereinafter specified.. electric dumbwaiters of the Burdett-Rowntree Manufacturing Co. type.

DUTY

....of the dumbwaiters to operate between the..........................floor and.......................... floor, approximately a distance of................feet, and to have ....stops with....................................door openings, with a capacity of................pounds at a speed of................ feet per minute. Size of car.. in depth, by..in width, by ....in height with door openings on...............sides with machine located....................................

....of the dumbwaiters to operate between the..........................floor and.......................... floor, approximately a distance of................feet, and to have ....stops with....................................door openings, with a capacity of........................pounds at a speed of........... feet per minute. Size of car.. in depth, by..in width, by ....in height, with door openings on...............sides with machine located....................................

....of the dumbwaiters to operate between the..........................floor and.......................... floor, approximately a distance of................feet, and to have ....stops with....................................door openings, with a capacity of........................pounds at a speed of........... feet per minute. Size of car.. in depth, by..in width, by ....in height, with door openings on...............sides with machine located....................................

Figure 1
Facsimile specifications for electric dumbwaiters in 1914 catalog of Burdett-Rowntree Manufacturing Co. (Hagley Museum and Library)

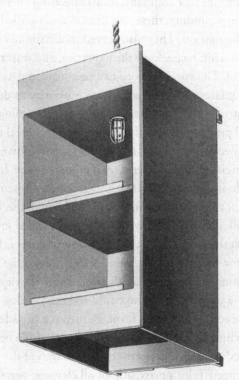

32   BURDETT-ROWNTREE MANUFACTURING CO.

TYPICAL BURDETT-ROWNTREE
DUMBWAITER CAR

Figure 2
"Typical 1914 Burdett-Rowntree dumbwaiter car," as shown in 1914 catalog of
Burdett-Rowntree Manufacturing Co. (Hagley Museum and Library)

specifications almost always included the name of the company itself, ensuring that a supervising engineer used only that manufacturer's products or services. But this official mention of a supplier's name should not be taken as the primary function of such guidelines. In standardizing communications between an elevator company, its architectural clients, and the people actually erecting a building, these specifications simplified the transmission of technical information. They also served to distribute credit in this competitive environment: to package the expertise, and thus the reputation, of the elevator firm. The specifications can be considered as a way of formalizing and thus guaranteeing the supplier's authority to define and control details of construction. This was a strategy for disseminating a particular set of technical practices without relinquishing technical expertise.[20]

In some respects, the elevator specifications endowed the individual architect or engineer with discretionary power. The building designer would select desired decorative features from a company's product lines to fill in some of the specification's blanks. Through the 1920s, for example, the Tyler Elevator Company of Cleveland offered lavishly illustrated hard-bound catalogs that displayed an array of colors, finishes, and fixtures for elevator cars, with facsimile specifications at the back.[21] An architect or a designing engineer could also determine the desired capacity for an elevator or a dumbwaiter, and could use the catalog to translate that decision into practical technical form. But from a broader perspective the inclusion of technical details in Tyler's catalog also can be seen to have put limits on clients' discretionary powers. The design of the facsimile specifications encouraged Tyler's provision of all elevator parts and Tyler's control of the installation itself. Architects using the catalog to prepare building specifications for contractors were instructed, in a section called "The Use of This Book," to select desired "Panels," "Transoms," "Jambs," "Door Closers," and other parts of the elevator entrance. Keeping all these elements together, as a single purchase made from Tyler, was emphasized as the point of the catalog:

To assure dependable, smooth-operating equipment, it is especially essential to have undivided responsibility: therefore, all of the parts which go to make up the complete elevator entrance should be included in one specification in one contract to one company.[22]

The facsimile specifications at the back of the catalog guided Tyler's clients into such an approach, discouraging in the process clients' invocation of outside (non-Tyler) technical help, parts made by other manufacturers, or their own discretion on technical matters. Through the publication of such specifications, Tyler could reduce the likelihood of both technical interference and commercial competition.

The published guidelines meant that architects not only did not *need* a profound understanding of the elevator technology they requested, but that Tyler did not wish its clients to pursue or exercise that understanding if they had it. Not only did these specifications make it legally necessary for the building's supervising engineer to use Tyler's products and services; once an order to Tyler was submitted, they actually made it unlikely that the architect or designing engineer, the supervising engineer, or the workers at the building site would take any technical initiative. Facsimile specifications like Tyler's performed a bundling of technical tasks that created something of a symbiotic relationship between a supplier of mechanical devices or services and its clients. We might briefly contrast their function with that of a simple bill of sale that recorded the exchange of money for product. Specifications prescribed a whole string of technical actions—parts to be used, personnel to be employed—to accompany that exchange.

Undoubtedly, securing control of its expertise in such ways was crucial to a company like Tyler. The geographic dispersal of the sites to which Tyler sent its products and services and the scale and complexity of construction operations threatened the safe, efficient application of the company's products, not least because the technical knowledge involved in the installation of elevators and dumbwaiters was complex and shifting.[23] The proliferation of city building codes for the operation of elevators after 1900 indicates one source of pressure on elevator companies and presents another understudied area in the history of technology and work.[24] Further, questions about how many cars, operating at what speed, would optimize a building's performance could not be answered without thought to safety, cost, and efficiency. Larger buildings might require faster elevator service, but their owners might also demand that less potentially rentable space be consumed by elevator equipment. To fulfill these complicated conditions, elevator companies constantly introduced new types of motors and safety brakes. All these technical challenges were exacerbated by the acute anxiety about passenger safety in elevators. *Engineering News-Record* published countless reports of accidents and infractions.[25] Specifications that encapsulated a firm's technical expertise and discouraged deviation and initiative by people purchasing technologies helped elevator companies and other building suppliers achieve a measure of reputational predictability in an extremely unpredictable environment.

If specifications helped elevator manufacturers and like-minded business owners to represent and reproduce their technical expertise, and thus to garner new customers, they also helped to regulate relations among the technical occupations that performed the work of building. When the Tyler Elevator Company noted that its catalog specifications guaranteed "undivided responsibility" for elevator installation, it meant not only that Tyler's personnel alone would oversee the work, but that responsibility would remain in the hands of technical experts—it would not be delegated to the legions of mechanics, electricians, or other technicians contracted to work on a building site. In 1900, electrical engineers, machinists, and other experienced personnel might reasonably have claimed some knowledge of elevator or dumbwaiter technology. As elevators and dumbwaiters became more popular, it is likely that building maintenance staffs also became familiar with their mechanical workings. In this context, Tyler's specifications served the professional ambitions of architects and engineers—two groups that, in 1900, were eager to distinguish their services from those of the skilled building trades. The relationship between the architects and engineers who designed buildings and the many types of practitioners who contracted to erect the buildings was shaped by a range of conflicting interests. Financial goals interwove with technical agendas, and both sets of concerns enveloped ideologies about the world of business, labor, and culture. Specifications could be crafted by suppliers to favor architects and engineers to the detriment of skilled tradespeople. Here we find a second set of "lateral" negotiations among accomplished technical practitioners, but now based in inter-occupation rather than inter-firm contestations.

Specifications for Refrigerators: Facing Competition among Skilled Occupations

The story of another popular new feature of early-twentieth-century buildings, refrigerators, illuminates this connection between technical specifications and the experiences of modern industrial workers. As did the proliferation of elevators, the arrival of refrigerators in American homes and institutions followed a broad set of technical and cultural developments at the end of the nineteenth century. Commercial ice making, particularly for the benefit of brewing and other food-processing operations, emerged as a viable industry in the 1860s. Cold-storage warehouses and refrigerated railroad cars soon followed, bringing the possibility of long-term storage and long-distance transport of many types of food and, for many Americans, a diet based on mass-produced and commercially distributed foods. The icebox, based on the large-scale production and home delivery of ice, helped this new variety of foodstuffs (fresh and packaged) find a place on the American table. A number of other changes, including shifts in the organization of middle-class households (fewer servants and unmarried female relations to contribute to food preparation) and fashions in diet (a greater variety and complexity to middle-class meals), paved the way for the popularity of mechanized refrigeration technologies. With the expansion of utilities providing gas, and then electricity, to homes, the possibility of high-profit, mass-marketing of home refrigerators was virtually ensured.[26] As Ruth Cowan and others have shown, General Electric, already a prosperous firm experienced at marketing electrical technologies, aggressively marketed the refrigerator after 1900 as a means of modernizing the American kitchen and promoting the consumption of electricity.[27]

The first refrigerators were expensive and not infrequently came into a new home as part of the kitchen design, entering many affluent homes not as a retail purchase made by consumers but under the auspices of the architect or engineer designing the home. Certainly institutions, such as hotels and hospitals, had kitchens designed and equipped by architects.[28] GE understood that architects constituted a significant market for its products and published facsimile specifications in its catalogs that architects could reproduce in their own plans and building specifications. Recognizing the architects' own concerns with occupational self-identity, GE designed its refrigerator specifications to appeal to a prestigious, almost scientific, standing for architects among the building occupations. By reinforcing architects' claims to a distinctly modern, technology-based body of knowledge, specifications aligned suppliers and architects in a mutually beneficial relationship, at the expense of tradespeople. GE's "sample specifications for architects, engineers and others," published in its catalogs of the 1920s and the 1930s, convey a hierarchical division of labor to the building site, with building designers holding a top position in that hierarchy.

In part, the specifications performed this function by implicitly invoking the architects' technical authority. One such feature echoes the Tyler Elevator Company's inclusion of highly detailed technical matter in its catalog specifications for architects. GE's 1931 specifications, as provided in "long form" in a catalog of home refrigerators, include such facts of refrigerator construction as the conductivity of cabinet insulation ("not greater than .31 Btu per degree F per hour per square foot per inch of thickness") and inner finish ("one blue ground coat and two white finish coats of vitreous porcelain enamel").[29] This information may seem intended to provide a degree of precision to architects' plans and contracts. Remember, however, that these refrigerators were factory made, and the details of their construction could not have been altered by the architects or anyone else on the building site. So why, we might ask, include such technical matter in catalog specifications? Why not merely mention the supplier's name and the appropriate model number? Like the elevator company, GE conveyed with such documents the depth of its technical expertise. The firm grounded its status in part on the general reputation of the mechanical refrigerator as a high-tech replacement for the icebox—as an ultra-modern appliance that could bring the best science had to offer to a woman's care of her family.[30] GE then offered that reputation to the architects who specified GE refrigerators. The GE specifications gave architects direct evidence of the company's own modernity and reliability, but also let architects "borrow" that credibility by inserting the very detailed, esoteric refrigerator specifications in their own plans.

GE's specifications bolstered the dominant position of architects among the construction trades in an explicit fashion too. The 1931 refrigerator catalog notes that, in using the company's products, "the architect has the assurance that he is specifying an electrical product that will give satisfaction. He eliminates unnecessary work. No plumbing or extra wiring is needed."[31] The intimation was of convenience and reliability, but these were to be achieved by the omission of plumbers and most electricians from the installation process. This is, of course, a case of the streamlining of labor so common to early-twentieth-century industries. But the design of refrigerators and the use of GE's specifications in building plans had occupational impacts of particular consequence to architects.[32] In the realm of home design, American architects were competing with professional plumbers, who in this period were also trying to align themselves with applications of esoteric knowledge. For some time, plumbers' professional associations had claimed an affinity with medical doctors and a vital role in city sanitation. With the new popular interest in "scientizing" the American kitchen and bathroom—replacing outdated technologies with modern appliances and waste systems—plumbers might have challenged architects as the preeminent designers of these spaces.[33] As we argued above, the attribution of technical authority is a fluid process, and in 1910 plumbers could reasonably have sought the exalted social status afforded to the "professions" through claims of scientific or other culturally esteemed types of knowledge. In retrospect, we know that plumbers achieved no such status, while Americans were willing to grant architects an enduring occupational prestige. That prestige was constituted of both the traditional cultural stewardship attributed to fine-arts enterprises and the celebrated modernity of cutting-edge technical achievements. In a sense, architecture capitalized on two apparently (but only apparently) contradictory sources of intellectual status—the rearguard and the forward-looking—and, in the marketplace of services, rose above the "merely" technical. A company like GE could contribute to the construction of this status hierarchy by wedding the use of its product to the presence of architects while excluding plumbers.

The success of architects in distinguishing their abilities from those of other occupations after 1900 was due, of course, to a range of professionalizing efforts, including aggressive licensing and educational initiatives.[34] This was a period of self-conscious and concerted organization for many professions in America, each contending, by 1910 or so, with the wholesale conversion of productive labor to systems of divided, largely de-skilled tasks. In many respects, technical reputations were themselves commodities, to be exchanged among practitioners or enterprises. Perhaps a company like GE enhanced its own reputation through its association with architects. At least one other refrigerator company of the day featured exterior views of elegant homes in its advertisements, rather than images of refrigerators alone.[35] Elevator companies frequently advertised their installations with pictures of skyscrapers, rather than of the elevators themselves.[36] These gestures by building supply companies suggest an intriguing confluence of interests. Commerce may have been competitive by nature, but a certain cultural consensus could bring a shared status to producers and their clients.

The third case of turn-of-the-century building supply we will discuss here looks at a portion of the industrial sector that denied any such consensus. On the work site, the relationship between producers of building materials and the large body of laborers who did the actual work of erecting buildings was characterized by a divergence of interest, if not by outright conflict. Here, specifications play another powerful, but very different, role in distribution the opportunities of industry.

Specifications for Prefabricated Concrete Reinforcement: "Skilled" vs. "Unskilled" Labor

As specifications for elevators and refrigerators demonstrate, the competitive circumstances of the building industry after 1900 generated a hierarchy of skills and opportunities of which materials suppliers were well aware. Their specifications followed from and reinforced those hierarchies, delegating tasks and assigning credit for technical know-how in the construction world. The control made possible by the issuance of specifications extended down through the interactions of elite professions and circles of competing suppliers to the "lowest" level of employment—that of the manual workers who carried and assembled the raw materials of building. Though we now understand that few jobs actually involve a complete absence of judgment or physical acumen, we can recognize that the world of construction work in the early part of this century did involve a great deal of simple physical activity. Mechanization of earth-moving, digging, and lifting equipment was on the rise, but many jobs still focused on individuals pushing wheelbarrows, pulling ropes, or wielding shovels. This was labor that required little or no training and that offered very low wages and minimal job security. As far as possible, building companies would employ this kind of labor, instead of the experienced, higher-paid workers (such as stonemasons) typical of nineteenth-century construction operations. Though this may seem a "natural" development for a modernizing industry, we can problematize the exclusion of skilled workers from buildings sites as an aspect of labor history, as we have problematized the ascendancy of successful construction occupations in this period. Specifications for building materials again contribute to this historical project. Here we consider specifications issued by companies that made prefabricated metal reinforcing for concrete, a medium of dramatically growing popularity in the new century.

Early-twentieth-century concrete construction was based on concerted efforts to simplify the process of erecting a building, doing away with inefficiencies inherent in such piecemeal operations as bricklaying and carpentry. The technological procedures involved in the erection of a reinforced-concrete building after 1900 may be divided into preparing the site (by excavation or other means), creating foundations for walls and columns, erecting wooden forms, placing iron or steel reinforcement in those forms, mixing concrete and then pouring it into the forms, removing the forms after the concrete has set, finishing the exposed surfaces, and installing doors, windows, roof coverings, and sprinkler systems and other plumbing. The drive to speed up and economize industrial construction addressed processes and the flow of materials at each of these junctures but confronted the fact that the technology involved operations of two types.[37] On one hand, concrete, a pourable medium, could be handled efficiently on a mass scale: in theory, forms, once erected, could be filled without interruption. On a well-organized project, pouring could continue on one portion of a building while another portion set. On the other hand, the erection of wooden forms and the placement of reinforcement could require slow, precise attention from costly skilled workers. Managers of construction enterprises sought means of translating the second type of operation into the first. For example, as one engineer summarized in 1906, the essence of economy in concrete was to be found in the duplication of forms and the elimination of architectural details that complicate the construction of forms.[38] Concrete construction was greatly expedited when, after 1900, the construction of forms and the assembly of reinforcing rods increasingly were taken over by outside suppliers. These auxiliary businesses, located off the construction site, mass produced materials that otherwise had to be individually fabricated in the course of building. Some intricate types of forms and reinforcement continued to be fabricated by workmen on the building site, but enough were standardized and mass produced to provide substantial economies for builders. These products were often called "systems" by their promoters.

Commercially produced reinforcement systems first appeared in the 1890s. They capitalized on the idea that reinforcement material could be bent and assembled by machine in quantity off the construction site. In 1906, *Cement Age* published a review of ten commercial systems of reinforcement; by 1914, *Sweet's Catalog* carried advertisements for systems of preassembled reinforcement from dozens of firms, some placed by prominent building firms that produced reinforcement as a sideline and some by specialized manufacturers. For example, the Clinton System of 1906 featured "wire cloth"—electrically welded metal fabric produced in 300-foot rolls.[39] Other firms offered "unit girder frames" that constituted preassembled reinforcement for entire beams or girders, ready to be set into place by three or four relatively unskilled workers. The Unit Concrete Steel Frame Company went so far as to provide sockets that fitted into the bottom of a form to ensure correct placement of the reinforcement unit.[40] The costs of purchasing this kind of fabricated reinforcement were offset by savings in labor on the construction site and by avoiding the use of more steel than was necessary. By the early 1910s, mass-produced steel reinforcement was so affordable that even large construction firms stopped making their own reinforcing rods.[41]

Many manufacturers of reinforcing systems promised high performance and reduced need for skilled labor to the builder who used their products. In 1912 the Trussed Concrete Steel Company offered a typical innovation of this type: "Hy-Rib" sheets of metal lathing, intended to provide reinforcement for thin concrete roofs, floors and ceilings. The essence of the "Hy-Rib" line (which consisted of metal sheets about a foot wide and 6–12 feet long) was to carry the benefits of factory production to the construction process (figures 3 and 4). By specifying Hy-Rib, the catalog indicated, the architect or engineer of a building reduced the need for skill or experience on the part of those placing the reinforcing. Sheets of Hy-Rib could be spliced together easily and could be easily positioned and secured. If any cutting had to be done on the construction site, a laborer could do it with a simple guillotine-style apparatus (figure 5).[42]

It was not only Hy-Rib sheeting itself that encouraged the use of low-skill labor on the construction site, but also the way in which it was marketed to building designers: through catalogs and prepared facsimile specifications. Because it was offered in standard sizes, use of Hy-Rib helped architects and engineers to rapidly design building plans. The specifications included in a 1912 Hy-Rib catalog required the architect only to write in the

THE THREE TYPES OF HY-RIB

4-RIB HY-RIB. Ribs 13/16 in. high; 3½ in. apart.

3-RIB HY-RIB. Ribs 13/16 in. high; 7 in. apart.

DEEP-RIB HY-RIB. Ribs 1½ in. high; 7 in. apart.

Type of Hy-Rib	Gauge Nos. U. S. Standard	Spacing of Ribs	Height of Ribs	Width of Sheets
4-Rib **Hy-Rib**	24, 26, or 28	3½″	$\frac{13}{16}$″	10½″
3-Rib **Hy-Rib**	24, 26, or 28	7″	$\frac{13}{16}$″	14″
Deep-Rib **Hy-Rib**	22, 24, or 26	7″	1½″	14″

Standard Lengths, 6, 8, 10, and 12 feet.

Intermediate and shorter lengths are cut without charge but any waste is charged to the purchaser. **Hy-Rib** sheets interlock at sides and ends. In ordering, no allowance need be made for sidelaps as these are provided in the **Hy-Rib.** Allow 2 inches for end laps where splice occurs over supports; otherwise, eight inches.

4-Rib and 3-Rib Hy-Rib is shipped in bundles of 16 sheets; **Deep-Rib Hy-Rib** in bundles of 8 sheets.

Hy-Rib is supplied either painted or unpainted.

5

Figure 3

"The three types of Hy-Rib" (steel reinforcing for concrete floors, wall, or ceilings), as shown in 1912 catalog of Trussed Concrete Steel Co. (Hagley Museum and Library)

Hy-Rib Building at Hudson Motor Car Co. Plant, Detroit, Mich.

Hy-Rib Sidings, Quincy Gas, Electric & Heating Co., Quincy, Ill.
Smith, Hinchman & Grylls, Archts.

110

Figure 4
Buildings erected with Hy-Rib reinforcing, as shown in 1912 catalog of Trussed
Concrete Steel Co. (Hagley Museum and Library)

Shearing a sheet of **Hy-Rib** with the **Hy-Rib Cutter.**

104

Figure 5
"Shearing a sheet of Hy-Rib with the Hy-Rib Cutter," from 1912 catalog of
Trussed Concrete Steel Co. Here a worker uses a simple guillotine-style machine
to trim a sheet of Hy-Rib at a construction site. (Hagley Museum and Library)

desired "type" and "gauge." As in the cases of elevators and refrigerators, we can think of this standardized product or procedure as redistributing technical expertise. The Trussed Concrete Steel Company had displaced the task of calculating how much reinforcement might be needed for a wall or floor of given thickness from the field (i.e., from multiple construction sites) onto its own premises (the Hy-Rib design department).[43] Similarly, the problem of correctly forming reinforcement was solved in a location remote from the building site. Through the use of the Hy-Rib system, the time it might take to make such a calculation was saved by the architect, and the motions required to shape individual pieces of reinforcement by hand were eliminated. Both tasks were also definitively removed from any workman on the construction site.

The tradespeople who might have previously expected jobs fabricating reinforcement were only too conscious of this reassignment of labor. As early as the 1880s, building trades had organized themselves around this very issue. The departure of opportunities to use their own judgment on the construction site, let alone find secure employment, spurred carpenters, masons, painters, and plasterers to resist the incursion of standardized, mass-produced construction materials.[44] By 1910, suppliers had refined methods of marketing and transmitting information about their products (such as catalog specifications); this further diminished the discretionary powers of building laborers. But resistance, in the form of union drives and strikes, persisted among the building trades, and the use of concrete presented a difficult challenge in the conflict between old-style and new-style industrial operations. Unlike building materials such as brick, stone, wood, and even steel, concrete had only a short history of use in commercial building. No ancient or solid brotherhood of concrete workers existed to defend craft traditions. Building firms that used concrete, therefore, encountered little organized resistance to their management techniques. We can appreciate the organizational appeal of concrete for building firms by noting that early in the twentieth century concrete was not always cheaper to use than wood or brick. Rather, it presented a welcome alternative to conventional methods that were increasingly fraught with labor conflict. As concrete's popularity grew, its price fell. Material considerations also helped: Concrete buildings are largely fireproof, their interiors are easy to keep clean, and its ingredients are readily procured in most parts of the United States. But, as the Hy-Rib specifications make clear, an important source of concrete's appeal was that it offered a division of labor—of technical authority—that was highly compatible with the tenets of modern business operation. As in the cases of elevators and refrigerators, specifications expressed ideologies of workplace organization and helped put those ideologies into action.[45]

Computer Network Standards and the Displacement of Labor

At first glance, the work setting of late-twentieth-century computer network providers might seem to have little in common with the early-twentieth-century building industry. However, standards performed similar functions as instruments for redistributing work, skill, and power in these two cases. White-collar occupations have received their share of attention from labor historians, including a number of works that focus specifically on the rationalization of computer programming, but most of these analyses treat standards as a means of de-skilling labor—a tool in the hands of managers intent on reducing the salaries and the autonomy of their technical employees.[46] To see the full range of labor implications of standards, it is necessary to look at the actual processes involved in putting standards to use. By uncovering the labor required of the people who implement, test, and deploy standardized products, this study demonstrates how standards can, in some instances, become a means of intensifying white-collar labor and an opportunity for self-made experts to assert authority.

Our examples here come from the development of the Internet in the United States.[47] Standards have been of paramount importance to the data communications industry.[48] Standardized functions and interfaces make it possible to combine individual components such as cables, circuit boards, and software programs into large communications systems. The alternative to standardization would be to have each network use its own techniques, with no expectation of being able to communicate across different types of networks. This was roughly the situation until the mid 1970s. Various government and academic computing centers in the United States and in other industrialized countries had built their own one-of-a-kind experimental networks, and computer manufacturers such as IBM supplied their customers with network software that worked only with the company's own line of computers. Connecting different networks, a process known as internetworking, was so difficult as to be almost unheard of. Standardization had been the norm for at least a century in international telecommunications (which was almost entirely the province of national monopolies), but agreeing on standards was much harder in the much younger and more competitive computer industry. By the early 1970s, however, computer professionals in the United States, France, the United Kingdom, Canada, Japan, and other countries were beginning to address the compatibility problem by developing common techniques for operating data networks.[49] Today it has become almost inconceivable to design a network that uses locally defined rather than nationally or globally standardized procedures.

We begin with an episode from the history of the Internet that illustrates how the introduction of new standards dramatically intensified the work of the researchers responsible for operating that network's computers. The design and implementation of a particular set of standard protocols, known as TCP/IP, was one of the technical innovations responsible for the success of the Internet. The TCP (later TCP/IP) protocol is the basis of communication on today's Internet and World Wide Web. The development of TCP/IP, like that of the Internet as a whole, was sponsored by the US government, and this standard was designed and implemented by computer scientists working as government contractors. Though the contributions of the individuals who designed the various Internet standards have been widely recognized, the skills and efforts of those who implemented the standards have gone largely unrecorded. Yet their software implementations eventually provided the basis for the commercialized off-the-shelf TCP/IP products that made possible the popularization of the Internet in the late 1980s. In this sense, the adoption of the TCP/IP standard captured the labor of these experts and passed it on to a future generation of computer owners and users.[50]

Unlike the tradesmen described above, skilled computer professionals have rarely viewed standardization as a labor issue. Instead, they have described the effects of standards in terms of technical performance, noting that standards increase the compatibility, efficiency, and reliability of data networks. Computer scientists have tended to regard the extra work they are required to do to support standardization as a necessary evil. A typical statement of this philosophy came from Jon Postel, a computer scientist at the University of Southern California who helped create and test software standards for the Internet: "You can't have internetworking without standards. It is not a question of more or less work, but rather a question of existence vs. non-existence."[51] Yet the computer experts who built the Internet faced arduous and unavoidable labor when it came to putting the new TCP standard in place.

Consider the work required to adopt a set of "protocols," as the specifications for networking standards are called. Different protocols define how various network activities, such as setting up a connection between two computers or transferring a data file, should be performed. In order for a network to operate properly, all the computers connected to it must share a core set of protocols. In the early one-of-a-kind network projects, computer scientists had to design and implement their own network protocols. Implementing a protocol meant creating hardware interfaces and/or software programs that performed the functions specified by the standard. (The physical labor involved in managing computer systems—building interfaces, installing circuit boards, laying cables—is often overlooked, and this reminds us once again that distinctions between "manual" and "intellectual" occupations say more about social status than about actual work processes.) Once standard protocols became available, it was no longer necessary to design protocols from scratch, and the required hardware could usually be bought off the shelf; however, in the early years it was still necessary to write software that would perform the functions specified by the protocols. Indeed, this task was often more difficult than before, because the standard protocols tended to be more complex (for reasons discussed below).

Internet standards also have a "quality control" aspect: the testing of TCP/IP software implementations to ensure that they meet specifications. In the late 1970s and the early 1980s, network experts invented novel ways to demonstrate standards compliance, including competitive tests and high-profile trade shows. By making it easier for customers to evaluate, compare, and combine network products, verification activities became a means of transferring labor from the user to the producer of standardized products. (Here a typical user might be the administrator of a computer center, a paid programmer, or a university scientist or student using computer systems in a research project.) In addition to the effort required to install and operate the network's hardware and software, the process of selecting these products was itself labor intensive, demanding knowledge of local equipment configurations and of the features and limitations of the products on offer. When making a purchasing decision, the prospective user of a network system would want to be sure that all the components would work together, since a technical diagnosis of "incompatibility between network products" translated into a social cost of "more work for users."[52]

The adoption of standards by the computer industry, including mechanisms for verifying those standards, was one means of easing the burden on computer users—at the cost of creating more work for vendors. More generally, we argue that standardizing a technology such as network software is not merely a process of "rationalization" that alters labor conditions as a side effect; rather, standardization can sometimes be a means to the end of redistributing work, knowledge, and authority among producers or between producers and the users of their products. In examining aspects of the introduction of computer network standards, we will ask: Who bears the responsibility for meeting the labor demands generated by standards? How have standards activities been used to challenge or increase the authority of participants? In what ways do standards shift work from users to producers, and to what extent do these labor effects motivate the choice to adopt particular standards? Our examples of the labor issues and the symbolic importance surrounding Internet standards demonstrate how these protocols have been employed as social as well as technical instruments.

"I Survived the TCP Transition": Standardization as Labor

To historians familiar with the complexities of technological systems, it may seem self-evident that putting a standard into practice often requires large amounts of expert work. The people who work with the technology are sure to need new knowledge or training; machines, materials, and work practices may have to be realigned. But most accounts of the development of computer systems omit this part of the process entirely. They simply note how the standards came to be specified, as if implementation of the specification were an easy and automatic sequel—an assumption that reinforces the perception of standards as knowledge rather than practice.[53] One way to recover the labor dimension of standards—and of "knowledge work" more generally—is to observe how and why the introduction of standards creates conflicts among the people building a technological system. A striking example of this comes from the development of the Internet.

The Internet's predecessor, the ARPANET, had been developed between 1969 and 1972 by the US Department of Defense's Advanced Research Projects Agency. ARPA's role in the Department of Defense was to identify and develop cutting-edge technologies with possible military value, and the agency had been funding computer science at American universities since the early 1960s. The ARPANET was intended to provide data communications among the agency's many research sites. The ARPANET began as a single network, but in the early 1970s ARPA managers decided to connect it to some of the agency's newer experimental networks. The resulting set of interconnected networks become the Internet.

ARPA program managers Vinton Cerf and Robert Kahn devised a technical framework for internetworking that included a new protocol, called TCP, that would provide a common language for all the networks in the Internet. For their design, which has proved remarkably adaptable in the face of rapid growth, Cerf and Kahn have been celebrated in popular culture as "fathers of the Internet." Yet the introduction of TCP as a standard for the ARPANET was not necessarily welcomed by the programmers and system administrators who maintained the network's computers. In fact, though this episode is seldom discussed in the popular histories, the introduction of TCP as a standard for the ARPANET was a traumatic and disruptive experience for many members of the network community.

The ARPANET contractors were mostly university scientists and their graduate students, and ARPA's managers came from this same world. These managers interacted with their contractors as colleagues and preferred consensus-style decision making. Yet, although computer science researchers and their government funders may not fit our usual conception of "labor" and "management," work issues had been a cause of some tension from the beginning of the project. The ARPA contractors were skilled professionals who wished to focus their efforts on projects that met their own interests, but they were ultimately beholden to their financial backers. (During the 1960s, the Department of Defense's spending on computing research was far greater than the universities' budgets for computer science.) When the idea of building the ARPANET had been announced at a 1967 meeting of ARPA contractors, many of those present had been critical of the idea, in part because they did not want to commit their efforts to the formidable task of building the network; the threat of losing their ARPA funding helped persuade them to join the project. During the course of the network-building effort, ARPA's managers occasionally felt it necessary to break out of their collegial mode of management and prod reluctant contractors to do the work necessary to get their computers on the network.[54]

The adoption of TCP as a Department of Defense standard in 1980 once again brought this tension out into the open. The ARPANET had an existing protocol called the Network Control Program (NCP), which had been providing satisfactory service since the early 1970s. But NCP was specifically designed for the ARPANET and would not adapt well to the diverse set of networks that were to be included in the new Internet. So in the early 1970s Cerf and Kahn began designing a new standard, called the Transmission Control Protocol. (Subsequently some of the functions of TCP were split off into an Internet Protocol, and the pair became known as TCP/IP.) Experimental versions of TCP were tested in the late 1970s, and in 1980 the Office of the Secretary of Defense formally adopted the ARPA protocols as military standards. Cerf and Kahn were ready to switch the ARPANET from NCP to TCP, a move they saw as serving ARPA's overall plan to create a multi-network data communications system. But many of the agency's contractors at the various network sites were hesitant to adopt TCP; they were satisfied with NCP, and since most of them were not planning to use ARPA's other networks they saw no immediate reason to adopt a new protocol. Managers of computer systems were naturally reluctant to do the work of implementing a complex new standard when they did not expect to benefit from it.

In this situation, the ARPA research community's usual consensus mode broke down and underlying power relations came to the fore. Official memoranda from the Department of Defense reiterated that the new protocol would become the ARPANET standard. Defense managers set a timetable for the transition and threatened sanctions (denial of access to the network) for noncompliance. This series of measures forced the reluctant contractors to implement the new standard in a relatively short period of time. Though the computer scientists had no real way to reject the standard if they hoped to continue receiving Defense funding, newsletters circulated by Defense administrators during this process repeatedly chided contractors for being slow to implement the new protocols, hinting at passive resistance from their expert labor force.

The contractors' resistance to the new standard was based on a realistic appraisal of the magnitude of the work that would be required of them. Each computer had to have TCP software provided for it; since TCP performed a number of complex functions, writing this software was a challenging task that might take even an experienced programmer 18 months or more. In addition to implementing TCP, the computer system managers had to rewrite all their network applications (such as file transfer, electronic mail, and remote log-in programs) to work with the new procedures and data formats required by TCP, and they had to replace their old network interface hardware. Programmers and system managers at the various sites recalled that it required an enormous and protracted effort for them to meet the deadline for the adoption of TCP, which had been set for January 1, 1983. One participant recalled that "the transition from NCP to TCP was done in a great rush, occupying virtually everyone's time 100% in the year 1982. . . . It was a major painful ordeal." Another contractor agreed that there had been a "mad rush at the end of 1982." The official ARPANET newsletter warned contractors in September 1982, "If you have NOT implemented TCP/IP, the end of the world is near!" Dan Lynch, a computer systems manager at SRI (Stanford Research International), recalled: "Dozens of us system managers found ourselves on a New Year's Eve trying to pull off this massive cutover. We had been working on it for over a year. There were hundreds of programs at hundreds of sites that had to be developed and debugged." Lynch made up buttons that read "I Survived the TCP Transition" and passed them out to his colleagues.[55]

Although the adoption of TCP was a technical milestone that cleared the way for the rapid expansion of the Internet in the 1980s and the 1990s, participants' accounts remind us that it also represented a significant appropriation of labor. This incident illustrates how a focus on standards can highlight hidden labor conflicts. ARPA, an organization that normally relied on informal, consensus-based management, suddenly reverted to top-down control *for the very reason* that standards, by their nature, require system-wide conformity and often impose significant labor costs. Ironically, the high professional status of ARPA's academic computer scientists and their students, which freed them from hourly wage labor, also made the intensification of their work invisible: long nights spent in the machine room did not produce such obvious signs as higher labor costs or protest actions.

We have noted that the adoption of TCP required significant amounts of skilled labor because of the complexity of the protocol. The reason TCP was much more complex than its predecessor was that the change in protocols was explicitly designed to displace labor and responsibility from one area of the system to another. The ARPANET can be thought of as having two parts: a set of "host" computers at the various research sites and a communications network that connected these sites. With the old NCP protocol, the communications network did most of the work of setting up and managing a connection between host computers at different sites; the hosts themselves had a relatively small role. (An analogy would be the telephone system, where the machines at the endpoints—telephone sets—are simple devices, whereas the communications network that connects them is quite complex.) But, unlike the ARPANET, ARPA's other experimental networks were not able to support the level of internal complexity demanded by the NCP design. So Cerf and Kahn determined that the new protocol, TCP, should minimize the demands made on the communications network by transferring most of the work to the endpoints of the network—the host computers at ARPA's research sites.

This brings us to our second observation: that the redistribution of work within the ARPANET system was fundamental, not incidental, to the design of the TCP standard. It is true that "work" in this case refers primarily to the functions performed by computers; however, the protocol designers were also conscious that tasks done by a machine translated into work that had to be done by the people maintaining that machine. This is a point that critics of TCP made clear. For instance, proponents of a rival networking standard called X.25 claimed that their system was superior to TCP because it required much less effort from individual computer operators.[56] The ARPANET community was aware of this displacement of labor from the center to the periphery of the system, but it had little choice in the matter. With ARPA holding the purse strings, those who wanted the privilege of working with expensive computing machines were ultimately willing to make the effort required to switch to TCP, however much they may have grumbled.

Excelling at "Bake Offs": Testing as the Performance of Authority
We have seen that implementing standards can involve significant labor. Verifying that an implementation actually conforms to the relevant standard is also a challenging task, particularly in the case of complex networking software. Starting in the late 1970s, members of the Internet community began organizing large-scale protocol testing events to address this problem. The evolution of test procedures reveals how standards were bound up with authority in the workplace. For ARPA contractors developing the first TCP implementations, the interpretation of standards tests depended in part on the credibility of the participants. Later, when the protocols were commercialized, computer experts would be able to translate their knowledge of protocol standards into professional power.

Networking protocols are so complex that it is almost impossible to verify an implementation directly. Demonstrating mathematically that a software program will behave in a particular way is notoriously difficult, and this sort of testing has rarely been attempted in the history of software.[57] In theory, an implementation can be tested against a benchmark program, but in the early stages of protocol development such programs are not yet available. Therefore, protocol developers have tended to adopt an empirical method, testing all the available implementations against one another. This provides an opportunity to observe each program under a wide range of conditions as it interacts with the other programs, no two of which are identical. A program that passes all the tests is assumed to comply with the specification.

This method of testing obviously requires that the individuals developing protocol software coordinate their efforts. Two who took on major organizational and leadership roles in establishing such tests for the Internet protocols were Jon Postel and Dan Lynch. In the 1980s both were ARPA contractors at the University of Southern California's Information Sciences Institute; Postel was a researcher at ISI and Lynch was director of the computer center. As early as 1978, a few years after TCP was first specified, Postel had begun organizing the sites that had implemented TCP to run a series of tests on the software. Some of these tests could be done in isolation, but most involved exchanging data with programs at other sites, which were reached through the Internet. The tests were designed to check various features of the protocol and to simulate potentially problematic situations. Each site would report its results to the test coordinator (usually Postel), who would draw up a chart that showed, for every possible pairing of implementations, whether or not their interaction had been successful.

Postel dubbed these tests "Bake Offs," in a lighthearted reference to cooking competitions. Though the analogy has feminine connotations, the rhetoric surrounding the Bake Offs reflected the more masculine values of competition and aggression, turning the genteel culinary contest into a pie-throwing match. Each site participating in the Bake Off was awarded points if its software correctly performed various actions. But participants did not simply compare themselves against a positive standard; they also performed what amounted to destructive tests of one another's software. The written procedures for the Bake Off invoked a boxing metaphor: progressively difficult levels of tests as the "featherweight," "middleweight," and "heavyweight" divisions.[58] Many of these tests involved benign operations such as initializing a connection, sending a message to the computer at the other end, and closing the connection. But others were adversarial, referring to the software at the other end of the connection as an "opponent" that the tester should try to "knock out." For instance, the rules awarded 30 points for causing the other party's software to crash (a knockout). Participants could choose to send "Kamikaze packets" or "nastygrams," which were messages designed to cause trouble by combining unusual options to create a situation that the "opposing" programmer might not have anticipated. Points were also awarded to those who could demonstrate that another participant had implemented the standard incorrectly.

Functionally, the adversarial aspects of the Bake Off provided a way to test the software's performance under severe conditions. Symbolically, they also reinforced the competitive culture of the computer programmers, and they illustrate how authority was constructed in this culture. Success was relative, and credit could be earned through destructive as well as constructive means. Since the measure of an implementation's performance was relative rather than absolute, test results had to be defended rhetorically. Postel noted that test results were not definitive or self-explanatory but, rather, provided a basis for participants to claim that they had met the standard: "The only way to determine if an implementation was 'correct' was to test it against other implementations and argue that the results showed your own implementation to have done the right thing."[59] Presumably, the level of authority participants carried within the community colored the reception of such claims. Reflecting the expectation that test results were open to interpretation and dispute, the rules for the Bake Off even awarded "10 points for the best excuse."[60] The incentive for making such excuses was not only to save face but also to save labor, since if the software failed the test there would be strong pressure on the programmers to rewrite it.

More radically, participants who failed to meet the standard could argue that the specification itself was at fault and should be changed. (Again, the participant's credibility probably influenced whether the fault was judged to be in the implementation or in the standard itself.) In this way, Bake Off–style testing did not just measure a product against a static body of knowledge represented by the standard; it also functioned as a way to generate new knowledge about how protocols did and should behave, providing feedback between the processes of designing and implementing the evolving standard.

"Trust, but Verify": Verification as the Transfer of Labor

As numerous commercial versions of TCP/IP became available in the early 1980s, entrepreneurial standards experts saw an opportunity to develop large-scale, commercially run testing activities. During these verification events, vendors would undertake to publicly demonstrate that their products conformed to the standard. Rather than remaining behind the doors of the manufacturer's workplace, therefore, some aspects of product testing moved out into a new, symbolically "neutral," highly visible arena. Public verification activities shifted the labor of evaluating network products from potential buyers to vendors and third-party experts.

The 1983 changeover from NCP to TCP, though traumatic for many, also presented opportunities for ambitious systems managers. One of these was Dan Lynch at ISI, the biggest and most heavily used ARPANET site. Lynch volunteered to coordinate a large-scale revision of software programs during the transition to the TCP standard, and this experience established him as an authority on protocol testing.[61] Like the standards experts at early-twentieth-century building sites, Lynch set out to use his expertise as a basis for professional advancement; also like them, he envisioned himself occupying a middle ground between the university-based scientists who specified the standards and the companies that tried to apply them. Noting that vendors of computer products had difficulty interpreting the arcane technical documents created by ARPA's academic researchers, Lynch believed he could make his fortune by bridging this gap. He formed a company and offered to help computer vendors understand TCP/IP and market it to their customers.

Lynch's proposed vehicle for promoting TCP/IP was a trade show, which he called the "TCP/IP Interoperability Conference" or simply "Interop." Like the TCP Bake Offs, Interop was intended to demonstrate that different implementations of the Internet protocols could interoperate successfully. Conference staffers would link computers running various TCP/IP products into a local network called the InteropNet, which would also be attached to the Internet. People attending the trade show could log into computers on the InteropNet and test the network products by trying to communicate with other machines at the show or elsewhere on the Internet. The vendors would cover the costs of the trade show, and prospective customers would pay to attend, generating profits for Lynch's company. The first Interop was held in Santa Clara, California, in September 1988, with about 50 companies participating. It was so popular and profitable that Lynch decided to make it an annual event. The second Interop featured 100 companies, the third 200. By 1994 there were 500 exhibitors and many thousands of attendees, and the show's scope had expanded to test a variety of protocols besides TCP/IP. Eventually the organizers were running two Interops per year in the United States and annual shows in six other countries.[62]

To reinforce the image of Interop as a source of authoritative information about standardized products, Lynch and his successors used rhetorical strategies that presented the event as an opportunity for users to gain reliable, unmediated knowledge of networking products. Typical Interop press releases referred to the InteropNet as "a real-world scenario" and claimed that the trade show "gives exhibitors and attendees the opportunity to personally experience the melding of different technologies and equipment." The engineer in charge of one year's InteropNet claimed: "Our network marks the place where talk and hype move to the sidelines, and performance takes center stage."[63] There is, however, a certain paradox in offering a "hands-on" demonstration of networking, since the flow of electronic information is impossible to observe directly. Lynch himself was aware of this problem when he began the Interops. To make the experience seem more "real" to attendees, he made a point of keeping the infrastructure visible—cables and network switches were exposed to view, allowing attendees to see how the various computers were connected. A skeptical user who wanted to make sure the whole thing was not a hoax or a simulation could perform an impromptu test, such as unplugging a cable and noting the effect of this on the network. Once Interop had gained some credibility, however, the symbolism of the visible hardware was no longer needed. Interop did not necessarily free potential buyers from having to take standardization claims on faith; rather, it gave them the option of placing their trust in the Interop staffers—who were billed as disinterested experts—and the event itself, rather than the vendors.

The Interop phenomenon illustrates two ways in which technical standards can function as a means for redistributing power and authority among skilled workers.

On the one hand, Lynch was able to create a lucrative and respected position for himself based on his credibility as a standards arbiter. His credibility rested not only on his personal expertise but also on his assumption of a position of neutrality with respect to the commercial interests and, above all, on the fact that he invited potential buyers to test the products themselves. By appealing to the users' expertise, Lynch solidified his own authority, since every attendee who tried the InteropNet and was satisfied with its performance confirmed Lynch's judgment in championing TCP/IP. Like the early-twentieth-century architects who were able to "borrow" an aura of expertise from General Electric by adopting its refrigerator specifications, Lynch was able to profit by his association with a well-regarded standard.

On the other hand, the Interops shifted a burden from their target audience: the owners and operators of computer systems, who were potential buyers of TCP/IP products. The insight for labor history here is that these "consumers" of network products were workers in their own right. The users of network products did not simply *consume* hardware and software; they also *produced* network services in their own workplaces, which required both skill and labor. The purchaser of a network product had to perform tasks that could include evaluating the performance of the product and its compatibility with the user's existing systems; installing, testing, and debugging software and hardware; establishing new systems-maintenance routines; and retraining network administrators. The appeal of Interops, we argue, was that they redistributed some of this labor from consumers to producers. By encouraging companies to create interoperable products (so that they would have something to demonstrate at the show), Interops promoted standardization, which relieved computer users of the need to cope with incompatible systems. Interops also guaranteed potential buyers that specific products would work as advertised. For the price of admission, attendees could quickly and easily obtain the information they needed to make purchasing decisions; the conference fee was an exchange for the labor they would otherwise have had to expend on evaluating products and making them work together. Interops altered the balance of power between buyers and sellers by providing consumers with reliable knowledge about the worth of the products on offer.

Much of the work displaced from customers now fell instead on the suppliers, who had to ensure that their products met the standards required by the demonstrations; companies also paid to participate in Interop and contributed equipment and skilled personnel to help design and build the InteropNet. From the vendors' point of view, Interop justified this investment by serving several purposes. First, it provided a laboratory for testing their products; such testing was mostly done in advance of the show as volunteers from various companies put together the InteropNet. (This resulted in a secondary displacement of labor *within* the companies making network products: the engineers did extra work to prepare for Interop, but because of their efforts the company's products were more likely to meet customers' expectations—which meant less work down the line for the customer service department.) Second and most important, Interop was a marketing opportunity, in which the authority of Lynch and his successors as neutral experts—as well as the evidence of the customer's own eyes—served to bolster the vendor's claims. Finally, in addition to demonstrating the merits of individual products, Interops promoted TCP/IP as an industry standard against the alternatives of adopting some other standard or having no common standard at all.[64] In the early years of TCP/IP, potential customers who had been using products from IBM or other sources had to be convinced this new protocol would work at all. Lynch recalled seeing two engineers from Ford taking their manager to observe a demonstration of TCP/IP. One of the engineers pointed to the display and said, "See, it works! *Now* will you sign the purchase order?" The boss agreed.[65] Every user who switched to the TCP/IP standard was a potential customer for a company's particular TCP/IP products. As with the earlier case of construction specifications, the status of the product and that of the producer rose together.

Bake Offs and Interops highlight the *performative* nature of standards. As public spectacles, these events implicitly asserted the importance of standardization itself. In addition, such practices helped establish particular standards within the industry. Widespread industry adoption of TCP/IP was not based solely on a "rational" assessment by companies of their technical requirements; it was actively promoted by third parties, such as Lynch, who not only believed in the technical superiority of TCP/IP but also saw standards activities as a way to advance their own careers. Public tests also focused consumers' attention selectively on certain aspects of the standard—in this case, interoperability—at the expense of other possible concerns. Alternative demonstration methods could have been devised that emphasized different features of network software, such as speed or reliability; but the Interop demonstration perpetuated the idea that compatibility between different vendors' products should be the main selling point. At the level of the individual product, the very definition of "standard" became contextualized rather than absolute: the measure of a product's compliance was how well it interacted with other products and with users.

Because standardized products embody labor (the work of specifying, implementing, and testing the standard), standards can serve as a medium for transferring labor between different types of workers or between producers and consumers. Protocol standards, for instance, free the user from the labor of fitting incompatible components together, while verification activities free consumers from the labor of evaluating vendors' claims. In some cases, this displacement of labor is simply a by-product of the standards effort, as with the work involved in the 1983 adoption of TCP for the ARPANET; in other cases, it is the primary goal of standardization. In either case, examining the labor implications of product standards can alert historians to previously unsuspected changes in the distribution of responsibility, skill, and authority in the workplace.

Conclusions

By now it should be apparent that standards and specifications, whether for concrete or computer code, are neither simplifying nor uniform in their effects. Some guidelines do eliminate work or intellectual complexity in industrial enterprises, but many have just the opposite effect, displacing or creating labor as they increase the productivity of a technology or a company. What all standards and specifications do have in common is their dynamic nature. We expect such technical protocols to bring fixity to material activity, but closer examination reveals the conditions brought by industrial standards to be relatively short-lived and by no means universal. It is far more useful to conceive of standards not as stable or stabilizing entities but as *mediums for exchange*: a non-economic means of negotiation among the many actors seeking to produce, use, or profit from industrial technologies. Standards convey not only control of materials and techniques but also opportunities for control of markets and workplaces, whether by administrative or reputational means. At times, standards and specifications lend that power to those who create them (elevator companies, protocol designers); at other times, they bolster the credibility and authority of those who use the guidelines (architects, computer owners). Most interesting for us are the moments when the interests of the two groups either coincide or radically diverge: these are the points when standards and specifications reveal their currency in the world of commercial exchange.

We have hinted at the significant difference between prescribing a technical practice (as the standard and specifications writers clearly intend) and enforcing the desired practice. How can we know if people actually follow directions or use conventional designs in their work? This is a complex historical problem that involves tracking some of the truly hidden features of industrial operation: What happens when a contractor finds he has bought too little reinforcement for a building that is already behind schedule? What happens when a programmer prefers to find a "good excuse" rather than rewrite a piece of software? These questions highlight one of the most intriguing features of standards and specifications: their intention to wed word and deed. Written standards must be defended by actions (new workplace organizations, verification events). In turn, making practical use of a standard often involves further rhetorical strategies (arguing that a protocol should be redesigned, writing an article that calls attention to accidents involving your competitors' elevators). It is in such escalating attempts to regulate the performance of activities that the conflict and much of the significance of standards lies.

Notes

1. For an overview of the economic functions of standards, see Samuel Krislov, *How Nations Choose Product Standards and Standards Change Nations* (University of Pittsburgh Press, 1997). Considerations of standards that address the nature of industrial labor are discussed below, but it should be noted that David Noble's *America by Design* (Oxford University Press, 1977) remains the most comprehensive of such works. A number of now-classic community and company studies have offered a basis for our study by describing processes of industrial standardization (if not the history of written standards and specifications themselves). Among these are the following: Merritt Roe Smith, *Harpers Ferry Armory and the New Technology* (Cornell University Press, 1977); David Hounshell, *From the American System to Mass Production, 1800 to 1932* (Johns Hopkins University Press, 1984), and Mary Blewett, *Men, Women, and Work* (University of Illinois Press, 1988). More recently, Ken Alder (*Engineering the Revolution*, Princeton University Press, 1997) has probed the political and social origins and impacts of standardization with particular acuity.

2. See Philip Scranton, "None-Too-Porous Boundaries: Labor History and the History of Technology," *Technology and Culture* 29 (1988): 722–743; Sean Wilentz, *Chants Democratic* (Oxford University Press, 1984); Judith McGaw, *Most Wonderful Machine* (Princeton University Press, 1987); Larry D. Lankton, *Cradle to Grave* (Oxford University Press, 1991); Jacqueline Hall et al., *Like a Family* (University of North Carolina Press, 1987). Scranton's article surveys progress in this interdisciplinary effort since the 1960s.

3. Ruth Schwartz Cowan, "The Consumption Junction," in *The Social Construction of Technological Systems*, ed. W. Bijker et al. (MIT Press, 1987): 261–280; Lizabeth Cohen, *Making a New Deal* (Cambridge University Press, 1990). See also Ronald Edsforth, *Class Conflict and Cultural Consensus* (Rutgers University Press, 1987); Richard Wightman Fox and T. J. Jackson Lears, eds., *The Culture of Consumption* (Pantheon, 1986).

4. Historians of science offer additional important models here. Simon Schaffer, Theodore Porter, and Graeme Gooday locate elaborate and self-conscious institutional and occupational agendas in apparently mundane "by-products" of scientific inquiry such as scientific constants, methods of calibration, and statistical systems—see Schaffer, "Accurate Measurement is an English Science," Porter, "Precision and Trust: Early Victorian Insurance and the Politics of Calculation," and Gooday, "The Morals of Energy Metering: Constructing and Deconstructing the Precision of the Victorian Electrical Engineer's Ammeter and Voltmeter," all in

The Value of Precision, ed. M. Wise (Princeton University Press, 1995); see also Porter, *Trust in Numbers* (Princeton University Press, 1995).

5. Scranton, "None-Too-Porous Boundaries."

6. Thomas P. Hughes, "The Evolution of Large Technological Systems" and John Law, "Technology and Heterogeneous Engineering: The Case of Portuguese Expansion," both in *The Social Construction of Technological Systems*, ed. Bijker et al.

7. Like Lorraine Daston and Peter Galison's work on the shifting definition of "objectivity" over the last two centuries ("The Image of Objectivity," *Representations* 40, 1992: 81–128), our study finds that technical authority, like scientific certitude, has a social basis. The content of standards reveals what those in authority wish to advance as "best practice." We argue that, like laboratory standards, industrial standards are historically contingent and exhibit tensions between the globalizing programs of those who create standards and the local conditions experienced by those who use them. On the historical connection of moral virtue and the reliance on faculties of judgment in science, see Peter Galison, "Judgment Against Objectivity," in *Picturing Science, Presenting Art*, ed. P. Galison and C. Jones (Routledge, 1998). See also Alex Soojung-Kim Pang, "Visual Representation and Post-Constructivist History of Medicine," *Historical Studies in the Physical and Biological Sciences* 28 (1997): 139–171.

8. David E. Nye, *Image Worlds* (MIT Press, 1985).

9. Stefan Timmermans and Marc Berg, "Standardization in Action: Achieving Local Universality through Medical Protocols," *Social Studies of Science* 47 (1997): 273–305; Linda F. Hogle, "Standardization Across Non-Standard Domains: The Case of Organ Procurement," *Science, Technology and Human Values* 20 (1995): 482–500; Geoffrey C. Bowker and Susan Leigh Star, *Sorting Things Out* (MIT Press, 1999); Warwick Anderson, "The Reasoning of the Strongest: The Polemics of Skill and Science in Medical Diagnosis," *Social Studies of Science* 22 (1992): 653–684; Karen Rader, "'The Mouse People': Murine Genetics Work at the Bussey Institution, 1909–1936," *Journal of the History of Biology* 31 (1998): 327–354.

10. Daniel J. Hauer, "Specifications," in *Handbook of Building Construction*, volume 2, ed. G. Hool and N. Johnson (McGraw-Hill, 1920), pp. 1075–1076; Richard Shelton Kirby, *The Elements of Specification Writing* (Wiley, 1935), pp. 1–8, 80–106; John C. Ostrup, *Standard Specifications for Structure Steel—Timber—Concrete—Reinforced Concrete* (McGraw-Hill, 1911). See also Krislov, *How Nations Choose Product Standards and Standards Change Nations*.

11. Historians have recently turned their attention in greater numbers toward the function of written materials in the operation of industry; two exceptionally helpful works are JoAnne Yates's *Control through Communication* (Johns Hopkins University Press, 1989) and Geoffrey Bowker's *Science on the Run* (MIT Press, 1994).

12. For an overview of this effort, see Scranton, "None-Too-Porous Boundaries." In addition to Noble's *America by Design*, foundational works in this area include

the following: Harry Braverman, *Labor and Monopoly Capital* (Monthly Review Press, 1974); David Montgomery, *Workers' Control in America* (Cambridge University Press, 1979); Daniel Nelson, *Managers and Workers* (University of Wisconsin Press, 1975); David Gordon, Richard Edwards, and Michael Reich, *Segmented Work, Divided Workers* (Cambridge University Press, 1982). Many company and community studies have augmented this work; one of these is Stephen Meyer, *The Five Dollar Day* (SUNY Press, 1981).

13. Amy Slaton, "'As Near as Practicable': Precision, Ambiguity, and the Social Features of Industrial Quality Control," *Technology and Culture* (forthcoming).

14. E. M. Haas, "Standardization of Buildings," *Railway Age* 64 (February 8, 1918), pp. 287–298; H. Ward Jandl et al., *Yesterday's Houses of Tomorrow* (Preservation Press, 1991); *Twentieth-Century Building Materials*, ed. T. Jester (McGraw-Hill, 1995); Cecil Elliott, *Technics and Architecture* (MIT Press, 1992), pp. 82–83, 192–193.

15. The idea that the routinization of construction work might actually benefit building workers was expressed by some analysts. Their premise was that lower costs for building operations would translate into more buildings being erected, and thus into more jobs for construction workers. See Adolph J. Ackerman and Charles Locker, *Construction Planning and Plant* (McGraw-Hill, 1940), pp. 365–367.

16. Gwendolyn Wright, *Moralism and the Model Home* (University of Chicago Press, 1980), pp. 192–190, 198; Marc Silver, *Under Construction* (SUNY Press, 1986), pp. 1–13; William Haber, *Industrial Relations in the Building Industry* (Harvard University Press, 1971 [1930]), pp. 36, 47.

17. Sarah Bradford Landau and Carl W. Condit, *Rise of the New York Skyscraper, 1865–1913* (Yale University Press, 1996), pp. 35–36; H. P. Bates, G. H. Cheesman, and W. W. Lighthipe, "Elevators," in *Handbook of Building Construction*, volume 2, ed. Hool and Johnson.

18. A particularly interesting narrative about the construction of an industrial reputation is offered in chapter 3 of Bowker's *Science on the Run*. See also Nye's *Image Worlds*. The literature on how scientific and other "expert" professions achieve social credibility has grown steadily since the appearance of the following books: Steven Shapin and Simon Schaffer, *Leviathan and the Air-Pump* (Princeton University Press, 1985); Andrew Abbott, *The System of Professions* (University of Chicago Press, 1988); T. Haskell, ed., *The Authority of Experts* (Indiana University Press, 1984); R. Walters, ed., *Scientific Authority and Twentieth-Century America* (Johns Hopkins University Press, 1997).

19. For detailed descriptions of the organization of commercial building projects in this period, see *Handbook of Building Construction*, volume 2, ed. Hool and Johnson; Ray J. Reigeluth, *Safety and Economy in Heavy Construction* (McGraw-Hill, 1933); Amy Slaton, *Reinforced Concrete and the Modernization of American Building* (Johns Hopkins University Press, 2001).

20. Catalogs of the Tyler Elevator Company, Cleveland (ca. 1927), the Burdett-Rowntree Manufacturing Company, Chicago (1914), and other catalogs in the Trade Catalog Collection of the Hagley Museum and Library.

21. Catalog of the Tyler Elevator Company, Cleveland (ca. 1927), in Trade Catalog Collection of Hagley Museum and Library.

22. Ibid., p. 6.

23. Landau and Condit, *Rise of the New York Skyscraper*, p. 256.

24. See for example, John Preston Comer, *New York City Building Control, 1800–1941* (Columbia University Press, 1942), pp. 58, 72; Edward Van Winkle, "Elevator Car-Safety Appliances and the Building Code," *Engineering News* 51 (1903), p. 354.

25. Interestingly, many were authored by employees of elevator firms, clearly posting attacks on competitors, and reports often included evidence of extensive safety tests conducted by the author's own firm. Typical exchanges of this type appear in the following articles: "Air Cushion for Elevator Shafts in High Buildings," *Engineering News* 42 (1899), pp. 274–275; "A High Drop Test of an Elevator Safety Cushion," *Engineering News* 48 (1902), pp. 295–296; "A Device for Testing an Elevator Test Cushion," *Engineering News* 48 (1902), p. 316.

26. Steward T. Smith, "Mechanical Refrigeration," in *Handbook of Building Construction*, volume 2, ed. Hool and Johnson; Ruth Schwartz Cowan, "How the Refrigerator Got Its Hum," in *The Social Shaping of Technology*, ed. D. MacKenzie and J. Wajcman (Open University Press, 1985), p. 204; Sigfried Giedion, *Mechanization Takes Command* (Oxford University Press, 1969), pp. 600–605; Ellen Lupton and J. Abbott Miller, *The Bathroom, the Kitchen, and the Aesthetics of Waste* (Princeton University Press, 1992), pp. 11–13.

27. Cowan, "How the Refrigerator Got Its Hum," pp. 207–210.

28. See, for example, Arthur C. Davis, "The St. Regis," *Architectural Record* 15 (1904), pp. 554, 615.

29. "General Electric Refrigerators for Residences and Apartment Homes," catalog produced by General Electric Company: Electric Refrigeration Department, ca. 1931, in Trade Catalog Collection of Hagley Museum and Library.

30. Lupton and Miller, *The Bathroom, the Kitchen, and the Aesthetics of Waste*, pp. 48–64. See also additional GE catalogs in the Hagley collection.

31. "General Electric Refrigerators for Residences and Apartment Homes."

32. For a representative discussion of attitudes toward architects held by building suppliers, see C. A. Crane, "The Relation between Engineers and Contractors," *Proceedings of the American Concrete Institute* 13 (1917): 130–142.

33. A telling quote appears on p. 49 of Andrew Young's article "The Relations of the Plumbers and the Physicians" (*Public Health* 17, 1903): "Plumbing is no longer a mere trade. Its importance and value in relation to health, and its requirements regarding scientific knowledge, have elevated it to a profession."

34. Wright, *Moralism and the Model Home*, pp. 178–185, 200, 212–213.

35. Again, cultural, technical, and occupational trends were mutually reinforcing. The domestic refrigerator carried intimations of gentility, in particular embodying a blend of values of privacy and material accumulation. Efficiency was important.

Many upscale hotels and apartment buildings of the first years of the century maintained centralized kitchens and laundries in which the domestic needs of individual families could be addressed in large-scale, factory-like operations. Yet the extreme valuation of privacy prevailed: as dumbwaiters discretely carried meals from a central kitchen to each apartment's private dining rooms in such buildings, so individual refrigerators became the norm in each family's kitchen, though far more expensive and inefficient than would be a shared, institutional refrigerator. Early refrigerators had mechanical equipment in housing that could be separated from the cold cabinet itself; many apartment buildings so emphasized private ownership of appliances that a single refrigeration plant in a basement connected to cabinets in individual apartments. (Gwendolyn Wright, *Building the Dream* (Pantheon, 1981), pp. 138, 147)

36. See, e.g., advertisements by the Otis Elevator Company in *Architectural Record* from the 1910s and the 1920s.

37. Amy E. Slaton, Origins of a Modern Form: The Reinforced Concrete Factory Building in America, 1900–1930, Ph.D. dissertation, University of Pennsylvania, 1995, pp. 192–261.

38. Ross F. Tucker, "The Progress and Logical Design of Reinforced Concrete," *Concrete Age* 3 (1906), p. 333.

39. Advertisement for Clinton Wire Cloth Company, *Sweet's Catalog* 1906, p. 96. Some reinforcement makers were clearly less sophisticated than the large concerns. The Hinchman-Renton System, for instance, offered reinforcing made from "ordinary barbed wire." But even the simplest product lines offered purchasers economies of scale based on replacing individually assembled reinforcement with mass-produced assemblies. The Hinchman-Renton Company was based in Denver, which may be the reason they considered barbed wire to be "inexpensive and readily obtained in any quantity" (Walter Mueller, "Reinforced Concrete Construction," *Concrete Age* 3 (1906), p. 325).

40. Mueller, "Reinforced Concrete Construction," pp. 323–324; advertisement for Unit Concrete Steel Frame Company, *Sweet's Catalog*, 1906, p. 127.

41. The Aberthaw Construction Company, for example, had purchased patent rights to Ernest Ransome's twisted steel reinforcement designs in 1896 and manufactured reinforcing from them for its own use and for sale. When the patent rights expired, probably about 1915, Aberthaw found it uneconomical to produce its own rods given the many "deformed" rods now on the market, especially those of the Kahn Company. See The Story of Aberthaw (unpublished manuscript in archives of Aberthaw Construction Company, North Billerica, Massachusetts).

42. Trussed Concrete Steel Company, Detroit, Michigan, "Hy-Rib: Its Application in Roofs—Floors—Walls—Sidings—Partitions—Ceilings—Furring," 1912, in the Trade Catalog Collection of the Hagley Museum and Library. The Trussed Concrete Steel Company was owned by Julius Kahn, brother of the celebrated factory designer Albert Kahn. Julius Kahn built a vast enterprise on his innovative reinforcing system of "trussed steel bars": rolled steel bars of diamond cross-section with bent up "wings" attached to either side. The wings countered the shearing

forces found in concrete beams, adding 20–30 percent to the strength of a beam. Other reinforcing systems for walls and flat slabs that varied slightly from Hy-Rib products are described in Mueller, "Reinforced Concrete Construction" (pp. 321–332) and in A.J. Widmer, "Reinforced Concrete Construction," *25th Annual Report of the Illinois Society of Engineers and Surveyors*, 1915 (p. 148).

43. The Trussed Concrete Steel Company had many divisions, including The Truscon Laboratories of Detroit, and made a point of delineating the expertise of its own employees from that of its clients. A 1924 Truscon Maintenance Data Book, intended for "maintenance engineers, building owners, and others interested in the maintenance and profitable upkeep of buildings and equipment," gathered miscellaneous specifications and general advice for the upkeep of industrial plants. It featured illustrations of white-coated scientists in the company's laboratories, and celebrated the work of its "graduate engineers" on the problems of "well directed, intelligent maintenance."

44. See Harry C. Bates, *Bricklayers' Century of Craftsmanship* (Bricklayers, Masons and Plasterers International Union of America, 1955).

45. These patterns extend beyond the realm of traditional manufacturing into many other modernizing work settings of the twentieth century. Warwick Anderson ("The Reasoning of the Strongest") documented a similar attempt to distribute technical authority through the creation of written protocols. In his study of computerized diagnosis systems, Anderson treated "'craft' and 'scientific' representations of diagnosis symmetrically, as discursive resources used in the hospital context to legitimate the divergent competencies of . . . two occupational subgroups" (ibid., p. 655).

46. Philip Kraft, "The Routinization of Computer Programming," *Sociology of Work and Occupations* 6, no. 2 (1979): 139–155; Philip Kraft and Steven Dubnoff, "Job Content, Fragmentation, and Control in Computer Software Work," *Industrial Relations* 25, no. 2 (1986): 184–196; Joan Greenbaum, *In the Name of Efficiency* (Temple University Press, 1979); Juliet Webster, *Shaping Women's Work* (Longman, 1996); Roslyn L. Feldberg and Evelyn Nakano Glenn, "Technology and Work Degradation: Effects of Office Automation on Women Clerical Workers," In *Machina Ex Dea*, ed. J. Rothschild (Pergamon, 1983).

47. For a detailed description of the development of the Internet, see Janet Abbate, *Inventing the Internet* (MIT Press, 1999).

48. For an overview of a wide range of standards concerns for the data communications industry, see *Standards Policy for Information Infrastructure*, ed. B. Kahin and J. Abbate (MIT Press, 1995).

49. On compatibility issues in computing and attempts to create international standards for data networks, see chapters 2 and 5 of Abbate, *Inventing the Internet*.

50. The TCP/IP implementations used in the ARPANET did not go directly into the market: the US government (ARPA) sponsored computer manufacturers to develop commercial TCP/IP products for their computer lines and also paid for UNIX versions of the protocols. But the commercial and UNIX developers were able to build on the prior groundbreaking work of the nonprofit ARPANET sites, and in many cases the same people were involved.

51. Jon Postel, email to Janet Abbate, July 29, 1998.

52. For an account of the difficulties faced by users of early networks, see chapter 3 of Abbate, *Inventing the Internet*.

53. See, e.g., Katie Hafner and Matthew Lyon, *Where Wizards Stay Up Late* (Simon & Schuster, 1996); Arthur L. Norberg and Judy E. O'Neill, *Transforming Computer Technology* (Johns Hopkins University Press, 1996); Peter Salus, *Casting the Net* (Addison-Wesley, 1995).

54. For a discussion of tensions within the ARPANET community, see chapter 2 of Abbate, *Inventing the Internet*.

55. These accounts are drawn from documents archived at the library of Bolt, Beranek and Newman, a computer firm and one of the main ARPANET contractors. Eighteen months was the implementation time reported by a programmer at BBN, whose staff had already had considerable experience writing TCP software (John Sax, email to Alex McKenzie, May 20, 1991). The quotations are from the following, respectively: Mark Crisping, message to newsgroup comp.protocols.tcp-ip, June 21, 1991; Alex McKenzie, message to comp.protocols.tcp-ip, June 24, 1991; *ARPANET Newsletter* no. 16, September 30, 1982; Dan Lynch, message to comp.protocols.tcp-ip, June 23, 1991.

56. Abbate, *Inventing the Internet*, chapter 5.

57. "Any kind of formal verification [i.e., proving mathematically that a program does what it is supposed to] is a great deal more work than the bakeoff style of testing. So from the developer's point of view, getting things working together in a bakeoff is by far preferable to formal verification." (Jon Postel, email to Janet Abbate, July 29, 1998)

58. ARPANET Request For Comments (RFC) #1025, pp. 2–3.

59. Ibid., p. 1.

60. Ibid., p. 4.

61. Most of the information in this section is from an interview with Dan Lynch by Janet Abbate on July 23, 1998.

62. See the Interop web page (http: //www.interop.com).

63. Interop web site, 1998 press release (http://www.interop.com/Press_Releases/ni_042798.html).

64. Both of these were real possibilities. Many computer manufacturers had developed their own proprietary protocols that they could have used instead of TCP/IP; there was also an international standard called OSI that was a serious rival to the ARPA protocols.

65. Interview with Dan Lynch by Janet Abbate, July 23, 1998.

Engineering Politics, Technological Fundamentalism, and German Power Technology, 1900–1936

Edmund N. Todd

Standard histories of electrification describe technology's moving almost inexorably from small, independent stations through block, central, and overland stations to a national system.[1] However, Thomas P. Hughes argues that there were regional and national differences in the development of electric power systems. He and other historians of technology no longer give technology an independent role in history.[2] Nevertheless, specialists in other kinds of history continue to present technology as autonomous, as did many engineers and other commentators from the 1880s to the 1930s. But even successful engineers did not identify the same trends. Georg Klingenberg promoted giant electric power stations and a national system in the 1910s, Arthur Koepchen defended regional systems in the 1920s, and Wilhelm Stiel recommended independent power stations for textile mills in the 1930s.[3] They had to struggle to create their technological systems. Though it is understandable that these engineers would represent their efforts in terms acceptable early in the twentieth century, it is curious that historians would ignore a rich body of literature concerning technology, one of their central historical forces. Instead of culture lagging behind technology, as William F. Ogburn claimed, we see a lag in historiography.[4]

Klingenberg, Koepchen, and Stiel provide a focus for investigating representations of technology and technological change. Like their contemporaries, they presented technology as an exogenous factor driving social, legal, and political developments. Technology was changing Germany from an agricultural to an industrial nation, promoting urbanization, and driving a transformation in the scale of corporations. Klingenberg, Koepchen, and Stiel participated in those changes. Klingenberg (1870–1925), after receiving a doctorate in Rostock and qualifying to teach at the Technical

University of Berlin, taught power-plant design from 1898 to 1909. In 1902, he became director of power-plant construction and operations at the electrical manufacturing firm Allgemeine Elektrizitätsgesellschaft (AEG).[5] Koepchen (1878–1954) graduated from the Technical University in Karlsruhe and worked for 2 years in the electro-technical division of Felten & Guilleaume in Mülheim, near Cologne. In 1906, he was hired as technical director for a subsidiary of the Rheinisch-Westfälisches Elektrizitätswerk (RWE), based in Essen. In 1914, he became an acting member of the RWE's Vorstand (managing committee); in 1917 he became a full member.[6] During the 1920s, with a doctorate in engineering, Stiel (1878–1936) directed a department responsible for textiles, paper, cellulose, and leather at the electrical manufacturing firm Siemens-Schuckert in Berlin.[7]

These individuals can be called technological fundamentalists. First, they sought to make technology fundamental to decisions. Using the determinist language of their time, they presented their views as straightforward, factual ways of resolving problems. Widely accepted, determinism was a "social fact."[8] Second, unlike "realist," "determinist," or even "technocrat," the word "fundamentalist" stresses the role of belief. Technological fundamentalists believed they had a "revealed truth" that identified a real trend toward a proper future that could be rationally achieved. "Technocratic" faith in "realism" and "determinism" made them much like religious fundamentalists, who believe that they have the correct "God's-eye view" of the world.[9]

Faith alone was not enough. Although believing that technology was on their side,[10] the three engineers did not promote utopian schemes. To implement their visions of proper change, they had to deal with various levels of government, different corporations, and each other. They avoided "party politics," but they engaged in "engineering politics" as they sought to make their visions of proper change "technically and politically feasible."[11] To create a national system, Klingenberg sought support from the Reich against the "political" positions of his opponents. To build his regional system, Koepchen lined up allies among representatives from cities, counties (Landkreise), and corporations against the "political" actions of Reich officials, electrical manufacturers, and other opponents. To create independent stations for textile mills, Stiel opposed "political" systems deployed by directors of power companies. They made local concessions without giving up their fundamentalism. With dire consequences, these three "statesmen of technology" refused to legitimize the negotiations and compromises—a politics of contending interests—in which they participated.[12]

Historiographic Artifacts

Change in science and technology does not explain change in general. Laboratories and shop floors involve contingency and choice. Theory does not always guide experiment, and science does not always guide technology. Causal lines often run in the other direction.[13] Nor are disciplines all the same. Peter Galison refers to them as subcultures, the members of which, given appropriate contexts, may work out their differences locally without giving up disciplinary commitments. Knowledge and skills often work only in institutions, laboratories, or technological systems in which members of subcultures have adjusted theory, apparatus, and analysis to one another. For instance, not until the 1930s did electric power systems in Germany work well enough to support an academic discipline of power-system economics.[14] Thus, rather than assuming that science and technology are autonomous, it may be fruitful to investigate "distributions of knowledge" in social systems and to follow productive actors around to see what problems they face and how they resolve them.[15] As Hughes has demonstrated, those building technological systems need to resolve technical, social, political, economic, and even cultural problems.[16]

Because scientists and engineers struggle to create order and meaning,[17] nature and technology do not mysteriously arrive to provide a God's-eye "view from nowhere."[18] Nevertheless, scientists and engineers often write "shadow histories" that present science and technology as nonpolitical, higher forms of rationality. As they reconstruct histories leading to preferred futures, they hide and even forget the social effort that was necessary to shift knowledge and social structure. Even if recognized as historically inaccurate, shadow histories allow authors to move on to topics that they wish to investigate thoroughly.[19] However, scientists and engineers seem to believe their historical reconstructions.[20] By using their versions of history to pass on values, establish a sense of time, and project a future, scientists and engineers create meaning and impose order on themselves and the world in which they live.[21] Thus, history provides a terrain in which those fighting to make development follow their preferred paths retroactively seek to establish trends supporting their projections of an appropriate future.[22] Promoting a God's-eye view makes technology fundamental in these histories.[23]

Klingenberg, Koepchen, and Stiel wrote shadow histories identifying "proper" lines of development. For Klingenberg, the logic of history had led to large power stations and would lead to a national grid. Combining large stations would create steady use of equipment and reduce the cost of installed capacity as engineers resolved remaining technical problems.[24] For Koepchen, technology had driven development from "local stations," not to Klingenberg's national grid, but to the RWE's regional, "long-distance supply" system. Koepchen deployed history to buttress his argument: the "development of this large integrated system [was] beginning rather than ending." Although some small stations were seeking autonomy and expanding, "in the long run economic rationality [would] succeed." Combining North German thermal plants with South German hydroelectricity would produce "a rational balance" and allow "complete use of the varying availability of water power energy."[25] For Stiel, the logic of technological development was leading to independent stations, not to a national grid or regional systems. Whereas steam had earlier promoted centralization, electricity was decentralizing textile mills, which would generate their own power and control their own independent technological systems.[26]

Leading historians also present technology as driving change. Their shadow histories may set up topics that they find more interesting. For instance, Hans Mommsen notes the important impact of "rapid modernization," but he focuses on "political decision-making" and "the structure of social, economic, and political interests."[27] Others give technology a larger role. Mainly concerned with the impact of cycles of boom and bust produced by the "Imperial German growth machine," Hans-Ulrich Wehler writes that "innovations drove their number regularly upward" as part of a self-creating and self-implementing secular trend. For instance, he presents electrical technology as growing inexorably from small to large.[28] Disagreeing with much of Wehler's interpretation of German history, David Blackbourn may be more willing to recognize contingency and choice, except in technology.[29] Presenting technology as growing bigger, faster, more rationalized, Blackbourn notes that "by the beginning of the twentieth century, a recognizably modern corporate structure had established itself in industry," although more in "electrics" than in textiles. One could be pessimistic or optimistic, but "the trend was general" during this "classic era of mass production."[30]

Historians of professions also use shadow histories that simplify and shorten narratives. For instance, several legal historians treat technology— for them a vague part of industrialization—as shaping Germany. They evaluate the professional and legal adjustments made in response to industrialization. Kenneth F. Ledford notes that after 1871 lawyers promoted procedural reform as an "Archimedean point" in order to make themselves member of a "General Estate." Michael John investigates the development of a new civil code. He argues that industrialists and merchants were more interested in legal unity than in the specific content of the new civil code, passed in 1896.[31] Curiously, many historians of engineers use similar shadow histories. They investigate career structures and professional organizations, rather than looking at engineers thinking or acting on shop floors or inside design bureaus, laboratories, and technological systems. They analyze engineers who discussed political, social, and economic change and who struggled to improve engineering education, status, and income. These studies present technology as a "black box," neither opened nor investigated. In these shadow histories, technology happened to engineers, who, like others, struggled to adjust.[32]

Shadow histories of technology use machine images without analyzing their uses. They also draw on other images. Historians of the Kaiserreich and Weimar Germany often turn to Max Weber and his "Iron Cage" of rationalization, associated with capitalism and bureaucracy.[33] Blackbourn argues that bureaucracy was "the true common element" linking aspects of Wilhelmine Germany together and notes the prevalence of "machine" images. According to Blackbourn, Weber had a "double-sided perspective" in which, although "the bureaucratic machine was rational, essential, a motor for good," it "also threatened to confine humans in an 'iron cage'." Weber admired "American dynamism" and promoted "capitalism and technical rationality," but he believed that rationalization was replacing an "enchanted world" full of meaning with "the grey forces of regimentation." Thus, Blackbourn ties together pessimism and optimism, irrationalism and rationalism, and subjectivity and objectivity from the Kaiserreich to the Third Reich.[34]

In the guise of the "Iron Cage," shadow histories of technology become not figures of speech shortening narratives but artifacts requiring explanations and responses. Jeffrey Herf separates politics from technology and then investigates how "German thinkers who rejected Enlightenment reason" combined "irrationality" (Nazism) with "the most obvious manifestation of means-ends rationality" (technology).[35] Rolf Sieferle explores "conservative" responses to "technological civilization as an objective process."[36] These shadow histories support "cultural pessimists" who believed that technology was coercively restructuring Germany and that political action was useless.[37] A way out of the "Iron Cage" might then be necessary. Lawrence Scaff sees "a series of continuing assaults on the sociocultural world by means-specific instrumentalities that have their true sources and rationale in the production of technologies and in economic modes of action." Technology and technical means have been replacing values, value conflicts, and politics, but Scaff wants a politics "beyond both instrumental rationality and aesthetic experience."[38] John P. McCormick follows a similar path in reviving Carl Schmitt's criticism of turning politics into a technology.[39] Without looking inside technological change, these scholars reify machine images. Herf wants to live in the "Iron Cage"; Scaff and McCormick seek to escape it.

Fundamentalism in Context

Both historians and engineers have closed off investigations of technology and history as contested terrain. Klingenberg, Koepchen, and Stiel had good reasons for doing so. In presenting their quite different "ends and choices" as determined by technology (which was fundamental), Klingenberg, Koepchen, and Stiel themselves used a Weberian language suggesting the "pure Platonic interest of the technologist."[40] They based their thinking, acting, and representing on the belief that technological change followed its own logic. This was also an appropriate rhetorical strategy. They chose modes of representation that were useful in promoting change—social, political, and technical—from the 1890s to the 1930s. Presenting technology as transcendent—as following a natural logic—made processes of genesis, justification, and implementation of technological change politically and technically feasible.[41] The "Iron Cage" cast a deep shadow.

A determinist view of technological change was widely shared in the Kaiserreich and in the Weimar Republic. Many Germans believed that technology was rationalizing, leveling, and homogenizing German "culture." Pessimists thought nothing could be done to reverse the trend, while others sought ways to preserve "culture."[42] Enthusiasm was common.[43] A few intellectuals interpreted technological change as consistent with "conservative" agendas.[44] Werner Siemens announced that "research and invention [were] leading humans to higher cultural levels,"[45] while Hugo Stinnes sought ways of scaling up and integrating technological systems. Landräte (county commissioners), powerful local members of the Prussian bureaucracy, promoted regional infrastructural development to prevent mayors and industrialists from pursuing narrow interests that encouraged urbanization and industrial concentration.[46]

Many scholars in the Kaiserreich drew boundaries between the technical and the political.[47] They faced a common problem. Representatives of sciences and technologies that have policy relevance need most acutely to appear above politics. In seeking to mechanize choice, they develop methods of quantification and sets of rules, protocols, and procedures to protect "experts" from the outside. As Theodore Porter notes, "not Science, but politics, demands narrow rigor."[48] For example, after 1871, the Verein für Sozialpolitik (Association for Social Policy) tied political economy closely to policy development. In addition, German Social Democrats claimed that sociology meant socialism. In a "value-judgment fight" that broke out in 1909, Max Weber and others distanced themselves from the association and from socialism by asserting that facts and values were separate issues. Whereas sociology (science) could establish the facts, only politics could establish what ought to be.[49] In jurisprudence, "statute positivists" focused on law that had been properly created. They separated the formal system from politics and applications. In his attack on statute positivism, published in 1911, Hans Kelsen promoted a theory of "pure" law that separated "is" from "ought."[50] Private lawyers thought that proper procedures would make them members of a general estate representing general interests.[51] State officials presented themselves as neutral, nonpolitical rule followers.[52]

Dissatisfied with the party and interest-group politics of the 1890s, which they did not think dealt adequately with problems of industrial, economic, social, and cultural change, some middle-class theorists also suggested an aesthetics based on Sachlichkeit (objectivity or matter-of-factness). In 1907, stressing "functionalism" and "efficiency," they established a new direction in industrial design by establishing the Werkbund. Scientific and technical progress would replace a lost faith in political progress. A sachlich (realistic) appraisal of machines and materials would provide proper design, while functionalism in architecture and engineering would help integrate society. Klingenberg promoted similar views. He argued that architecture should reflect "the purpose of the building" and allow one to imagine that "inside are housed heavy masses with solid foundations." He rejected ornament—"friezes"—that could "never suit the form of the machines."[53]

During the 1920s, as Germans constructed a new state and suffered through inflation, stabilization, then depression, politics fell even further into disrepute. A common figure of speech equated politics with the Kuhhandel (shady business) found in the traditional marketplace. Attacks on science and technology increased. Although leading physicists lost faith in reason and causality, other Germans retained their faith in Sachlichkeit.[54] Presenting technology as fundamental allowed proponents of different ways of organizing state and society to appear as nonpolitical agents of "proper" change. Many industrialists, trade union officials, engineers, and politicians supported economic and industrial "rationalization" based on American technology and production. They envisioned quite different impacts, and each group believed that political machinations might divert the natural logic of rationalization.[55] Promoting "technocracy," one group of engineers sought to separate technology from capital, so that technology would no longer narrowly serve profit.[56] Of course, some scientists and engineers found "professional dignity" not in controlling the uses of their expertise but simply in making things work.[57]

Scholars believed that following procedures and rules circumvented values and politics. German historians of the "historicist" school envisioned scientists as using "abstract, classificatory methods" to study nature, which was "the scene of the eternally recurring, of phenomena themselves devoid of conscious purpose." Historicists wanted to use "intuitive understanding" to study "unique and unduplicable human acts, filled with volition and intent."[58] Karl Mannheim believed that following rules placed one beyond social influences.[59] Hans Kelsen created a vision of "pure law" that placed sovereignty in the legal system's "logical unity," whereas Carl Schmitt identified the head of state as the sovereign "source of 'objective' decisions—those 'above' parties."[60] Schmitt thought that technology was penetrating the legal profession by filling it with technicians mechanically applying rules. Thus, he assumed that technological change was a neutral, nonpolitical process. A form without content, technology involved only efficiency and control. These scholars did not investigate what lawyers, bureaucrats, engineers, and scientists actually did. Daily activities, choices, and negotiations were irrelevant, perhaps because scholars thought that they all contributed to the same process—rationalization.[61]

Using Fundamentalism

Engineers also participated in German culture. They promoted technological fundamentalism as a new "way of life,"[62] as they sought to make themselves significant bearers of German culture.[63] Stressing technology, nationalism, and industrialization, leaders of the Verein Deutscher Ingenieure (Association of German Engineers) defined development of the engineering profession as benefiting German society and the nation. Thus, the VDI promoted educational reform, safety, and norms to raise the status of engineers and promote economic growth. VDI engineers supported Taylorism to overcome narrow profit concerns of economic liberalism and presented themselves as neutral mediators between capital and labor. VDI leaders avoided the key political issues of the day that might have fractured its heterogeneous membership and instead supported technical and scientific progress to solve general problems and promote the general good. Their way of resolving social and economic issues was to treat them in a "scientific," not a "political" manner. Facts would determine choice and exclude interest conflict. Klingenberg joined his AEG colleagues Wichard von Moellendorff and Walter Rathenau in representing these claims as maximizing rationalization and standardization with the help of Taylorism. In 1927, expressing the fundamentalist view, a VDI curator stressed that technology had "its own necessities."[64]

Klingenberg agreed. Beginning in 1909, he promoted large power stations and a national grid,[65] describing them more fully in 1913. He argued that issues of "electrical policy" could be resolved only after considering "the most essential economic questions," which involved generating and distributing costs. He argued that "the economically most advantageous" station could be designed because "for each level of load there [was] a particular organization of machines and a particular distribution of load to them that [would] produce the smallest costs." Each plant was a "unified machine" with an "unequivocal managerial direction." Ignoring exceptional conditions, he planned only to deal with "fundamental considerations."[66] After establishing principles, he discussed generating equipment, boilers, coal transport and storage, ash removal, switching equipment, plant location, and architecture. His ideas became the basis for developing the field of power-system economics, which provided generally accepted methods by the mid 1930s.[67]

Koepchen, making technology fundamental, promoted a different vision of the proper end of technological change: regional systems. In 1920, he argued that electric power should be reorganized according to "general economic principles" independent of "all private-enterprise, communal or quite political special interests." However, "technical and economic requirements" would produce eight regional systems (including his power company, the RWE), not a national grid.[68] In 1930, Koepchen argued that "the task of technology" was to use "a minimum of national income to achieve the economic optimum." Thus, "technological progress" required a safe, economical, and "steady adjustment of all operations to the actual state of the technology." For Koepchen, the RWE's efforts to tie Ruhr coal to Alpine hydroelectricity provided a model of rational, fact-based, nonpolitical, technology-driven development.[69]

Stiel also made technology fundamental in promoting independent power stations of an even smaller scale. He argued for the economic benefits of combining heating and electric generation in paper mills. Georg Siemens writes that Stiel was "a scientist by nature, who had studied the technological foundations" of the textile industry.[70] Stiel wished to provide "the most appropriate energy supply for factories and their individual machines" in order to maximize productivity and quality. Setting up an on-site "textile power station" created the lowest possible costs for facilities and operations by "tightly matching the needs of the individual factory." Powering each machine with an electric motor would improve factory operations, increase production, and improve the quality of work. He preferred alternating current because of its "convenient transforming of tension" and because "cheap, simple, practical" squirrel-cage motors could be used. Closely involved in implementing textile electrification, Stiel could not ignore contingency and choice. The "real operating situation" after construction had to be considered as well as "expected changes in operations in the foreseeable future." Design had to be "flexible" enough to allow "operational changes and expansions without impairing economy."[71]

Klingenberg, Koepchen, and Stiel made technology fundamental to three different systems.[72] They could not mechanize choice. In typical fundamentalist fashion, they stressed the use of data and experience to bridge the gap between abstract system and implementation.[73] Discerning individuals would recognize the truth. Klingenberg asserted that "the most practical design, [concerning] the relative position of boiler, machine, and switch rooms and the kind of coal storage, [was] to be determined case by case according to [his] criteria." Stressing experience (Erfahrung) allowed him to state what was best without discussion and to resolve disputes without appearing to participate in them.[74] For Koepchen, the RWE exemplified the proper adjustment of theory to local conditions, but the RWE often had to wait while others came to understand the right way.[75] Stiel argued that electrifying textile manufacturing was possible only "if the special technological conditions of the textile industry [were] taken into careful consideration." In addition, the choice of self-generation or drawing energy from outside depended "entirely on the special conditions of individual cases." Although representatives of public power companies had argued that they could not take electricity from independent power stations at acceptable rates, experience showed otherwise.[76]

Klingenberg, Koepchen, and Stiel tapped into a consensus—right and left, pessimists and optimists—that technological change had one logic, both pervasive and irresistible. However, facts, logic, experience, and "pure Platonic interests" were not enough to achieve their different ends. Nor are they for historians. To avoid charges of subjectivism and relativism, Wehler and other historians of the Kaiserreich seek objective structures in society.[77] Making technology fundamental helps to create "objectivity" by constructing a realm of causal factors seemingly independent of human choice.[78] Nevertheless, historians promote contending interpretations. They may turn to "academic politics."[79] Geoff Eley notes that Jürgen Kocka, in his survey of the nineteenth-century German bourgeoisie, "pretends that Blackbourn and Eley did not exist," although they had been arguing against Kocka and Hans-Ulrich Wehler since the early 1980s.[80] In a fine display of academic rhetoric, Wehler accuses three widely cited authors of an "astounding ignorance of historiographical development, academic-political naiveté, and theoretical Philistinism."[81] History remains a contested ground for historians, even when they present technological change as an objective process creating change in other realms. In addition, shadow histories of technology ignore the political efforts necessary to build different systems.

Engineering Politics

Technological fundamentalism provided useful political rhetoric. German university professors, claiming to be nonpolitical because they separated facts and values, cloaked their role as state officials serving conservative states.[82] In addition, separating facts and values allowed these "mandarins" to promote themselves.[83] Herbert Mehrtens describes the "caste politics" pursued by German mathematicians, who promoted and protected their field during the 1920s and the 1930s but who presented their efforts as nonpolitical.[84] Thus, claiming value neutrality and transcendence, university professors, scientists, and engineers served special interests.[85] Behavior could undermine rhetoric,[86] but technological fundamentalism remained unchallenged, as it does among some historians. Although Klingenberg, Koepchen, and Stiel promoted their contending solutions in a neutral language of technical necessity, they still had to find allies to build their different systems. Although opposition to party politics was widespread, negotiation and compromise were necessary to establish consensus among contending interests.[87] They did not resolve their fundamental differences, but each of them participated in engineering politics to implement his plans.

Although Klingenberg made technology fundamental,[88] he was well aware of the political efforts that went into building power systems. In Germany, local governments were sovereign entities in matters of local infrastructure.[89] Towns and counties (Landkreise) could establish their own companies or could join mixed public-private corporations that combined local governments with private capital.[90] Thus, in order to build power stations and distribution systems, Klingenberg and the AEG worked with local governments.[91] South Africa provided different allies. Slick maneuvering allowed the AEG to penetrate the British empire after 1906. In order to sell equipment and technology in South Africa, the AEG purchased a majority of stock in a power company. Outmaneuvering the city of Johannesburg, the AEG used its power company, renamed the Victoria Falls and Transvaal Power Company, to build its own system. Allied with gold-mine owners, who sought to resolve labor problems and had close connections to the state in the Transvaal, the AEG gained rights of way and circumvented opposition on the part of coal-mine owners. Close enough together for existing technology to connect them in a grid, gold mines purchased more than 95 percent of the energy generated. By 1923, they were consuming more electricity than London, Birmingham, and Sheffield combined.[92]

In Germany, only an alliance with a higher level of government would allow Klingenberg and the AEG to circumvent local politics. In 1913, the AEG sought unsuccessfully to form an alliance with the Reich government to create a national system.[93] In the unique political situation provided by World War I, Klingenberg designed and directed construction of the Golpa power station, south of Berlin, which went on line in December 1915. Built for the Reich, it did not include local governments.[94] In addition, Klingenberg envisioned a system of large power stations connected over a net of 100-kilovolt power lines which would produce a savings of "national wealth" (Nationalvermögen) that, he believed, could not be ignored. Klingenberg described how Germany would have to be reorganized to suit his vision of a proper electric supply system. Political boundaries, individual companies, towns, and counties could no longer interfere in proper development.[95] Klingenberg's vision gained support among engineers and historians. However, neither Klingenberg nor the Reich government was able to build a political coalition powerful enough to bypass local, provincial, and state governments. Neither these political problems nor severe technical problems could be resolved before World War I, or even in the 1930s.[96]

Local politics was integral to successful power politics. During World War I, the AEG lost power systems to the province of Brandenburg, the city of Berlin, and the Reich.[97] While Klingenberg sought support for a national grid, another group of technological fundamentalists were wiring together towns and counties, and eventually provinces and states into regional coalitions that could protect themselves. Like the AEG, the RWE had to deal with local governments that exercised sovereignty over infrastructure. In 1908, local officials forced the RWE out of Westphalia. To expand in the Rhineland, the RWE had already become a mixed company that grew by exchanging RWE stock for town and county systems. Public officials joined the RWE's board and provided the RWE with influence necessary for local expansion and for dealing with state officials concerned with regional administration, railroads, and canals.[98] During World War I, Reich funding allowed the RWE to expand dramatically. After World War I, the RWE's coalition of big business, towns, and counties successfully avoided Reich control and socialization.[99] Astute political maneuvering, not the inexorable logic of history or technology, had made the RWE one of the largest power companies in Germany.

During the 1920s, the RWE added states to its coalition in order to connect Ruhr coal and Alpine hydroelectric stations. The expanding RWE coalition fought state officials, who were constructing state power systems. In this "Electrical War," each side defended its own system as being driven by technology and reason. Unable to defeat each other, the RWE and Prussia signed the "North-German Electrical Peace" in 1927, although they continued to line up allies among local governments in the Saarland. Then, three state-controlled power companies—Prussian, Reich, and Bavarian—formed a corporation to unify electrification. The RWE and western allies formed a counter-organization, but negotiations produced compromise. The reconstitution of the eastern group to include the RWE and its allies and the "Real Electrical Peace" of 1929 between the RWE and Prussia resolved political conflict by demarcating territories within which companies consolidated themselves politically and technologically. In 1930, the RWE's connection between the Ruhr and the Alps began operating. After warding off one further attack from Berlin, the RWE devoted the 1930s to making its expanded system economically viable. Expansion had created significant overcapacities that would be utilized later in the 1930s by tying the RWE to National Socialist goals.[100]

Promoting a regional, west German system, Koepchen provided the main obstacle, political and technological, for engineers like Klingenberg who wanted a national grid. In addition, Koepchen railed against those who, without understanding the proper path of development, might build generating systems in individual factories.[101] However, another technological fundamentalist, Stiel, argued that rational technological development would lead textile mills to exactly the outcome ridiculed by Koepchen. Statesmen of technology like Klingenberg and Koepchen dealt with government officials and with one another as they built systems, while Stiel, head of the textile department at Siemens-Schuckert, persuaded mill managers and owners to turn to individual electric drive. Stiel also compared self-generation with outside power sources. He presented power companies as political, and plant design as an objective, nonpolitical process that would benefit from self-generation. Often needing a source of steam, textile mills could also generate electricity, which would also free them from labor problems affecting outside sources. In addition, power companies might regularize their own systems to the detriment of the mills drawing current from them. For instance, they might seek to get amperes and volts into phase (expressed as $\cos\phi$) at the expense of mills. Thus, Stiel referred to the "$\cos\phi$ politics of power companies."[102]

Presenting textile electrification as a neutral and rational process, Stiel ignored significant political characteristics of textile regions. Textile production took place in areas that Gary Herrigel has termed "decentralized realms," such as Elberfeld and Barmen in the Wupper valley, Mönchengladbach and Krefeld on the left bank of the Rhine, and Thüringen. Instead of large, hierarchically organized corporations like Siemens, the AEG, and the RWE, which characterized "autarchic regions" around Berlin and in the Ruhr, decentralized realms had small firms that relied on significant infrastructural support provided by local government. During the 1920s and the early 1930s, decentralized realms fought to maintain their independence from the national government, which, it was feared, might promote big business. In addition, members of the "old middle class" suffered severe economic setbacks. National Socialists sought and gained support from these increasingly politicized groups, as well as from larger, concentrated industries. Many historians have denigrated the culture of small business and of artisan and household production as an anti-modern impediment to proper capitalist production in increasingly large factories.[103] Stiel supported small producers without discussing their political values.

Decentralization continued, and after World War II it provided a basis for flexible production.[104]

Consequences

Representing large corporations, Klingenberg, Koepchen, and Stiel sought to build different technological systems. They made technology fundamental to their different projects, and they represented engineering as apolitical. Each believed that technology was on his side and would lead, except for political opposition, to a proper future rationally arrived at. They rejected alternatives by referring to their experiences and to shadow histories that provided bases for projecting their visions of the future. Instead of conflict, negotiation, contingency, and construction, they wrote about farseeing, insightful, all-knowing engineers and managers who recognized how things should properly be done and who heroically overcame significant obstacles to build technological systems. Technological progress was a figure of speech in their narratives, rather than part of a process to be analyzed. However, only success could fully support claims of transcendence. Implementing visions for development, which by definition did not yet exist and which differed from one another, required technical and political finesse. They negotiated with others as they sought to create a world in which their assertions would hold true. Klingenberg, Koepchen, and Stiel were not engineers who had to become political. Rather, they had to engineer politics to be good engineers.

The three engineers factored "engineering politics" out of their rhetoric because they and other Germans opposed politics. Technological fundamentalism may not have provided a good basis for compromise.[105] Perhaps more important, hiding engineering politics meant that commitment to objectivity (Sachlichkeit) undermined democratic processes.[106] In the Weimar Republic, very little worked to promote democracy, negotiation, or pluralism as organizational necessities. Separating politics from technology—values from facts—meant that infrastructural development was not presented or seen as a successful political process. Instead, success was presented as the triumph of clear reasoning and facts over politics. Hence, the process of infrastructural development did not help legitimize politics in the Weimar Republic. Legitimizing choices in terms of technological necessity implied that ending politics could be a viable and constructive step. During the Weimar Republic, lost faith in political processes eventually benefited National Socialism. A "catch-all-party of protest," the Nazis promised an end to politics and drew on support from scientists, engineers, techno-phobes, and many others, "united above all by a profound contempt for the existing political and economic system."[107]

Though groups opposing urban, technological, and scientific develop-ments may have helped the National Socialists to power, proponents of tech-nological fundamentalism made the Third Reich viable. This is the sad story of scientists and engineers in the Weimar Republic and the Third Reich. Far from being constitutionally incapable of working under Hitler, scientists and engineers preserved their castes and systems. Electrical engineers kept power systems working during World War II by learning how to interconnect them from Austria to the North Sea. Under the guise of technical and economic efficiency, and even professional dignity, "nonpolitical" scientists and engi-neers found it quite possible to serve a variety of masters, from the Kaiserreich to the Federal Republic. In addition, their "nonpolitical," "value-neutral" posture allowed them to deny responsibility.[108]

How the agents of change acted, as well as how they thought, should be investigated. This strategy involves following the actors around to find out what they did and thought as they tried to create knowledge and shift soci-ety. This approach treats actors as aware and calculating but not all-know-ing or all-powerful members of a social system that is a "distribution of knowledge."[109] Klingenberg, Koepchen, and Stiel were knowing partici-pants in the construction of knowledge. Each recognized that opponents had to be shifted and new social systems constructed as they struggled to find ways of building systems. Each understood aspects of German society, engineering, and politics. Each responded to developments beyond his con-trol, just as other people responded to developments promoted by engineers like Klingenberg, Koepchen, and Stiel.

Technology did not force them to think and act in a particular way. Nevertheless, identifying changes in science and technology as contingent historical processes may offend scholars who present themselves, their cat-egories, and their thoughts as fundamentally correct. Their belief in their own objectivity allows them to decide that some actors' categories are time-lessly true. They may choose sides in past arguments and then make one side capable of "ice-cold reasoning."[110] They may claim that ability for themselves. That technological fundamentalism continues to find life has to do with how people think and act and thereby constitute institutions. "Shadow histories" are, thus, "not attempts at accurate representation, but rather attempts to forge a moral identity."[111] Forging a moral identity may be laudable, but so is creating accurate representations. Historians who treat science and technology as fundamental and exogenous reify shadows cast by Weber's "Iron Cage" and provide a basis for claims that the solution to too much objectivity is more subjectivity.[112] Presenting technology as a coercively restructuring and rationalizing agent may be useful for setting up discussions of other topics, but envisioning technology as "congealed politics"[113] and asking about the material components of social and political systems may be more useful than investigating shadow impacts of reified shadows.

Acknowledgments

At SHOT (1993) and ICOHTECH (1996) conferences, I presented parts of this essay. Members of Yale's History of Medicine workshop made valuable suggestions in 1997. I would like to thank Mike Allen, Hans-Joachim Braun, Hanko Dobi, Robert Glen, Gabrielle Hecht, Sharon Traweek, Frank Trommler, and several anonymous referees for suggestions that shaped this text.

Notes

1. Rudolf von Miller, "Ein Halbjahrhundert deutsche Stromversorgung aus öffentlichen Elektrizitätswerken," *Beiträge zur Geschichte der Technik und Industrie* 25 (1936): 111–125; Georg Boll, *Entstehung und Entwicklung des Verbundbetriebs in der deutschen Elektrizitätswirtschaft bis zum europäischen Verbund* (Verlags- und Wirtschaftgesellschaft der Elektrizitätswerke m.b.H., 1969).

2. Thomas P. Hughes, *Networks of Power* (Johns Hopkins University Press, 1983); Merritt Roe Smith and Leo Marx, eds., *Does Technology Drive History?* (MIT Press, 1994).

3. Georg Klingenberg, *Bau großer Elektrizitätswerke* (Springer, 1913); A. Koepchen, "Zur Sozialisierung der Elektrizitätswirtschaft. Ein Gutachten zum Entwurf des Gesetzes," *Elektrotechnische Zeitschrift* 41 (1920): 481–485; A. Koepchen, "Das RWE in der deutschen Elektrizitätswirtschaft (Vortrag, gehalten

im Haus der Technik in Essen am 28. März 1930, von A. Koepchen, Essen)," in *RWE, Bericht über das Geschäftsjahr 1929/30*, pp. 3–12; Wilhelm Stiel, *Elektrobetrieb in der Textilindustrie* (Hirzel, 1930).

4. William F. Ogburn (*On Culture and Social Change*, ed. O. Dudley Duncan, University of Chicago Press, 1964, pp. xv–xvii and 23–32) argues that different parts of society could lag behind others, but that from the 1890s into the 1920s society had to adjust to technological change.

5. Norbert Gilson, "Die Vision der Einheit als Strategie der Krisenbewältigung? Georg Klingenbergs Konzeption für die Energieversorgung in Deutschland zu Beginn des 20. Jahrhunderts," in *Der Optimismus der Ingenieure*, ed. H.-L. Dienel (Franz Steiner, 1998), pp. 57–76; Norbert Gilson, *Konzepte von Elektrizitätsversorgung und Elektrizitätswirtschaft* (GNT, 1994), p. 411; C. M., "Georg Klingenberg," *Zeitschrift des Vereins Deutscher Ingenieure* 29 (1925): 1613–1618.

6. Gilson, *Konzepte*, p. 411; Ernst Henke, *Das RWE nach seinen Geschäftsberichten, 1898–1948* (RWE, 1948), p. 29; Helmut Maier, "Arthur Koepchen (1878–1954)," in *Ingenieure im Ruhrgebiet*, ed. W. Weber (Aschendorff, 1999), pp. 184–223.

7. "W. Stiel," *Elektrotechnische Zeitschrift* 57 (1936): 350; Georg Siemens, *History of the House of Siemens*, volume 2 (Karl Alber, 1957), p. 124; Stiel, *Elektrobetrieb*.

8. Frank Trommler, "The Avant-Garde and Technology: Toward Technological Fundamentalism in Turn-of-the-Century Europe," *Science in Context* 8 (1995): 397–416. Paul Rabinow argues ("Representations Are Social Facts: Modernity and Post-Modernity in Anthropology," in *Essays on the Anthropology of Reason*, Princeton University Press, 1996, p. 31) that "different historical conceptions of truth and falsity . . . are historical and social facts."

9. Richard Rorty (*Achieving Our Country*, Harvard University Press, 1998, p. 34), attributes the term "God's-eye view" to Hilary Putnam. Rorty (*Philosophy and Social Hope*, Penguin, 1999, pp. 155–157) notes that "scientific realism" and "religious fundamentalism" both provide absolutes by appealing to "objectivity as fidelity to something nonhuman."

10. Bruno Latour (*Science in Action*, Harvard University Press, 1987, pp. 94–100) discusses claims to have nature on one's side in disputes.

11. Robert W. Smith ("The Biggest Kind of Big Science: Astronomers and the Space Telescope," in *Big Science*, ed. P. Galison and B. Hevly, Stanford University Press, 1992) discusses making the telescope "technically and politically feasible."

12. Sharon Traweek (*Beamtimes and Lifetimes*, Harvard University Press, 1988) uses the term "statesmen of science" for those senior physicists who deal with Washington. Because they do, they are seen as tainted by politics. Lynn Hunt (*Politics, Culture, and Class in the French Revolution*, University of California Press, 1984, p. 229) notes that the rhetorical framework developed during the 1790s did not support development of "liberal politics, politics as the representation of interests."

13. For various approaches to science and technology, see *Handbook of Science and Technology Studies*, ed. S. Jasanoff et al. (Sage, 1995); Bruno Latour and Steve

Woolgar, *Laboratory Life*, second edition (Princeton University Press, 1986); Timothy Lenoir, *Instituting Science* (Stanford University Press, 1997); Wiebe E. Bijker, Thomas P. Hughes, and Trevor Pinch, eds., *The Social Construction of Technological Systems* (MIT Press, 1987); Wiebe E. Bijker and John Law, eds., *Shaping Technology/Building Society* (MIT Press, 1992). Walter G. Vincenti (*What Engineers Know and How They Know It*, Johns Hopkins University Press, 1990) and Nathan Rosenberg ("How Exogenous is Science," in *Inside the Black Box*, Cambridge University Press, 1982) note that engineers resolve many problems before scientists can explain what engineers are doing. Edward W. Constant II ("Reliable Knowledge and Unreliable Stuff: On the Practical Role of Rational Beliefs," *Technology and Culture* 40 (1999): 324–357) notes that engineers develop reliable knowledge that continues to work even if stuff breaks down.

14. Peter Galison, *Image and Logic* (University of Chicago Press, 1997). Ian Hacking ("The Self-Vindication of the Laboratory Sciences," in *Science as Practice and Culture*, ed. A. Pickering, University of Chicago Press, 1992) argues that ideas, machines, and data are adjusted to one another to create a mature laboratory science. Gilson (*Konzepte*) describes the development of theory, apparatus, and analysis necessary for power systems economics to become a viable specialty.

15. Barry Barnes (*The Nature of Power*, University of Illinois Press, 1988) treats social systems as distributions of knowledge. Robert E. Kohler (*Lords of the Fly*, University of Chicago Press, 1994) brackets epistemological questions and stresses practice as providing a fruitful way of understanding productivity. See also Bruno Latour, *Science in Action* (Harvard University Press, 1987) and *We Have Never Been Modern* (Harvard University Press, 1993).

16. Thomas P. Hughes, *Networks of Power* (Johns Hopkins University Press, 1983) and *Rescuing Prometheus* (Pantheon Books, 1998). Whereas Hughes ("Walther Rathenau: System Builder," in *Ein Mann vieler Eigenschaften*, ed. T. Buddensieg et al., Klaus Wagenbach, 1990) presents Rathenau as a systems thinker seeking to resolve technical, political, financial, and other problems in order to deploy electric power systems. Hans Dieter Hellige ("Walther Rathenau: Ein Kritiker der Moderne als Organisator des Kapitalismus: Entgegnung auf T.P. Hughes' systemhistorische Rathenau-Interpretation," in *Ein Mann vieler Eigenschaften*, ed. Buddensieg et al.) argues that Rathenau was not a systems thinker focussing narrowly on technology. Mikael Hård ("German Regulation: The Integration of Modern Technology into National Culture," in *The Intellectual Appropriation of Technology*, ed. M. Hård and A. Jamison, MIT Press, 1998, p. 51) stresses "the political dimensions of [Rathenau's] vision."

17. See Latour and Woolgar, *Laboratory Life*; Donald MacKenzie, "From Kwagalein to Armageddon? Testing and the Social Construction of Missile Accuracy," in *The Uses of Experiment*, ed. D. Gooding et al. (Cambridge University Press, 1989); W. Henry Lambright, "The Political Construction of Space Satellite Technology," *Science, Technology, and Human Values* 19 (1994): 47–69.

18. For discussions of objectivity as a "view from nowhere" see Thomas Nagel, *The View from Nowhere* (Oxford University Press, 1986); Richard Rorty, *Truth*

and Progress, volume 3 (Cambridge University Press, 1998), pp. 98–121; Theodore M. Porter, "Objectivity as Standardization: The Rhetoric of Impersonality in Measurement, Statistics, and Cost-Benefit Analysis," in *Rethinking Objectivity*, ed. A. Megill (Duke University Press, 1994).

19. For discussions of the uses of history and shadow history, see Rorty, *Truth and Progress*, pp. 274–275, 278, and Richard A. Watson, "Shadow History in Philosophy," *Journal of the History of Philosophy* 31 (1993): 95–109. For discussions of the different attitudes of scientists and historians toward history, see Paul Forman, "Independence, Not Transcendence, for the Historian of Science," *Isis* 82 (1991): 71–86, and Charles E. Rosenberg, "Woods or Trees? Ideas and Actors in the History of Science," *Isis* 79 (1988): 565–570.

20. See *The Historiography of Contemporary Science and Technology*, ed. T. Söderqvist (Harwood Academic, 1997), and Alfred I. Tauber's review of that book (*Science, Technology, and Human Values* 24 (1999): 384–401).

21. See Sharon Traweek, *Beamtimes and Lifetimes*, pp. 11–12, 77–78, 86, 93, 151, and "Big Science and Colonialist Discourse: Building High-Energy Physics in Japan," in *Big Science*, ed. P. Galison and B. Hevly (Stanford University Press, 1992), pp. 100–128, esp. p. 101. In contrast, Wolfgang J. Mommsen (*Imperial Germany 1867–1918*, Arnold, 1995, pp. 119–140) treats culture as an independent realm that influences people.

22. Latour and Woolgar (*Laboratory Life*) describe how one group reinterpreted an earlier paper to show that they had done work claimed by another group. Steve Woolgar ("Some Remarks About Positionism: A Reply to Collins and Yearley," in *Science as Practice and Culture*, ed. A. Pickering, University of Chicago Press, pp. 327–342, esp. 328–329) provides a history of the social studies of science in which his views are more up-to-date than those of his critics. Fritz Ringer (*Max Weber's Methodology*, Harvard University Press, 1997, pp. 146–147) presents his approach as new and that of several sociologists of science as old. Harro van Lente (*Promising Technology*, WMW, 1993) focuses on expectations in guiding change. Bruno Latour (*Aramis or the Love of Technology*, Harvard University Press, 1996) notes fictional aspects of technological projects that do not yet exist. For discussions of imagined futures, see Joseph J. Corn and Brian Horrigan, *Yesterday's Tomorrows* (Smithsonian Institution, 1984), and *Imagining Tomorrow*, ed. J. Corn (MIT Press, 1986).

23. Thomas J. Misa discusses the role of historians in helping machines make history in the following articles: "How Machines Make History, and How Historians (and Others) Help Them to Do So," *Science, Technology, and Human Values* 13 (1988): 308–331; "Retrieving Sociotechnical Change from Technological Determinism," in *Does Technology Drive History?* ed. M. Smith and L. Marx.

24. Klingenberg, *Bau großer Elektrizitätswerke*, pp. 57–73.

25. Koepchen, "Das RWE," pp. 3–8.

26. Stiel, *Elektrobetrieb*, pp. v–vi, 1–3, 12–19, 27–35.

27. Hans Mommsen, *The Rise and Fall of Weimar Democracy* (University of North Carolina Press, 1996), pp. viii, xi.

28. For a general discussion of his approach, see Hans-Ulrich Wehler, *Deutsche Gesellschaftsgeschichte*, volume 1: *Vom Feudalismus des Alten Reiches bis zur Defensiven Modernisierung der Reformära, 1700–1815* (Beck, 1987), pp. 6–31. For the quotations, see Hans-Ulrich Wehler, *Deutsche Gesellschaftsgeschichte*, volume 3: *Von der "Deutschen Doppelrevolution" bis zum Beginn des Ersten Weltkrieges, 1849–1914* (Beck, 1995), pp. 558, 571, 609, 613. On electrification, see ibid., pp. 523–524, 601, 607, 612, 617–618, 627. Wehler presents himself as all-knowing, while "the objective tendencies of growth were for contemporaries not easily, nor quickly recognized" (ibid., p. 571). He promotes his interpretation with seemingly exhaustive footnotes, but he does not evaluate different points of view. Those he recognizes, he condemns. In support of his view of electrification, Wehler (*Deutsche Gesellschaftsgeschichte*, volume 3, p. 1391) cites Hughes (*Networks of Power*), who does not, however, share Wehler's view of technology or of electrification.

29. David Blackbourn and Geoff Eley's *The Peculiarities of German History* (Oxford University Press, 1984) and the earlier German version began a long discussion with Wehler and his allies about the proper way to approach Imperial Germany.

30. David Blackbourn (*The Long Nineteenth Century*, Oxford University Press, 1998, pp. xix, xxii, 313–330, esp. pp. 323–325) criticizes "academic enthusiasts for contemporary 'flexible specialization' in small firms." He cites C. F. Sabel and J. Zeitlin ("Historical Alternatives to Mass Production," *Past and Present* 108 (1985), pp. 133–176), who argue, however, that large-scale industry did not follow naturally from technological change but had to be achieved. For a more recent development of this theme, see Gary Herrigel, *Industrial Constructions* (Cambridge University Press, 1996). David A. Hounshell (*From the American System to Mass Production, 1800–1932*, Johns Hopkins University Press, 1984) traces the development of the elements of mass production that Henry Ford brought together in 1913. Wolfgang König and Wolfhard Weber (*Netzwerke, Stahl und Strom*, Ullstein, 1990, pp. 427–441) discuss "rationalization and mass production" but treat Americans, especially Henry Ford and Frederick Taylor, as leading the way. Thus, mass production may have had a classic era in Germany prior to World War I as an idea, but not as an actuality.

31. See Kenneth F. Ledford, *From General Estate to Special Interest* (Cambridge University Press, 1996); Michael John, *Politics and the Law in Late Nineteenth-Century Germany* (Oxford University Press, 1989); Konrad H. Jarausch, *The Unfree Professions* (Oxford University Press, 1990). Bruno Latour ("Give Me a Laboratory and I Will Raise The World," in *Science Observed*, ed. K. Knorr Cetina and M. Mulkay, Sage, 1983) notes the technical and political effort necessary to make the Archimedean figure of speech plausible in the case of Louis Pasteur's laboratory. Sheila Jasanoff (*Science at the Bar*, Harvard University Press, 1995) investigates the interaction of experts from these different realms.

32. Gert Hortleder, *Das Gesellschaftsbild des Ingenieurs* (Suhrkamp, 1970); Kees Gispen, *New Profession, Old Order* (Cambridge University Press, 1989); Jarausch,

The Unfree Professions. In contrast, Gary Lee Downey and Juan C. Lucena, noting the centrality of knowledge and labor in engineering, write that "virtually all inquiries into engineering find some conceptual position for the content of engineering knowledge" ("Engineering Studies," in *Handbook of Science and Technology Studies*, ed. S. Jasanoff et al., Sage, 1995, p. 169).

33. Max Weber (*The Protestant Ethic and the Spirit of Capitalism*, Scribner, 1930, pp. 181–182) develops this figure of speech, according to which "victorious capitalism . . . rests on mechanical foundations." The "Iron Cage" forms the central image for at least two studies: Arthur Mitzman, *The Iron Cage* (Knopf, 1970); Lawrence A. Scaff, *Fleeing the Iron Cage* (University of California Press, 1989).

34. Blackbourn, *The Long Nineteenth Century*, pp. 384–385, 397–399.

35. Jeffrey Herf, *Reactionary Modernism* (Cambridge University Press, 1984), pp. 1–2. Herf investigates a number of "literati" and gives Heinrich Hardensett prominence among several engineers (pp. 179–187). Stefan Willeke ("Die Technokratiebewegung in Deutschland zwischen den Weltkriegen," *Technikgeschichte* 62 (1995): 221–246) notes that Hardensett was so committed to his belief in the transcendence of technology that he eventually opposed the Nazi regime, which, he believed, was not pursuing technological goals properly.

36. Rolf Peter Sieferle (*Die Konservative Revolution*, Fischer, 1995, pp. 27, 206) notes that völkisch (pure German) thinkers thought universalism was the "expression of the interest or point of view of a foreign people."

37. For a discussion of this process in the US, see Leo Marx, "The Idea of 'Technology' and Postmodern Pessimism," in *Does Technology Drive History?* ed. M. Smith and L. Marx (MIT Press, 1994).

38. Scaff, *Fleeing the Iron Cage*, 233, 166, 239. Jürgen Kocka ("Max Webers Bedeutung für die Geschichtswissenschaft," in *Max Weber, der Historiker*, ed. J. Kocka, Vandenhoeck & Ruprecht, 1986, pp. 13–27, esp. 15) suggests that Weber wanted to protect politics from science. Scaff writes about unintended consequences that are quite different from those investigated by Edward Tenner (*Why Things Bite Back*, Vintage, 1996), who sees people constructing complex systems that have unintended and unforeseen consequences that can be understood case by case rather than as abstractions interacting with other abstractions and hence, as Scaff would have it, beyond control.

39. John P. McCormick (*Carl Schmitt's Critique of Liberalism*, Cambridge University Press, 1997) presents technology in a determinist form. He finds "an intensifying fundamentalism" in "attempts to stake secure positions against the rapidly changing socioeconomic landscape in the supposedly timeless entities of family, nation, and faith" (p. 313). He should add technology to those entities.

40. Scaff (*Fleeing the Iron Cage*, p. 29 and note 54) uses these phrases but suggests "the actual domination of human ends and choices . . . by *techne* (or technical means)." Scaff does not envision the iron cage as an "absolutist ideology" (pp. 175, 231), perhaps because he believes that it is simply true. This is an example of fundamentalism.

41. Latour (*Science in Action*) demonstrates that genesis and justification can both be investigated and that treating genesis as coming out of the blue ignores a good deal of work that goes into creating new ideas. Nevertheless, Ringer (*Max Weber's Methodology*) would like to separate genesis and justification.

42. Fritz Ringer (*The Decline of the German Mandarins*, Harvard University Press, 1969) describes the reaction of academics to perceived change.

43. Blackbourn, *The Long Nineteenth Century*, pp. 271–273, 279–283, 394–396.

44. Sieferle (*Die Konservative Revolution*) discusses five individuals concerned about the impact of technology.

45. Werner Siemens, "Das naturwissenschaftliche Zeitalter," in *Tageblatt der 59. Versammlung deutscher Naturforscher und Ärzte zu Berlin vom 18.-24. September 1886* (Berlin, 1886), pp. 92–96, as quoted in Gerhard A. Ritter, *Großforschung und Staat in Deutschland* (Beck, 1992), p. 7.

46. Edmund N. Todd, "Electrification in the Ruhr circa 1900," *Osiris 5* (1989): 243–259; Todd, "Prussian Landräte and Modern Technology: Electricity as a Source of Power in the Ruhr, 1900–1915," *ICON 2* (1996): 83–107.

47. Separating the political from the technical has a long history. According to Steven Shapin and Simon Schaffer (*Leviathan and the Air-Pump*, Princeton University Press, 1985), Robert Boyle publicized experiment as a way around the sectarian struggles of the English Civil War. Hunt (*Politics, Culture, and Class in the French Revolution*, p. 229) notes that Bonaparte presented himself as standing "above party and factions" and ready "to rid the nation of such unseemly political machinations."

48. Traweek (*Beamtimes and Lifetimes*) shows that high-energy physicists depend on face-to-face communication and oral exchanges of information. Individual character and skill matter. Theodore M. Porter (*Trust in Numbers*, Princeton University Press, 1995, pp. 221–223, 230) stresses this aspect of Traweek's argument. Julie Johnson-McGrath ("Speaking for the Dead: Forensic Pathologists and Criminal Justice in the United States," *Science, Technology, and Human Values* 20, no. 4, autumn 1995: 438–459) describes forensic pathologists' seeking independence from local politics by establishing proper procedure.

49. Robert N. Proctor, *Value-Free Science? Purity and Power in Modern Knowledge* (Harvard University Press, 1991), pp. 85–98. Ringer (*Max Weber's Methodology*) argues that facts and values are logically separate. Recognizing the logical distinction, Loren Graham (*Between Science and Values*, Columbia University Press, 1981) investigates several scientists who sought to combine science and values. Bruce Bimber (*The Politics of Expertise in Congress*, State University of New York Press, 1996) investigates neutrality as an institutional strategy.

50. Peter C. Caldwell, *Popular Sovereignty and the Crisis of German Constitutional Law* (Duke University Press, 1997), pp. 4–5, 45–52. Caldwell notes that, because courts changed significantly during the twentieth century, "Accounts of the major Weimar theorists of constitutional law, not court decisions, provide the continuity between Weimar constitutionalism and that of the Federal Republic" (ibid., p. 11).

In electric power, applications appear more lasting than theoretical statements constructed to demonstrate paths that must be followed.

51. Ledford, *From General Estate to Special Interest*.

52. Todd ("Prussian Landräte and Modern Technology") discusses literature concerning bureaucrats and roles played by state officials in constructing electric power systems.

53. Trommler, "The Avant-Garde and Technology"; Klingenberg, *Bau*, pp. 51–54.

54. Paul Forman, "Weimar Culture, Causality, and Quantum Theory, 1918–1927: Adaptation by German Physicists and Mathematicians to a Hostile Intellectual Environment," *Historical Studies in the Physical Sciences* 3 (1971): 1–115. Gerald Holton (*Science and Anti-Science*, Harvard University Press, 1993, pp. 180–184) draws a direct connection between "anti-science" and the rise of Nazism in Germany. Whereas many scholars have traced the origins of Nazism to middle-class cultural pessimism, Frank Trommler ("The Creation of a Culture of Sachlichkeit," in *Society, Culture, and the State in Germany, 1870–1930*, ed. G. Eley, University of Michigan Press, 1996) argues that Sachlichkeit characterized middle-class groups even more.

55. Mary Nolan (*Visions of Modernity*, Oxford University Press, 1994) describes debates concerning American technology and production. Participants seem to have treated "rationalization" as a neutral process.

56. Willeke, "Die Technokratiebewegung in Deutschland zwischen den Weltkriegen."

57. Monika Renneberg and Mark Walker (*Science, Technology, and National Socialism*, Cambridge University Press, 1994, pp. 4–17) identify "making things work" as one aspect of technocracy. They note that representatives of competing segments of the Nazi system met at the Wannsee conference, but identified "scientific, technological and bureaucratic rationality and efficiency [as] the way to solve their and Germany's problems, in this case the 'Final Solution of the Jewish Question.'" Primo Levi ("A Conversation with Primo Levi by Philip Roth," in Levi, *Survival in Auschwitz*, Collier Books, 1986, p. 179) describes an Italian bricklayer who "hated Germans" but built "straight and solid" walls "out of professional dignity."

58. Georg G. Iggers, *The German Conception of History*, revised edition (Wesleyan University Press, 1983), pp. 5, 14, 271. Iggers (pp. 292–293) does not find the same level of decency in some German academic rhetoric that Shapin and Schaffer, *Leviathan and the Air-Pump* find in Robert Boyle's recommendation that opponents be treated as possible converts.

59. David Bloor (*Wittgenstein, Rules and Institutions*, Routledge, 1997, pp. 142–144) suggests that Mannheim thought following rules removed social influence whereas Wittgenstein saw "rules, and the so-called unfolding of meaning, as themselves social processes."

60. Caldwell, *Popular Sovereignty and the Crisis of German Constitutional Law*, pp. 9, 137.

61. Caldwell (*Popular Sovereignty and the Crisis of German Constitutional Law*) and McCormick (*Carl Schmitt's Critique of Liberalism*) also treat science and technology as neutral and nonpolitical.

62. For discussions of this concept, see Shapin and Schaffer, *Leviathan and the Air-Pump*; and Paul Rabinow, "Severing the Ties: Fragmentation and Dignity in Late Modernity," in his *Essays on the Anthropology of Reason*.

63. See Burkhard Dietz, Michael Fessner, and Helmut Maier, eds., *Technische Intelligenz und "Kulturfaktor Technik"* (Waxmann, 1996); Hård, "German Regulation."

64. Lothar Burchardt, "Standespolitik, Sachverstand und Gemeinwohl: Technisch-wissenschaftliche Gemeinschaftsarbeit 1890 bis 1918," in *Technik, Ingenieure und Gesellschaft*, ed. K.-H. Ludwig and W. König (VDI, 1981), pp. 167–234, esp. pp. 194–195, 209, 215–216; Wolfgang König, "Die Ingenieure und der VDI als Großverein in der wilhelminischen Gesellschaft 1900 bis 1918," ibid., pp. 235–287, esp. pp. 252, 257, 260; Erwin Viefhaus, "Ingenieure in der Weimarer Republik: Bildungs-, Berufs- und Gesellschaftspolitik 1918 bis 1933," ibid., pp. 289–346, esp. 296–299 and quotation on p. 334. See also Gispen, *New Profession, Old Order*; and Dietz et al., *Technische Intelligenz und "Kulturfaktor Technik."*

65. Georg Klingenberg, "Elektrizitätsgroßwirtschaft unter staatlicher Mitwirkung," *Elektrotechnische Zeitschrift* 29 (1909): 118; "Elektrizitätswerke und Überlandszentralen," *Elektrotechnische Zeitschrift* 34 (1913): 315–317; "Elektrische Großwirtschaft unter staatlicher Mitwirkung," *Elektrotechnische Zeitschrift* 37 (1916): 314–317, 328–333, 343–348, 714–716.

66. Klingenberg, *Bau*, pp. iv, 1.

67. Klingenberg described two power-generating systems in the last two-thirds of his book, and in 1920 he described a third (*Bau großer Elektrizitätswerke*, volume 3: *Das Kraftwerk Golpa*, Springer, 1920). See also Gilson, *Konzepte.*

68. Koepchen, "Zur Sozialisierung," pp. 481, 483–484.

69. Koepchen, "Das RWE," pp. 3–4.

70. "W. Stiel," *Elektrotechnische Zeitschrift* 57 (1936), p. 350; Siemens, *History of the House of Siemens*, volume 2, pp. 124–125.

71. Stiel, *Elektrobetrieb*, pp. 12, 13, 19, 27–28. See Stiel's comparison of shaft transmission, electric group drive, and individual drive (pp. 115–143). Stiel devoted a quarter of his book to energy supply and auxiliary operations and three quarters to setting up electrical systems and motors in a variety of textile mills. The more closely historians study shop floors, the more they see contingency and choice—see Misa, "How Machines Make History" and "Retrieving Sociotechnical Change."

72. Nolan (*Visions of Modernity*, pp. 40, 245) states that few speculated about alternatives in Weimar.

73. Rules do not exist for applying rules. Institutions provide a guide. See Bloor, *Wittgenstein, Rules and Institutions*; and Mary Douglas, *How Institutions Think* (Syracuse University Press, 1986).

74. Klingenberg, *Bau*, pp. 8, 38, 47–50, esp. p. 50; *Das Kraftwerk Golpa*, pp. 78–79.

75. Koepchen, "Das RWE," pp. 5, 9.

76. Stiel, *Elektrobetrieb*, pp. v, 12, 17, 27, 79.

77. Richard J. Evans (*Rereading German History*, Routledge, 1997, pp. 38, 81) notes that they rely on "American neo-Weberian sociology of the 1950s." Iggers (*The German Conception of History*, p. 277) notes that "what confronts the historian is a reality which responds to his questions and which may not be distorted. In this sense, Wehler and Kocka . . . assumed that the possibility exists to formulate real, not merely ideal, types by which historical reality can be measured."

78. See Misa, "How Machines Make History"; Misa, "Retrieving Sociotechnical Change"; Langdon Winner, *Autonomous Technology* (MIT Press, 1977). Mary Douglas ("Environments at Risk," in *Science in Context*, ed. B. Barnes and D. Edge, MIT Press, 1982) argues that members of groups use accepted conceptions of "time, money, God, and nature" to persuade each other to behave "properly." David Edge ("Reinventing the Wheel," in *Handbook of Science and Technology Studies*, ed. S. Jasanoff et al., Sage, 1995, p. 18) notes the "tenacity" of "the 'old,' positivistic image of science, as an abstract, timeless search for irrefutable facts—ending the pain of uncertainty, the burden of dilemma and choice, separable from 'society,' and leading inexorably to technical innovations for the good of all." Richard J. Evans (*Death in Hamburg*, Oxford University Press, 1987, pp. 503–504) demonstrates that tenacity. He suggests that Barry Barnes is wrong in seeking to apply T. S. Kuhn's ideas to sociology. Barnes does no such thing. Barnes (*T.S. Kuhn and Social Science*, Columbia University Press, 1982, p. 115) states that "how far inference and evaluation in science relate to contingencies in the narrow scientific setting, and how far to macrosociological factors, is a straightforward empirical matter."

79. Rabinow ("Representations Are Social Facts," pp. 47–51) discusses "academic politics" as integral to knowledge construction.

80. Geoff Eley, "German History and the Contradictions of Modernity: The Bourgeoisie, the State, and the Mastery of Reform," in *Society, Culture, and the State in Germany, 1870–1930*, ed. G. Eley (University of Michigan Press, 1996), p. 83, note 34. Sharon Traweek (*Beamtimes and Lifetimes*, Harvard University Press, 1988) discusses novices, training, moities, and ritual avoidance in high-energy physics, a community of discerning individuals.

81. Wehler, *Deutsche Gesellschaftsgeschichte*, volume 3, p. 1383, note 2.

82. Proctor, *Value-Free Science*, pp. 85–98.

83. Ringer, *The Decline of the German Mandarins*.

84. Herbert Mehrtens, "The Social System of Mathematics and National Socialism: A Survey," in Renneberg, *Science, Technology, and National Socialism*, pp. 291–311; Herbert Mehrtens, "Irresponsible Purity: The Political and Moral Structure of Mathematical Sciences in the National Socialist State," ibid., pp. 324–138.

85. Forman ("Independence," pp. 73–77, identifies this as a lack of responsibility, as does Mehrtens ("Irresponsible Purity"). Dealing with a similar issue, Trommler ("Creation of a Culture of Sachlichkeit," p. 484) writes that "once morality is transformed into Sachlichkeit, politics deteriorates into the worst form of ideology."

86. Ledford (*From General Estate to Special Interest*) demonstrates that lawyers building their claims to social position as an estate serving the public were increasingly recognized as serving their own interests.

87. Mommsen (*Rise and Fall of Weimar Democracy*, pp. 191–196) notes that the parliamentary system in Prussia remained viable during the 1920s in part because organized interests fought at the national level. He notes that "responsible party leaders" recognized "the need for political compromise" (p. 195). For a discussion of building political coalitions as an aspect of constructing technological systems, see Lambright, "Political Construction of Space Satellite Technology."

88. On the founding of Märkisches Elektricitätswerk in 1909 and the development of the Victoria Falls and Transvaal Power Company soon thereafter, see Klingenberg, *Bau*, pp. 74–75 and 116–119. On the establishment of the Golpa power station south of Berlin during World War I, see Klingenberg, *Das Kraftwerk Golpa*, pp. 1–7, esp. p. 6.

89. Wolfgang R. Krabbe (*Die deutsche Stadt im 19. und 20. Jahrhundert*, Vandenhoeck & Ruprecht, 1989, p. 92) makes this point.

90. Richard Passow discusses a variety of mixed corporations in *Die gemischt privaten und öffentlichen Unternehmungen auf dem Gebiete der Elektrizitäts- und Gasversorgung und des Straßenbahnwesens* (Gustav Fischer, 1912).

91. *Das Märkische Elektricitätswerk* (Berlin: 1934).

92. On the development of Klingenberg's system in South Africa, see Renfrew Christie, *Electricity, Industry and Class in South Africa* (State University of New York Press, 1984), pp. 5–49.

93. Walther Rathenau of AEG was also promoting a national system: see Helga Nussbaum, "Versuche zur Reichsgesetzlichen Regelung der deutschen Elektrizitätswirtschaft und zu ihrer Überführung in Reichseigentum, 1909 bis 1914," *Jahrbuch für Wirtschaftsgeschichte* (1968), part 2, pp. 117–203. Hård ("German Regulation," pp. 46–51) stresses Rathenau's Sachlichkeit and the power industry's reliance on "scientific technology" as a basis for Rathenau's promotion of a national reorganization led by the state to assimilate modern technology. Trommler ("The Creation of a Culture of Sachlichkeit") notes that the new aesthetics did have a politics: conservatives recognized claims of efficiency and materialism as potent attacks on the old order, while Social Democrats opposed attempts to construct alternatives to their own materialism.

94. Boll, *Entstehung*, pp. 28–30; Klingenberg, *Das Kraftwerk Golpa*, pp. 3–6.

95. Klingenberg, *Bau*, pp. 71–73.

96. Boll (*Entstehung*, pp. 77–102) describes the technical problems between 1933 and 1945. Gilson (*Konzepte*, p. 189) questions Klingenberg's figures.

97. *Das Märkische Elektricitätswerk*; Hughes, *Networks of Power*, pp. 198–200, 289.

98. Todd, "Electrification in the Ruhr circa 1900"; Todd, "Prussian Landräte."

99. Koepchen, "Zur Sozialisierung," pp. 483–484; Koepchen, "Das RWE," pp. 3–8. Koepchen fit the RWE into what Gerald D. Feldman (*The Great Disorder*, Oxford University Press, 1993, pp. 272–305) calls "the Gospel according to Stinnes." Hugo Stinnes directed the expansion of the RWE from 1902 until 1924.

100. Koepchen, "Das RWE," pp. 10–12. For the views of Prussian officials, see Jaques, "Preußen und das Reich in der deutschen Elektrizitätswirtschaft," *Elektrotechnische Zeitschrift* 47 (7 October 1926): 1160–1163. For evaluations of these developments, see Thomas Herzig, *Geschichte der Elektrizitätsversorgung des Saarlandes unter besonderer Berücksichtigung der Vereinigten Saar - Elektrizitäts-AG* (Kommissionsverlag, Minerva-Verlag Thinnes & Nolte OHG, 1987), pp. 137–155; Norbert Gilson, "Der Irrtum als Basis des Erfolgs. Das RWE und die Durchsetzung des ökonomischen Kalküls der Verbundwirtschaft bis in die 1930er Jahre," in *Elektrizitätswirtschaft zwischen Umwelt, Technik und Politik*, ed. H. Maier (TU Bergakademie Freiberg, 1999), pp. 51–88; Helmut Maier, "'Nationalwirtschaflicher Musterknabe' ohne Fortune. Entwicklungen der Engergiepolitik und des RWE im 'Dritten Reich'," ibid.

101. A. Koepchen, "Westdeutsche Elektrizitätswirtschaft, Vortrag gehalten am 3. Februar 1936," *RWE, Bericht über das Geschäftsjahr 1935/36*, pp. 27–35.

102. Stiel, *Elektrobetrieb*, pp. v, vi, 1–3, 12–19, and 27–35, esp. 27.

103. Herrigel (*Industrial Constructions*, pp. 1–32, 61–62, 123–139) describes the extensive political and economic problems faced by areas supporting decentralized production in the 1920s and the 1930s. On economic problems and Nazi political campaigns, see Thomas Childers, *The Nazi Voter* (University of North Carolina Press, 1983), pp. 64–71, 79–80, 142–144, 151–159, 211–224.

104. In writing about the engineering profession, Jarausch (*The Unfree Professions*, pp. 178–181) finds "mass rationalization" dramatically increasing production during World War II. Nevertheless, Richard James Overy (*Why the Allies Won*, Norton, 1995) finds German officers supporting production of a wide variety of types of military equipment, quite unlike the Soviet concentration on mass producing a very few models of required equipment. On postwar developments in decentralized regions, see Herrigel, *Industrial Constructions*.

105. Johnson-McGrath ("Speaking for the Dead," p. 455) makes this point about positivism. Ian Buruma ("Japan: In the Spirit World," *New York Review of Books* 43, no. 10, p. 33) notes: "Syncretic religions are more likely to foster tolerance than universalist faiths based on dogma. When beliefs are negotiable, they are less likely to be imposed."

106. Nolan (*Visions of Modernity*, pp. 179–205) notes that right-wing efforts to reshape workers were presented as nonpolitical scientific ways to increase productivity in firms envisioned as autonomous from society and politics. Themselves technological determinists, Social Democrats, Communists, and Trade Union officials were unable to develop successful counter programs.

107. Childers, *The Nazi Voter*, p. 268.

108. Articles in Renneberg and Walker, *Science, Technology and National Socialism* and in Dietz et al., *Technische Intelligenz*, demonstrate many continuities. By making technology fundamental, Boll (*Entstehung*) presents constructing power systems during the Third Reich as heroic.

109. Barnes develops this approach in *The Nature of Power*. Ringer (*Max Weber's Methodology*) denigrates the "'strong' relativist program put forward by Barry Barnes and David Bloor" (p. 147) and hopes to maintain the concept of "ideology" and "false consciousness" (pp. 142–149).

110. Wolfgang J. Mommsen, *Max Weber and German Politics, 1890–1920*, second edition (University of Chicago Press, 1984), p. 434; see also Ringer, *Max Weber's Methodology*, pp. 142–149.

111. Rorty, *Achieving Our Country*, p. 13.

112. Loren R. Graham (*The Ghost of the Executed Engineer*, Harvard University Press, 1993, p. 99) refutes the suggestion that the Soviet Union represented too much scientific objectivity in the form of "arrogant, absolutist reason" (hence the need to trust subjectivity). Graham stresses the problems that follow from making projects politically but not technically feasible.

113. Hughes, *Rescuing Prometheus*, p. 197.

Modernity, the Holocaust, and Machines without History

Michael Thad Allen

Walk into a bookstore and pick up any book whose author loosely fits the label "public intellectual." Many will fit under the broad category of what is now popularly known as the "Culture Wars": authors such as Alan Bloom, Jean-François Lyotard, and Richard Rorty. Flip through the index to the H's. Chances are that you will find a few pages on "Hitler" or "Holocaust," and if you read them you will find that, despite the bewildering diversity and vituperation, the unique atrocities of Nazism and "modernity" occupy a prominent place in these authors' historical imaginations. For example, Francis Fukuyama urges us to embrace the shining truth of modernity, which he defines as the progressive historical implementation of the Anglo-American Enlightenment, because a modern United States vanquished National Socialism.[1] Rorty, on the other hand, admonishes us to seek *not* the truth because it is authoritarian and leads down a slippery slope to the gas chambers.[2] Not to be outdone, M. Stanton Evans, who shares Rorty's disquiet over "the ideologies of the modern era," informs us that "Hitlerian concepts of unfettered power, genocide against defenseless people, and eugenic efforts to direct the growth of populations . . . resulted from the idea that man, severed from his connection to the absolute, had been assimilated entirely to the world of nature."[3] To rescue the West from this distress, he invites us to seek salvation in Jesus Christ. All perceive the modernity or anti-modernity of the Nazis as a climacteric of twentieth-century history. At stake is not so much an accurate portrayal of the Holocaust, but how that past can be fashioned as either a mortal weapon or impenetrable armor for competing claims to know where World History should be taking humankind.

In the face of this overwhelming multiplicity, let me state unequivocally what "the modern," "modernity," and "modernization" mean in the present meditation.

First, any claim to modernity or to "modernize" places a stake upon futurity. It seeks to fix the preconditions that must be fulfilled in order to make that future possible and identifies the patterns of development of these preconditions in history. Thus, to understand "the modern" requires more than any one idea. It requires comprehending the structure of a debate. This debate constantly identifies false starts as backward historical developments. To know what is modern, in contrast, means to hold the key to a new era, to claim the right to know what is best for the full flowering of human history and, no less, to herald the creation of a "new man." The primary casualty of this debate is reasonable discussion of first principles, which "modernity" sacrifices from the outset. To claim that anything is modern is to claim that it is desirable not because it necessarily represents the good, the true, or the beautiful but because it represents historical destiny.

Second, to say that "modernity" constitutes a debate does not imply that all is relative or that what is being debated is a matter of indifference. "Modernity" is about the power of industrial technology to transform civilization. As Reinhardt Koselleck pointed out long ago, the historical imagination has been preoccupied with modernity since the industrial revolution. Because the industrial revolution in tandem with the French Revolution inaugurated more than 200 years of continuous, rapid change, our ability to assimilate the past into the present has become a deep-felt need which Koselleck attributes to the startling pace of technological change: ". . . the more a particular time is experienced as a new temporality, as 'modernity,' the more demands made on the future increase. . . . this is certain to be an effect of the technical-industrial modification of a world that forces upon its inhabitants ever briefer intervals of time in which to gather new experiences and adapt to changes induced at an accelerating pace."[4] Again, what is at stake is our future. Because of the ongoing industrial transformation of society, nothing can be taken for granted, "the center does not hold," and therefore, in order to prepare as well as possible to muddle through an uncertain tomorrow or even to know who we are, citizens of modernity constantly stare into the past in order to discern patterns which supposedly propel us forward in space and time (even in futility—like Walter Benjamin's Angel of History). Most often, this means an attempt to take charge of the patterns of technological change and claim that they will define our destiny.

It should be no surprise, in this light, that the special role of science, technology, and modernity in the Holocaust plays such a large role in the "Culture Wars" to which I alluded above.[5] At the most superficial level, modern technology defines the uniqueness of the Holocaust. Fatigued and blinkered as we are, the simple fact of genocide—even large-scale genocide in the heart of Europe—is less likely to shock us than it shocked those who believed unequivocally in Western progress at the middle of the twentieth century. Knowing more of the collectivization campaigns of the Soviet Union, we are also less likely to count the sheer numbers of dead as a distinctive sign of evil. In addition, it now seems that quite primitive regimes such as Serbia and Rwanda are capable of piling up bodies on a vast scale. What was truly surprising at the end of the twentieth century was that anyone ever thought this kind of barbarity would ever disappear from human experience.[6] On the other hand, only the Germans strove to use "assembly-line" methods. Only the gas chambers enlisted the services of the most advanced chemical industry in the world and the officious punctuality of the Reichsbahn. Only the concentration camps statistically managed murder, and only they built economies of scale and scope into their extermination systems.[7]

The National Socialists unmistakably turned to modern technology and science as their means; at the same time, they took Germany's historical prowess with modern technics as evidence that their achievements could deliver them to the "End of History," its full unfolding. A brief survey of the *Journal of the Four Year Plan* allows some insight into what the Nazis meant by "modern world" and what technology had to do with it. The *Journal*, founded by Hermann Göring, showcased the Third Reich's industrial policy on glossy folio-size pages replete with statements of Nazi futurity. Göring proclaimed in the first issue that National Socialism had divined the secret to a reborn economy and promised to "fulfill tasks of world-political and world-historical proportions."[8] Economics Minister Walther Funk similarly claimed a mandate to direct "tasks which have never before existed in a highly qualified modern national economy."[9] A lesser official declared "We are National Socialists and that gives us duties and tasks for all future times"[10]; likewise, staff writers lauded the Frontarbeiter (Front Workers) organization as a crucible in which a new man would be forged.[11] Claims to modernity also took the form of specific artifacts and technological systems, as when Jakob Werlin wrote of the centralized, hierarchical management of the Volkswagenwerke and its rigorous insistence on norms, standardization, and mechanized production.[12]

The Nazis' millenarian modernism has caused great discomfort, for arguments that Hitler had a coherent modern social vision seem to lend themselves to apology. How can we fault the Nazis if they wanted a "modern" society just as everybody else did? Some warn us to abandon the discussion of National Socialist modernity altogether. Hans Mommsen asks, in near total frustration, "Does anyone earnestly wish to make the fascistic understanding of politics, which touched upon mere mobilization for the transformation of visionary-utopian goals, into the foundation of a present day 'modern' society?"[13] The answer, of course, is that the Nazis did, which cuts to the quick of a central false dilemma of the modernity debate. Those who would identify modernity with historical destiny, as the necessary legislator of the norms of contemporary society implied by Mommsen, leave scant room for the rational judgment of first principles— on what primary grounds should we condemn the regime *whether it was modern or not*? Too often the "modernity" debate seeks to substitute historical destiny for substantive discussion and, in so doing, strives to discount the actual history of the Nazi regime. Ironically, calls to abandon the question of "modernity" and National Socialism is in many ways analogous to Marxist debates of the 1960s and the 1970s, which demonstrated a marked aversion to any discussion of the Nazis' robust anti-capitalism. But that debate never went away just because its implications were unpleasant to some historians and other academics.[14] Modernity and National Socialism shows little sign of doing so either. In view of Heinrich Himmler's enthusiasm for "a completely new, modern concentration camp for new times, capable of expansion at any moment" the question is not whether the Nazis were modern or whether we should discuss it.[15] The question is what Nazi modernity meant and what it implies for us in the present.

Modernity Is Not the Enlightenment, or Why Nazi Modernity Is Hidden

As I have already noted, a macabre fascination with technology looms large in debates about the Nazi genocide. What, then, connects modernity, technology, and the Holocaust? Saul Friedländer has commented upon the feeling of Rausch (inebriation or ecstasy), which, he suggests, seduced Germans into believing that the Holocaust was meet and right.[16] Rausch stresses human transcendence through existential experience, which invites initiates to "feel the power." Rausch, as described by Friedländer, derives from terror and confrontation with the incomprehensible, as, for instance, when genocide sprang all known moral boundaries, requiring "ordinary men" to act outside any traditionally accepted authority. Nazi leaders were quick to explain that only Hitler's Volksgemeinschaft could provide the medium for this Rausch, precisely because National Socialism had made them subject to a new morality and a new historical epoch. They were invited to "feel" its history-making acts, among them mass murder. This is clear in Himmler's 1943 speech at Posen, in which, amidst the butchery, he urged his men to feel proud, empowered, and justified in their transgressions, for Nazis were masters of their own morality. For this very reason they believed themselves to stand at the brink of historical destiny, defining modernity's final and highest expression.[17]

This same sense of intoxication also manifests itself in technological endeavor.[18] Even some who are not historians of technology, such as Modris Eksteins, have argued that new technology was "a means of escaping from the confines of reality, a way of liberating the imagination."[19] Technology has long been a "religion" in the West, and, as David Noble has recently observed, we should not forget that this *secular* deity manifests itself through history. Noble locates it first in medieval Christianity, in which "technology . . . became at the same time eschatology."[20] Eschatology appealed to the Nazi mentality, although Hitler's 1000-Year Reich aspired to a racial utopia entirely of human making. This too was quintessentially modern, a narrative of human teleology to be fulfilled in this world and not through divine salvation in the hereafter. This caused no insignificant amount of friction between organs of the National Socialist German Workers Party (NSDAP) and traditional institutions. For example, defense of the modernity of National Socialism can be found in the censorship bureau of Richard

Walther Darré's Reichsnährstand (Reich Agricultural Estate)—an institution often presented as exemplary of Nazi anti-modernity.[21] The Catholic magazine *Nature and Culture* had upset a certain censor, Dr. Böhmkamp, by arguing that "natural science can give no answer to the most important problems of our being and that science is absolutely worthless." The rejection of evolution in the name of the biblical creation myth seemed to bother Böhmkamp the most: "We have here the crassest confusion of the confessional church when it comes to modern scientific results, and we must see this as an attempt to topple the foundations of National Socialist ideology."[22] Here the author coupled modernity, based in "science," with the fundamentals of National Socialism as a secular faith. The specific "science" he had in mind was the "racial teaching" of eugenics.

Racial science has received increasing attention from historians, including Robert Proctor, Henry Friedlander, Götz Aly, and Michael Burleigh. Nevertheless, in disproportion to the stress placed upon the Nazis' "machinery of death," academic discourse has left the unique technology of the Holocaust almost completely untouched.[23] Instead, debates about modernization and National Socialism have entered a seemingly endless round, which comes full circle about once a decade. In 1972, Henry Ashby Turner, polemical as ever, suggested that the study of modernization in Germany could open the phenomenon of Nazism to more useful conclusions than previous analysis of capitalism, fascism, or totalitarianism. Nearly 15 years later, Thomas Nipperdey suggested roughly the same thing.[24] Almost 10 years after that, Hans Mommsen titled an article "Yet One More Time: National Socialism and Modernization."[25] It is little wonder that Gerald Feldman remarked disparagingly that "an invitation to consider any historical period from the perspective of modernization theory is a bit like being invited to climb a mountain in the fog."[26]

Technology has always played a crucial role in all definitions of modernization, and historians and sociologists by no means neglect it in their analyses. As Ian Kershaw has noted, modernization "implies long-term change spanning centuries and transforming 'traditional' society . . . into industrial class society with highly developed industrial technologies."[27] As Mary Nolan has shown, a preoccupation with modern technology was common, from right to left, among industrialists, engineers, managers, and union leaders of the 1920s.[28] And Ernst Jünger, with near exaltation, identified the same trends that Kershaw would summarize more than 50 years later: ". . . the technological engagement of industry, economy, agriculture, transportation, administration, science, public opinion—in short, each special substance of modern life in a self-enclosed and elastic space, inside of which a common character of power manifests itself."[29] The "common character of power" immanent within new technology and organization that Jünger celebrated filled the imaginations of ordinary Germans. The National Socialists were no exception.

The debate continued after 1945, as one might expect. Attempts to predicate historical destiny on "modernity" did not end with Hitler's suicide, and this has only added to the fog so wryly identified by Feldman. However, despite the prominence of modern technology in discourse of the 1920s, the 1930s, and the 1940s, after the Third Reich's demise the historical fact of National Socialism became a new linchpin in bitter struggles over the Enlightenment. This struggle has involved, essentially, two camps. Immediately after World War II, self-conscious moderns of the liberal left pointed to the Third Reich's supposed anti-modernity as the quintessential evil from which they proposed to rescue history. Later, so-called postmoderns quickly adopted the same strategy, but by declaring the Nazis to be modern they have used the Holocaust to discredit liberalism in general. The central theme common to these otherwise antinomic visions is a determined conflation of modernity with the Enlightenment, though this was wholly alien to the National Socialist movement and uncommon in the interwar period.

Very early on, champions of the Enlightenment advanced the erroneous conclusion that National Socialism was "anti-modern." Historians such as Henry Ashby Turner, joined by intellectual historians such as George Mosse and Fritz Stern, claimed that, essentially, National Socialism was a revolt of modernity's losers. Challenged by the progressive transformations inevitably caused by industrialization, these losers wished to "turn back the clock."[30] Hitler promised to wreak revenge on an isolated social group which reactionaries associated with everything repugnant about modernity: the Jews in modern urban areas, modern professions, and modern art movements who had profited from the liberal democracy and egalitarianism of the Weimar Republic.[31] Turner is the only historian of industry to have followed this trajectory, which Timothy Mason once pejoratively referred to as the fixation of intellectual historians only.[32] Turner is also one of the few to have tried to explain the obvious contradiction between the supposed anti-modernity of Nazi ideology and Germany's startling military and industrial might. In brief, he argued that the Nazis came to terms with modern industry out of practical necessity. Hitler needed modern industry to realize his military pursuit of an atavistic "Lebensraum."[33]

The sociologist Ralf Dahrendorf had adopted a similar interpretation of Nazism in the mid 1960s: "Hitler needed modernity, little as he liked it."[34] Yet Dahrendorf went beyond analysis of industry to argue that, whatever . the Nazis intended, they unwittingly destroyed the bastions of traditional society. After the war, traumatized as it was, German society could no longer resist modernization.[35] Thus liberal democracy, set equal to modernity, flowered in the Bundesrepublik despite continuities with the Nazi legacy. Several historians have pressed this point: If the Nazis achieved "modern" credentials, they certainly did not mean to; furthermore, those credentials are themselves dubious. So argued the social historian Hans Mommsen: the Nazis offered only a "vorgetäuschte Modernisierung" (feigned modernity)—a sham unity of the diffuse and disarrayed factions of Weimar.[36]

The postwar generation of scholars overwhelmingly associated "modernity" with the Enlightenment's legacy of liberal rationalism and with the ascendance of a politically active bourgeoisie in the cause of democratic government. They made no secret of their own first principles; however, they advanced them in a curious way, as if the historical destiny of the "modern" world entailed equality, liberal education, a society open to merit, and the rule of law as the natural accompaniment to ongoing industrialization. Their efforts to make this a reality in postwar German society should not be taken lightly. Many of those who had built the Third Reich were still alive and well in the 1950s and the 1960s, and the need to discredit their intellectual traditions was felt keenly and approached in deadly earnest. What better strategy than to label the Nazis as anti-modern—that is, retrograde troglodytes, or, as Wolfgang Sauer pithily called them, "the Philistine Underground"![37] The political stakes of this literature help explain the impact of David Schoenbaum's book *The Brown Revolution*, which argued that National Socialism created advantageous conditions for social mobility and political participation, despite its authoritarianism—perhaps because of it. His book makes no mention of the modern, modernity, or modernization, but Schoenbaum was drawn into the modernization debate nonetheless.

By the 1980s, many had begun to express uneasiness with modernity. Nipperdey, one of the few postwar historians to strike an ambivalent stance toward modernity, attempted a syncretism by claiming that, although the Nazis were anti-modern, they were also anti-tradition, and therefore they were "revolutionary anti-moderns."[38] Mostly younger historians have pursued this argument. Jeffrey Herf, who stands Nipperdey's point somewhat on its head, argues that the Nazis were "reactionary moderns."[39] Yet these authors still abide by the touchstone assumption that the liberal Enlightenment sets the paradigm for what it is to be "modern," something staunch National Socialists would have denied.

For the very reason that the Nazis embraced modernity, other historians have correspondingly condemned Enlightenment rationalism itself through guilt by association. At the risk of grossly oversimplifying, I suggest that Götz Aly, Susanne Heim, Zygmunt Bauman, Detlev Peukert, and Ronald Smelser all have uncovered contributions made to "modern" Germany by the Third Reich and all judge that legacy to be vexed and, at worst, malignant.[40] Peukert, for instance, argued that the Nazis showed the "shadow side" of modernity.[41] If portrayals of the Nazis as a reaction against modernity have explained Hitler's commitment to modern technology as a "compromise," we should not be surprised to find that critics of the anti-modern thesis place Nazi jubilation over modern technology at the center of their arguments.[42] To Zygmunt Bauman, a straight line runs from the logic of modern bureaucracy and technology to the killing machinery of the death camps: "The murderous compound was the work of a typically modern ambition of social design and engineering mixed with the typically modern concentration of power, resources, and managerial skill."[43] The Holocaust was social engineering with a vengeance that stemmed from "the Enlightenment" an "enthronement of the new deity, that of Nature, together with the legitimation of science as its only orthodox cult, and of scientists as its prophets and priests."[44] Partially in support of Bauman's ideas, Detlev Peukert's last essay warned of the "the genesis of the 'Final Solution' from the spirit of science."[45] Unlike immediate postwar scholarship, these historians and sociologists warn us that another event like the Holocaust may be exactly what we will get if an ascendant bourgeoisie, modern political economy, science, and industry continue to develop unrestrained.

Nevertheless, "modernity" remains more or less equated with the Enlightenment and liberalism. Alan Beyerchen has recently remarked that the Nazis "were united in their belief that they were fighting against modernity itself, which they called liberalism, democracy, urbanization, parliamentarianism, and other names they abhorred."[46] This last statement shows the current need to historicize modernity and the Holocaust, for there is no reason to believe that when the Nazis heaped scorn upon the legacy of liberalism they spoke in codes: they hated liberalism. When they mentioned modernity, as an architect did regarding the Hermann-Göring Werke of Salzgitter, they did so alternately with the gusto of crusaders and the dewy eyes of poets:

Here a model rises into the light that will serve as precedent for private industry for future building. Here the engineer, the scientist, and the architect have sat down together and have created in communal labor the ideal modern industrial works, as rational as it is beautiful, at once a symbol . . . of true industrial culture: Technology, Science, Art.[47]

It is difficult to discern abhorrence for "the modern" or its offspring, "modernization" or "modernity," in this statement, and the Nazis would have violently rejected the label of "reactionary" or "anti-modern" applied to them ahistorically after 1945. Sadly, having chosen to advocate the advance of "modernity" instead of championing the substantive issues at the heart of Enlightenment humanism directly, liberal scholars such as Mosse, Stern, Dahrendorf, like many a disillusioned spouse, find that modernity has been cheating on them. Despite insistence on a "one best way," the modern has shown little fidelity to liberal social and political reform.[48] Modernization, in other words, philanders, and this is nowhere more apparent than among the historians (predominantly on the new German right) who have set out to historicize modernity and National Socialism. Far from creating a fresh consensus, they have polarized debate all the more. They have maintained the positive valuation of modernity expressed by immediate postwar scholarship; on the other hand, they have reversed the long-standing criticism of Nazi "anti-modernity." To the extent that National Socialism made possible a modern Germany, so argues especially Rainer Zitelmann, those contributions should be judged good and even celebrated as part of German national identity.[49] Thus, above all, this "mountain in the fog" of books, special editions of journals, and review essays demonstrates precisely the point with which I began this essay: that modernity has always encompassed a debate far more than a singular body of consistent political philosophy or a coherent, predictable pathway of historical development.

The Machine in the Fog

Nevertheless, this is a debate about industrial technology. Because of a wide-ranging consensus after World War II that the Enlightenment and modern destiny are one and the same, the history of technology has receded from view, but it is present nonetheless in a curious form. As important as technology has been to debates about modernity, most scholars usually equate one and only one technological rationality and one and only one pathway of industrialization with Enlightenment rationalism. Machines appear as artifacts without history even to those (e.g., Herf, Koselleck, Kershaw, Eksteins) who acknowledge the central importance of technology in these debates. Undifferentiated, machines have their entrances and their exits upon the world stage, but no character. But which machines are modern? Here a more accurate history of technology can play a crucial role. For if, as I argue, the attempt to associate some kind of destiny with the change caused by industrialization defines the "modernity debate," then selecting a single technological trajectory as the harbinger of that destiny involves historical choices. This holds whether or not the selection is made consciously. Some industrial technologies acquired favored status as "modern" over the course of the 1920s and the 1930s; others, equally new and equally capable of invoking massive social change, economic growth, and organizational transformation failed to awaken interest and were in fact discredited as illegitimate pathways toward future development.

The Nazis were fascinated, for instance, with large-scale industry over shop production (despite the misimpression that Nazi ideology favored the old middle class). Gottfried Feder, an engineer and Hitler's early economic guru, is often taken as the quintessential Nazi "anti-modern" because he wished to safeguard small shopkeepers from large department stores. He was also a rabid anti-Semite who blamed the Jews for Germany's economic troubles during the 1920s. Yet when Feder wrote of the "moral personality" of the "true entrepreneur" in *The Program of the NSDAP*, he did not name butchers, bakers, and candlestick makers. He named inventors, "constantly concerned with innovations and improvements in factory and sales." And who did Feder name as the "most prominent and world renowned example of this kind of entrepreneur?" Henry Ford—hardly an inspiring figure for anyone seeking to "turn back the clock."[50]

Yet the 1930s were precisely the decade in which General Motors surpassed Ford by introducing more consumer-oriented marketing and more flexible approaches to serial production. This did not inspire Feder. The Four Year Plan, the Herman Göring-Werke, Volkswagen, SS concentration-camp industry, Albert Speer's war production ministry, and the rantings of Adolf Hitler displayed a consistent fascination with techniques of modern organization and standardized mass production—in short, for the capital-intensive, concentrated industry that was new to the twentieth century.

As defined by Thomas Parke Hughes, modern systems are characterized by centralized hierarchy, with smaller systems subordinate to larger, more encompassing systems through mechanisms of control both technical and social. Rather than conflate such systems with an eminent destiny compelled by an all-embracing Enlightenment rationality, however, one should remember that centrally controlled systems have themselves come under attack by industrialists and entrepreneurs as irrational. This is due to their inflexibility and, not least, their unprofitability.[51] As sociologists of technology such as Michael Piore and Charles Sabel and historians of technology such as Philip Scranton and Jonathan Zeitlin have shown, other, highly innovative alternatives to mass production presented themselves in the 1920s and the 1930s alongside assembly lines—namely, flexible, all-purpose machine tools and the mass marketing of customized goods through adaptable means. Nevertheless, specialized machine tools, rigid norms of design, and assembly lines captured Hitler's fancy, He latched onto them as an intimation of a future world, not the "old-fashioned" craft modes of production, the budding "flexible specialization," or the consumer's "freedom of expression." Remarking on traditional shop practice, Hitler at times ridiculed nostalgia for the supposed quality of non-mass-produced goods: "One wishes to believe that it [handmade goods] concerns something unachievable. That is a bluff. A modern giant press stamps me something to an exactness that is totally impossible with our craftsmen." Noting the productivity of Fordist automobile factories, Hitler continued: "The entirety is the pure work of automatic machines. . . ."[52]

Hitler's preoccupation for automation, I would argue, was largely if not wholly aesthetic, for he was no engineer. After all, even in the 1940s only a few of the world's companies found automated production profitable or even possible. Nevertheless, the spectacle of these few plants seemed the prophecy of a new order; they prompted the Swiss historian Sigfried Giedeon to write *Mechanization Takes Command.* There are few records of the same overwhelming experience elicited from other modes of production. Whether advocates (such as Aleksei Gastev) dedicated poetry to the grand machinery of mass production or whether critics (such as Aldous Huxley) conjured up unparalleled images of barbarism, the continuous flow of assembly lines stimulated the imagination like no other history of technological development in the early twentieth century. It seemed unique, something "man" had wrought for the first time, and it gripped the Nazis with its existential intoxication as much as with its promise of efficiency and output. Ernst Jünger described the mass-production aesthetic as a new mode of being. A new man had emerged, Jünger proclaimed, who did not strive for his essence in individual expression but who took delight in the massive release of power through collective organization and technological control: the grander the scale, the more intoxicating it was; the larger the influx of resources, the greater the complexity, the more naked the technology, the more titillating it was.

In industrial terms, this aesthetic meant a preference for productivism over consumption, which both Richard Overy and Mary Nolan have noted from quite different methodological directions.[53] And here I wish to define productivism precisely in its relation to Nazi modernity: it meant the tendency to view the purpose of the factory not as economic output in goods or services but as the production of identity. In this sense, National Socialism shared with other political movements of its day a central belief in work as the formative activity of modern collective identity. Whereas distribution and consumption carried the eye far from the factory system and tended to emphasize particular, prosaic uses of things, the aesthetic of productivism made the assembly line into a symphony of national destiny whose output could be enjoyed for its own sake. This would also seem to explain the prevalent fascination with Ford, who stressed extreme vertical integration and engineering prowess, over GM, which stressed marketing and sales. It did not hurt that Henry Ford was also an anti-Semite.

Again material drawn from the *Journal of the Four Year Plan* is instructive. First of all, in many of its articles liberal ideals of equality received a drubbing as old-fashioned and infantile:

> [The Führer] does not preach the childish and mocking lecture of the equality of man and man's equal claim on the distribution of the goods of this earth, but rather [the Führer] gives you back consciousness of your belonging to a great people. In the people [Volk] you are bound to decline or prosper and in the people each can achieve as much as he can according to how he applies his ability and achievements . . . so that ability and achievement is the only measure of progress . . . not—what a utopia!—to eliminate the inequality of man but the inequality of conditions under which they work.[54]

If enhanced production was not to serve the greater equality of human beings through the increased gratification of material consumption, what was the goal of industrial society? Another article assured readers that industry could now once again serve "the moral and cultural life of the nation." Rather than the mundane output of commodities to satisfy whimsy, the Four Year Plan set itself the tasks of manufacturing national identity: "The social question is first solved when all productive [schaffenden] German national comrades reawaken their proud and happy consciousness, fully worthy and responsible to partake of the nation's providence as well as its cultural and economic goods to the measure of their productive contribution."[55] This doctrine defined the worth of each individual according to his or her contribution to the national community, and in this light factories were supposed to produce more than commodities; their products were supposed to be Germany's strength, prowess, and sense of unity.

An article titled "The Future German Economic Order" argued that a centrally planned economy was not about prosperity but about national virtue. The author proposed in relatively straightforward terms what I define here as *productivism*:

> . . . the peacetime economy works for the higher value and power of human life and brings living human communities to their best development. Once again it is not economic goals that the economy has to obtain. It is not only the man as consumer and receiver of economic production . . . in much stronger measure the economy must work for man as producer and provider.[56]

Production was to manufacture the German spirit, a spirit forged as the maker rather than as the consumer of goods. At the heart of Nazi modernity was the dream of a perfect system—what Charles Maier has called the "society as factory." But it was to be a system whose overarching output was supposed to be culture—the New Order of National Socialism.

Modernity and the Holocaust

In the Third Reich, "the New Order" meant something quite specific: the transformation of Europe into Nazi Germany's "Lebensraum." Götz Aly and Susanne Heim speculate that the desire to "modernize" a Greater Regional Economy (Grossraumwirtschaft) in central Europe fueled the Nazis' drive to eliminate the Jews, and their book *Vordenker der Vernichtung* makes the logic of capitalist expansion in the East a primary evil. Aly and Heim have been criticized for overdrawing their conclusions and for bypassing ideological motivations among perpetrators. Nevertheless, there can be little doubt that the exclusion of Jews from German society intensified in step with the accelerated armaments drive and the economic growth of the Four Year Plan.[57] I would argue that the Nazis' New Order had more to do with productivism than with an exclusively capitalist "logic" (a logic that many came to associate with "modernity" only after World War II).

I would like to turn to two examples: the norms and standards of manufacturing intended for Aryan settlements and the "factories of extermination" designed to eliminate the European Jews. In each case the motivation to fulfill "destiny" with modern technology had as much to do with the regime's intentions as with anti-Semitism and racial supremacy. The eastern territories conquered by the Wehrmacht serve as an excellent illustration. The SS received a mandate directly from Hitler to "make the East German." The special SS office of the Reichskommissar für die Festigung deutschen Volkstums (Commissar for the Reinforcement of Germandom) drew up blueprints for model Aryan settlements, which were supposed to secure the racial and cultural dominance of Germany over the conquered lands. Himmler's expert for regional planning adamantly opposed bucolic nostalgia in this endeavor:

German agriculture has just begun to overcome the backwardness of the last decades. The technical and economic criteria for optimal production are still in a great deal of flux and represent a complex problem, but the organization of new farmsteads will exclude from the very beginning any romantic or far-fetched fantasies of "bucolic forms."[58]

Thus, the New Order in the East undeniably included dreams of a perfectly planned and integrated "modern" economy with an emphasis on technology (though this was by no means its only dream). For one thing, the SS proposed to use the captive labor of the concentration camps to mass produce building materials, furniture, and appliances for its Aryan settlements. These settlements, in turn, were supposed to convert the East into a "German" landscape. I will return to the means of production the SS wished to use. Here it is important to note that the SS's conception of modernization was also inherently racially motivated. "The Führer himself has set forth the great outlines for building up the eastern territories," wrote one manager of the SS's concentration-camp industries. He did not propose to turn back the clock to craft industrial forms; rather, he referred to the SS's wish to build up modern factories. In contrast, he criticized the "backwardness" of Polish society, in which "habitation worthy of human beings found absolutely no place."[59] In Himmler's New Order, modern industry represented a means of transforming a Polish and Jewish landscape into an Aryan one; at the same time, the presence of modern machines was a part of what was supposed to be Aryan about that landscape.

A mundane example of how the SS sought to manufacture its New Order, the SS company German Noble Furniture, serves to illustrate how these megalomaniac fantasies became embedded in the minutiae of technological detail. Through the lens of productivism, even seemingly banal manufacturing standards could acquire the aura of hallowed national symbols. Like any other artifacts of industrial societies, standards have a history that accompanies them into daily use, and few regimes have ever taken their cultural implications as seriously as Nazi industrial planners did.

German Noble Furniture was one of the SS's "settlement" industries, founded to supply the New Order. In the summer of 1940, the chief executive officer of all SS companies, Oswald Pohl, prepared to bid on Emil Gerstel AG of Prague, a highly valued furniture manufacturer. Emil Gerstel, the owner, was one among many Jews whom occupation authorities were forcing to sell out, and his factory had attracted the SS's interest because of its advanced manufacturing methods. In Prague the SS solicited the help of a well-positioned officer in private industry, Dr. Kurt May, son of the Stuttgart furniture manufacturer A. May. In 1932, at the age of 22, Kurt May had taken over the firm upon his father's death.[60] In the late 1930s, in the crusading spirit of an industrial modernizer, he appeared in occupied Czechoslovakia seeking new investments. A friend of May's described his methods as follows: "By means of special production of individual parts by contracting firms, planned mass production, and modern wood-saving methods, a price level was to be achieved [by May] within the means of the majority of buyers but which, at the same time, guaranteed sound and taste-ful work."[61] According to his contemporaries, "Dr. May was . . . concerned with the creation of culturally valuable household articles," and his vision included production norms for German culture.[62] His father had founded the Verband Deutscher Wohnkultur (Society for German Cultural Living), and May had himself founded the Deutsche Heimgestatlung (German Home Design Society).

At first, the attempted takeover of the Jewish firm brought Pohl and May into a fracas with the civil administration of occupied Czechoslovakia. In the autumn of 1940, Pohl had submitted a bid to Constantin Freiherr von Neurath, the regional governor of the former Czech lands that the Reich had not directly annexed (i.e., the Protektorat, encompassing Bohemia and Moravia). Neurath opposed the SS's goal of technological modernization because he feared the ensuing standardization of goods might lead to a "limitation of production to fixed models which threaten free artistic expression."[63] Thus Neurath was opposing the modern nature of the SS's business—its preference for mass production and for large-scale, central-ized organization. The brief conflict is therefore instructive.

First of all, Neurath was hardly representative of the activists in Hitler's movement. His anti-modernism cannot be equated with any dominant cur-rent of Nazi ideology. He had made his career as a staid civil servant of the German diplomatic corps. He was an old-fashioned conservative, and in the last days of the Weimar Republic he had served Chancellor Franz von Papen as Foreign Minister. When Hitler became Chancellor, in January of 1933, Neurath remained in his ministerial post. In fact, he remained there until February of 1938, when Hitler replaced him with Joachim von Ribbentrop. Because Neurath had left the government before the round of annexations that began with Austria, he was not closely associated with either the NSDAP or Hitler's expansionist policies. At times the foreign press even speculated that Neurath was "anti-Nazi" because of his aris-tocratic, "cultured" background, and some speculated that Hitler had

appointed him because this "outsider" might elicit some measure of coop-
eration from Czech citizens. From the start, however, much more dedicated
Nazis among the Sudeten-German nationalist movement challenged
Neurath's authority, as did others within his own administration.[64]

Because of Neurath's weak position, his efforts to champion a craft-based
economy cannot be considered typical of Nazi policy in the Sudetenland. It
is also important to note that Neurath lost, as anti-moderns repeatedly did
in the Third Reich. To characterize the entire Third Reich as "anti-modern" is out of all proportion to the weight such voices carried. In addition,
when "anti-moderns" raised objections, they never actually claimed to be
against modernity; unlike those Nazis who championed modernity, people
like Neurath avoided the word. One searches in vain for quotations such as
"We will destroy modernity!" or "I hate modernism!" Instead, Neurath
based his objections on typically smug hypocrisy and cultural pretension
to what rightly counted as "German" design and "artistic expression"—
the very rhetoric mobilized by the Third Reich's modern productivism.

The SS countered that its unified management within Hitler's "National
Community" could ensure that standardized production would ring true.
"If the entire field of design down to the level of the last detail of utility is
seen as a unity and focused on man," explained an engineer who designed
SS furniture at the time, "then the senseless contrast between 'soulless industrial wares' and the exclusive, spiritual quality of craftsmanship falls
away."[65] The SS, by its own admission, strove to enforce a homogeneity of
"cultural" taste, but declared that this very homogeneity expressed the
German will. Neurath remained unconvinced and continued to frustrate
the merger through 1940 and into 1941.

So began a case of true SS industrial espionage. May transferred his
German Home Design Society (an eingetragener Verein—nonprofit society) to the SS, which then founded a GmbH (limited corporation) of the
same name with a high-profile storefront in Berlin's Potsdamer Straße.
Meanwhile, May posed as a civilian entrepreneur and bid on the coveted
Gerstel factory. In Brünn, he also purchased a 75 percent interest in another
Jewish firm, D. Drucker AG. Several of his partners from Stuttgart simultaneously founded Deutsche Meisterwerk GmbH to subsume these industrial properties.[66] This gave all the dealings the façade of private
entrepreneurship beyond the SS's direct involvement. Pohl quickly began

negotiations with industrial authorities in the Protektorat to subsume Deutsche Meisterwerk as a subsidiary. Because Deutsche Meisterwerk was nominally German owned and run by private citizens, the transfer remained hidden from Neurath's direct scrutiny. The closure of all transactions took place in 1941, and thus Pohl deftly manipulated modern organization— that is, a national corporate bureaucracy—to overcome the merely regional scope of the Protektorat's authority. The SS created a new affiliate, Deutsche Edelmöbel (German Noble Furniture), to consolidate all furniture manufacturing, and May entered as CEO of "wood working industries."[67] This poised the SS to furnish its Aryan settlements as a direct supplier.

As furniture design and manufacture came on line for the SS, May proudly reported among his personal contributions "comprehensive new application of modern, specialized machines."[68] He also helped organize exhibits of household interiors, with the assistance of an engineer from the Industrial Arts School of May's native Stuttgart, Diploma Engineer Hermann Gretsch. Gretsch criticized the tastelessness of past ages and the slavish imitation of foreign styles, especially those motivated by a mindless liberal capitalism: ". . . if we find so many houses so ugly today, it is because they were made in the period of beloved mammon!" He then suggested that German values and race had to be crafted into material artifacts: "Race, heritage, tradition, and lifestyle are important, but designers completely forgot them. They have forgotten that they must also satisfy cultural needs."[69] In this light, such men had no problem seeing racial supremacy—in whatever form—as "modern"; it resided in the standards of their industrial production.

Gretsch called his own style "Agrarian Objectivity," characterizing it as a "timeless" aesthetic located in an imagined epoch before capitalist spoliation. He consciously coined the term as a direct attack upon left-leaning artistic movements associated with "New Objectivity," whose consummate representative was the Bauhaus of Dessau. The Bauhaus had sought to create homes and living spaces as "machines for living" and made ostentatious use of modern, mass-produced materials such as steel, concrete, and glass. Its designers strove to strip away flourishes that did not serve practical, functional needs; this is perhaps best represented by Marcel Breuer's famous chair made of bent steel tubing and flat-black leather.[70]

To those who would define "modernism" by the international style of architecture and design promoted by the Bauhaus, Gretsch and the SS's

Figure 1
A Marcel Breuer chair. (Bauhaus Archives)

German Home Design undoubtedly seem "anti-modern." Were they not simply reactionaries who longed to turn back the clock to an imaginary past? (Gretsch acknowledged his own nostalgia for the epoch of Biedermeier.) Gretsch tried to derive racially pure "German" design from what he considered the eternal virtues of peasants, whose "life-style was more objective than the so-called New Objectivity." In contrast: "The Farmer took it for granted that everything he designed had to be practical as well as beautiful. He obeyed the eternal laws of nature."[71] In Gretsch's living spaces, the woman's place was unmistakably in the kitchen, preferably with many children, and Gretsch took care to place a baby carriage in the "parents' room." Scenes of peasant life hung on the walls of his displays.[72] Again, this would seem to confirm those who view the Nazis as hopelessly backward-thinking retrogrades. In a survey conducted in 1929, for instance, Erich Fromm claimed to have discovered an overweening preference for conventional aesthetics among Nazis, including wall pictures of

Figure 2
Sketch of a kitchen by Diploma Engineer Hermann Gretsch, from *Planung und Aufbau im Osten* (Deutsche Landbuchhandlung, 1941).

dictators and generals as opposed to progressive, original works of art.[73] Gretsch's taste could not have proved Fromm wrong.

On the other hand, as Mechtild Rössler and Sabina Schleiermacher aptly note: "The history of the modern is not only the history of international developments in philosophy, architecture, art, and aesthetics, is not only the Vienna Circle, the German Werkbund, and the Bauhaus . . . modern also means a new culture of technology and science."[74] In fact, when it came to technology and industry, Gretsch shared with Bauhaus design an impulse to plan domestic spaces as "machines for living." The SS's "German design" confirms Rössler and Schleiermacher no less than Fromm. Gretsch did not object to the New Objectivity's spirited idealization of technology; he simply found Bauhaus designs ugly, pretentious, degenerate, and "un-German." He too called upon the SS to hammer out norms for functional dwellings: "The useful is always the beautiful and lends simple, unified principles of form."[75] Furthermore, simple design aided mass production: "We should not aim for short-term fashion 'hits'; we should aim for standard-ized designs that are universally viable. Only this will reduce the serial pro-duction expenses over long term production and thus make designer furniture profitable."[76]

Unlike the Bauhaus designers, however, Gretsch believed that traditional materials such as wood suited the "German spirit" better than tube steel. Kurt May, whose firms specialized in woodworking, intended to adapt modern production to the values advocated by Gretsch. Their efforts sug-gests that Nazi modernity had much less to do with fine distinctions of intel-lectual history than with technology, organization, and their fusion in production. As Frank Trommler has pointed out, the aesthetics of the Werkbund and the Bauhaus do not necessarily define modernity; rather, they often betray the attempts (which had failed by 1933) of high artists and architects to win the modernity debate—that is, to proclaim the aes-thetic destiny supposedly immanent in the ascendant technology and orga-nization of the twentieth century.[77]

Some may object (as do the critics of Aly and Heim) that the active imag-inations of SS planners seldom amounted to much more than memoranda filed away in Himmler's desk drawer. And true enough, May's Deutsche Edelmöbel did not produce the New Order; it merely filled minor subcon-tracts for military suppliers during the war. But the preoccupation with Nazi industrial norms hardly began and ended with the SS. The *Journal of the Four Year Plan* repeatedly advocated the adoption of German mass-pro-duction standards, and authors of articles in that journal also did so in sec-tors more suited to fluctuating demand, flexible consumer response, and open-ended production practices (such as construction, machine-tool build-ing firms, and textiles). Furniture design was, in fact, a case in point, for classic mass production has always proved a failure in this industry.[78] Here the SS pursued modern production *despite* its irrationality.

Nothing was more typical of this irrational modernization than the means the SS developed to murder the European Jews. Here too the regime sought to combine its fantasies of racial supremacy with modern technol-ogy. Contemporaries both outside and within Germany immediately rec-ognized the nature of the SS genocide as "modern," and it was precisely the modernity of the technology that horrified them. The *New York Times* reported on November 26, 1944, that "a new modern crematorium and gassing plant was inaugurated at Birkenau."[79] At Auschwitz the SS seems also to have bragged about the tightly coupled integration of gassing and cremation in one technological system—"the best ever done in this line," as an SS man instructed one prisoner.[80] An adjutant of one of the SS's highest-ranking generals wrote of his travels to Auschwitz: "The camp Auschwitz has a special task in the Jewish Question. Here the most modern means make it possible to carry out the order of the Führer [to exterminate the Jews] in the shortest period of time without a great deal of sensation."[81] He went on to describe in detail how the system operated, stressing the modern organization of steady-flow production. Outside Germany this seemed a new vision of hell; within the SS it often awakened awe and pride.

It is well known that most Jews perished by primitive means and that the crematoria at Auschwitz broke down repeatedly. What is less well known is that "modern" killing, and by this I mean an integrated system approximating mass production, started relatively late. The Nazi genocide had proceeded with large-scale gassings since the autumn and winter of 1941. By early 1942 the SS had erected several installations exclusively to kill victims with carbon monoxide exhaust. The concentration camps had built industrial-scale cremation ovens much earlier, in the late 1930s. Nevertheless, these early systems were hardly "modern." The Belzec death camp simply burned its victims in open pits, and Odilo Globocnik's[82] experiments with the prussic acid agent "Zyklon B" were amateurish. Rudolf Höss testified after the war that his assistant at Auschwitz had attempted to use Zyklon B as early as the autumn of 1941, on some Soviet prisoners of war. But the basement in which they had locked the prisoners was likely too cold to activate the gas pellets effectively.[83] After the decision to begin mass gassing on a larger scale at Auschwitz, the SS staff simply sealed the doors and windows of existing barracks with felt or paper strips. In another case, they used two old farmhouses confiscated from Poles, and a low-ranking SS man teetering on a ladder poured the poison through openings knocked in the walls. These were the ad hoc experiments of psychopaths, scarcely the cool, calculating work of "technocrats" or a faceless, Behemoth-like bureaucratic machine.[84]

Christopher Browning begins *Ordinary Men*, his chronicle of one SS shooting squad, with the poignant observation that the Holocaust was a short, intense wave of mass murder. In the spring of 1942 the Nazis had killed only about 20–25 percent of the Holocaust's victims, but by mid February of 1943 only 20–25 percent of their victims were still among the living.[85] Since the invasion of the Soviet Union in the summer of 1941, and through 1942, the SS's killing spree proceeded with rudimentary technological systems. The simple shooting of Jews beside mass graves continued in the East; in fact, it increased over the course of 1942 even as large-scale gas chambers came on line.[86] In other words, a crude mixture of ad hoc technology and outright barbaric killing more than sufficed.

The only killing systems that one might truly describe as modern industrial factories were Crematoria II and III in Birkenau. Here the engineers and architects of the Central Building Directorate of Auschwitz planned undressing rooms and gas chambers in the basement and long banks of cremation ovens on the ground floor designed to run from central furnaces, thus achieving economy of scale. The layout of the buildings accommodated a steady flow of work. Special slave-labor commandos transported the corpses from the cellar to the cremation ovens on an electric lift, and the SS had even undertaken plans for semi-automated transport systems to deliver coal to the furnaces and to remove ashes from collecting bins.[87]

But the Central Building Directorate of Auschwitz designed these killing factories very slowly, over the late autumn and the winter of 1942. Crematorium II at Birkenau, the first, would not begin full operation until the spring of 1943. Jean-Claude Pressac notes that from April to October of 1943 the crematoria at Birkenau ran for only about 2 months at full capacity. In other words, the Auschwitz camp complex (of which Birkenau was a part) vastly overbuilt its killing machinery.[88] A graph recording transports to Auschwitz (figure 3) illustrates this quite well. This graph is based on the careful analysis of Franciszek Piper. Although the numbers of victims in transports to Auschwitz are not equal to those killed in the gas chambers (some prisoners were shunted into slave labor), it is nevertheless a good measure of activity in the killing camp. After March of 1943, killing continued on a large scale, but by Auschwitz's standards this period was a rather long lull. In the spring of 1944, Hungarian Jews started to arrive in staggering numbers (more than 430,000). At the end of 1942, however, Hungary, a German ally, still refused to release its Jews to the Nazi genocide. It is highly unlikely that the staff of Auschwitz was planning its capacity for this huge influx (represented by the spike from April to August of

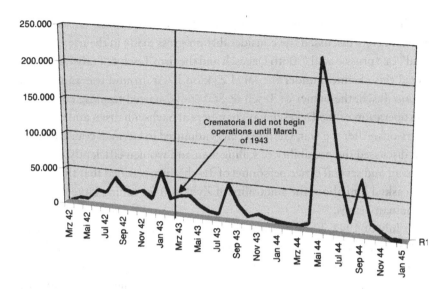

Figure 3
Graph of numbers of Jews transported to Auschwitz-Birkenau. (after F. Piper, *Die Zahl der Opfer von Auschwitz* (Oswiecim: State Archive Press, 1993)

1944) more than a year in advance. Thus, at the beginning of 1943, ironically, there never was a modern technology more lacking in justification as a means to an end.

This would seem to lend credence to the argument that the Holocaust represents the supreme danger inherent in modern industrial development, which has unleashed technological change without concern for the consequences to humanity. If the machinery of death at Auschwitz was not "technics out of control," what possibly could be?[89]

To answer this question, I would like to look more closely at how and why the idea of a Zyklon B gas chamber first entered Auschwitz. Rudolf Höss testified after the war in his famous confession that mass killings using gas began in the summer of 1941. We know this cannot be true, for the first experiments with Zyklon B began no earlier than September. Nevertheless, Höss's slip of memory is instructive. SS engineers did plan gas chambers for Zyklon B in early July of 1941, but they did so as part of an integrated and partially automated system for the fumigation of clothing. The chemical firm Degesch marketed machinery for this purpose, and at the beginning of July some advertising materials arrived at Auschwitz, including a copy of a scholarly article by the company's chief executive officer, Dr. Gerhard Peters. Peters discussed the considerable progress made in the use of "blue acid" or "prussic acid." Both Degesch and the firm Tesch & Stabenow marketed this chemical under the label Zyklon B. At around this same time, Bruno Tesch, the owner of Tesch & Stabenow, started training courses in the operation of the Degesch gas chambers at Sachsenhausen and other SS garrisons. After the war, Gerhard Peters admitted in German courts to having discussed the possibility of killing men and women efficiently with the poison, and several office personnel of Tesch's firm testified that the SS had also asked him about the suitability of Zyklon B for killing large numbers of human beings.[90]

In July of 1941, however, murder was not yet the purpose. The modern technology that Degesch was pushing involved clothing-fumigation chambers, each 10 cubic meters in volume, with specially designed ventilation systems and automatic devices for introducing Zyklon B pellets. Peters's article, still preserved in the files of the construction directorate of Auschwitz, begins by discussing the "advances in this particular year that have contributed considerably to the modernization of the well-known [gassing] process."[91] The chambers could be built in series in a factory-like arrangement: "The operation is organized so that an undressing room [for prisoners] lies on one side of the chambers [the "impure" side]. During the delousing of the clothing, the prisoners can bathe and submit to a general examination until they arrive at the other side [the "pure" side of the installation] where they receive their clothing in a dressing room."[92] The SS, Degesch, and Tesch & Stabenow conceived of these chambers as a means of managing prisoners as "human material" in a mass-production fashion.

Enthusiasm for such a "modern" system is unmistakable among the construction engineers and architects of Auschwitz. They had long planned a prisoner-reception building at Auschwitz, including a laundry and delousing station. The new SS chief of engineers who entered service in June of 1941, Hans Kammler, gave these buildings top priority. After the SS became aware of the Degesch system, however, the Central Building Directorate rushed to modify them. This happened as early as July 14, little more than a fortnight after Dr. Peter's advertising materials were received. "Without overall planning there can be no economic and timely installation of technically modern equipment," wrote one SS engineer in an attempt to spur Auschwitz's subcontractors to more concerted action on the project, and in the SS's eyes what counted as "modern" about the new Zyklon B reception building was its approximation of mass production.[93] As early as September of 1941, a report mentioned that blueprints had been drawn up so that "the floor plan is so organized in order to make possible an assembly-line operation [Fließbetrieb] such that the flow of work is never interrupted."[94]

This initial enthusiasm was not confined to Auschwitz. The leadership of the SS's corps of civil engineers in Berlin insisted that "the final means of all delousing installations . . . be geared toward the process with blue acid."[95] The SS believed it had found the "one best way" to handle sanitary conditions and insisted upon the Degesch's factory-style chambers, a new invention, as the standard for all concentration camps. This decision is peculiar, for these very gas chambers had already failed by the spring of 1942. The SS architect Fritz Ertl marked exclamation points in the margins of the above-mentioned letter from Berlin, as if to express his exasperation with the lack of comprehension of the real problems with delousing. Regional authorities were putting the camp's administration under pressure and complaining that civilian workers were catching diseases when they worked in proximity to Auschwitz. Epidemics were common there, as they were at other concentration camps, owing to the atrocious shortages of shelter, sanitation, food, and potable water. In fact, SS men had also reported sick with spotted fever. By December of 1942, the chief physician of Auschwitz, Eduard Wirths, had had to borrow mobile autoclaves from SS doctors in Kattowitz, and he reported that "the Cyklon-B fumigation has been completely abandoned, for it has been shown that the success of this process is not 100% certain."[96] The prussic acid only sublimated at relatively warm temperatures—hardly an "efficient" poison in the cold Polish winter. The reception building, constructed according to principles of an "assembly line," a building that had once been projected by Auschwitz chief of engineers to cost 2.7 million Reichsmarks, proved a waste of time and money.

We might therefore return now to the questions posed above: Does the technology of Auschwitz represent the greatest nightmare in which the dreams of reason have begotten monsters? Was the mechanized genocide "less an eruption of the irrational than an extreme form of technocratic reason?"[97]

First of all, the SS was scarcely indulging an unbridled enthusiasm for Enlightenment rationality when it chose to make a fetish of "assembly lines," the latest chemical novelties, and complex automated gadgetry. Zyklon B was a rather unusual choice. Chemical fumigants were a new technical product that had found their first widespread use after World War I. As such, they counted as modern, something Dr. Peters of Degesch labored to point out in his article. Paul Weindling has noted, however, that other chemical fumigants at this time were much more effective and much easier to handle. German authorities and German industry—not just the SS—had adopted fumigation with prussic acid at a time when others were adopting DDT. The dangers of DDT are well known, but it is hard to imagine its use in homicidal gas chambers. In contrast, as early as 1915 plaintiffs had brought suits before the German courts because the highly potent toxins in Zyklon B had led to wrongful death.[98] What could have been the attraction of Zyklon B? Weindling notes that it had been invented in Germany and thus seemed to represent the "inventive spirit" of that nation. In the case of the Degesch fumigation chambers, German firms had also developed delivery systems to approximate the methods of mass production.

The reason for the adoption of the Degesch-style chambers at Auschwitz thus seems to have been unbridled enthusiasm, not for technology generically, and far less for "reason," but for *modern* technology. The SS engineers' calculations leapt the bounds of mere "instrumental rationality" such that modern technology began to be an end it itself. We should not be surprised, therefore, to find that the search for "high-tech" solutions to epidemics continued at Auschwitz even after the massive failed investment in the Degesch chambers. Wirths soon reported that he was experimenting with radio waves to sterilize clothing. Needless to say, this too proved a dismal failure.[99] The most rational solution to the epidemics of Auschwitz, of course, would have been to halt the never-ceasing transport of new prisoners. The SS failed to do that because killing the Jews was a central focus of Nazi racial supremacy, not because of some dialectic of the Enlightenment.

A similar history of irrational enthusiasm for mass production accompanied the modern "factory of extermination" typified by Crematoria II and III at Birkenau, for which the Degesch chambers served as predecessor and model. Here the SS clearly intended to approximate principles of mass production with the continuous flow of "human material." SS architects and engineers designed these two buildings to facilitate forcing prisoners into the gas chambers at one end, transporting their corpses to the crematoria, and burning their bodies in a steady-flow progression. Nevertheless, the system constantly broke down, not unlike the Zyklon-B fumigation chambers. Because the SS insisted on cramming the cremation ovens with too many bodies (four at a time, sometimes more), they burned too hot and the firebrick soon caved in.[100] In the summer of 1944, when the Hungarian Jews started to arrive at Auschwitz in large transports, the SS reverted to ad hoc gas chambers in the storage rooms of Crematoria IV and V, where Zyklon B was simply shaken in through small windows in the side of the building. Here the SS engineers and architects never designed a steady-flow, linear progression of genocide into the building's layout. The SS also began once more to burn bodies in open pits instead of in the "modern crematoria," which were then largely defunct.[101] Thus Auschwitz ended its orgy of murder where it began: with psychopathic improvisation. The personnel who conceived and carried out these operations were not in thrall to the monsters of reason; they were simply monsters.

Conclusion

The barbarous utopia of the SS, which it tried to build into an "assembly line" for murder, was wholly in keeping with what I have termed *productivism*. That is, the National Socialist movement in general identified modernity with technology as the means of forming national identity in production. If social engineering was part of this, so too was the irrational belief that German national identity resided in a destiny tied to modern machines. Regarding the gas chambers and crematoria of Auschwitz, the leader of one mobile SS killing squad, Erich von dem Bach Zelewski, reportedly commented that the machinery of Auschwitz was "something the Russians could not accomplish: it reflected the German gift."[102] What better example of Hitler's productivism might we find: the Nazis consciously adopted a modern technological system in order to purge Europe of its Jews and thus sought to manufacture a racial empire in their own image? Productivism urged Hitler's true believers to seek the wellsprings of national identity and homogenous community in modern production, and this is exactly what killing the Jews of Europe was supposed to do: prepare a clean slate so that the National Socialists could manufacture their New Order, a utopia of Nazi values.

Degesch's factory-style Zyklon B chambers were the immediate predecessor and model for Crematoria I and II at Birkenau. That SS men after the war confused them often enough with the killing chambers is further evidence of this.[103] Allow me to repeat, however, that I am not arguing that modern technology "caused" the Holocaust or that anti-Semitism played no role; in fact, anti-Semites were among the most energetic advocates of modern technology in the Third Reich. But I do argue that modernity determined the unique means of the "Final Solution" at Auschwitz. In this sense the "postmoderns" have been right all along. But it is wrong to see this as the necessary teleology of Enlightenment rationality. The crassest irrationality and modernization have never contradicted each other, least of all in the minds of anti-Semites.

The history of technology reveals that a belief that the machine would "change the world" was central to Nazi modernity, but more than this a specific kind of technological system—the hierarchically governed, centralized, mass-production factory new to the early twentieth century. Wedded to Nazi productivism, the legitimate purpose of the state became the promotion of industrial growth as a symbol of national strength and unified identity rather than the wealth, peace, and freedom of citizens. National Socialism emphasized existential feeling—feelings like Rausch—through an inevitable fulfillment of destiny by modern technological means.

Hans Mommsen asks: "Does one earnestly wish to make the fascistic understanding of politics into the foundation of a present day 'modern' society?" One might answer with another question: What about the modernity of this morbid history could possibly bind us to do so? To reject the modernity of the Holocaust, however, one must reject the common assumption of Mommsen and many others that modernity is destiny. The SS's vision of modernity was partial, and cannot be taken as an overarching definition of *the* modern world. A multitude of values and attitudes have accompanied the rise of an industrial, scientific, and technological society, and the Nazis were undeniably selective. Yet still less did Nazi modernity conform to the many claims made in its name by postwar historiography, and a selective approach to "modernity" hardly distinguished National

Socialism. Rather, it seems a central element of the entire "modernity" debate. Fritz Stern, George Mosse, and Ralf Dahrendorf have all associated liberal democracy with "modernization." However much one sympathizes with their political goals rather than with Hitler's, this characterization of history is hardly less arbitrary. It is the Nazis' participation in the debate, the effort to present their dogma as historical destiny, that made them "modern."

We should beware of the moral paralysis that debates over modernity induce precisely because they render discussion of substantive principles impossible in the name of an "inevitable" or "inexorable" logic of industrial development. This paralysis was already evident at the time of Hitler's rise to power. Lyndall Urwick, Director of Geneva's International Management Institute, warned in 1933: "A machine technology points to the obvious economies of large-scale units of business control: amalgamations founder because there is widespread ignorance as to the methods of managing these aggregations." The new technics of the twentieth century in communication, production, and distribution seemed to outstrip civil institutions in the 1920s and the 1930s, and Urwick made his own passionate plea for the world's democracies to adapt more efficiently: "In country after country liberty of speech and of person are lost, because democratic institutions fail to devise an administrative structure adapted to the speed and complexity of social evolution."[104] In the same year, the Chicago Century of Progress exposition featured an Art Deco statue with the inscription "Science Finds—Industry Applies—Man Conforms."

In such a light, the ability to force the pace of modern organization and technology could too easily displace inquiry into the first principles of citizenship, the state, and political economy, which should define social change. Fascists would condemn democracy directly for institutionalizing an old-fashioned "cultural lag" behind the dynamic changes of technology and organization. Although Urwick deplored the Nazis, he believed as they did that "modernization" itself was inevitable. Once Hitler could pose as a master of technological success, even staunch resisters began to see him as the inevitable product of historical destiny. In the mid 1930s, when National Socialism achieved real economic growth, visions of modernity offered slim recourse to anyone who still wished to condemn the regime.

Notes

1. Francis Fukuyama, *The End of History and the Last Man* (Free Press, 1992), pp. 128–130.

2. Richard Rorty, *Truth and Progress* (Cambridge University Press, 1998), pp. 323–326. Here Nazism is condemned (as is Jürgen Habermas, ironically) for daring to posit universal theories of social justice. Rorty also attributes the poverty of metaphysics to the Enlightenment (pp. 317, 321), and argues that, rather than first principles, "utility" should be the standard of just government. For Jean-François Lyotard, as for Rorty, "Auschwitz" is the ultimate "differend," the hole in our understanding that gives the lie to any absolute statements of truth and ultimately to the Enlightenment itself. See Lyotard, *The Differend* (University of Minnesota Press, 1988), pp. 43, 56–57, and 89–110.

3. M. Stanton Evans, *The Theme Is Freedom* (Regnery, 1994), pp. 15, 52. Evans's argument is, in fact, the same as Rorty's: the Holocaust and fascism demonstrate the evil of metaphysical theories of first principles, especially those of the Enlightenment. Evans merely has a much different answer to this problem.

4. Reinhart Koselleck, *Futures Past* (MIT Press, 1985), p. xxiv.

5. Note the enthusiasm for information technology among many so-called postmoderns. See Jean-François Lyotard, *The Postmodern Condition* (University of Minnesota Press, 1979), pp. xxiv–xxv. For an example of technology and the Holocaust in the "Culture Wars," see p. 6 of Fukuyama, *The End of History*. See also the central themes of John Ralston Saul's *Voltaire's Bastards* (Vintage Books, 1993).

6. Omer Bartov, *Murder in Our Midst* (Oxford University Press, 1996), esp. pp. 1–11; Sven Lindquist, *"Exterminate All the Brutes"* (New Press, 1992).

7. For an array of positions in the "uniqueness" debate, see Alan Rosenbaum, *Is the Holocaust Unique?* (Westview, 1996). On technology's role in this uniqueness, see Christopher Browning, "Barbarous Utopia: The Terrible Uniqueness of the Nazi State," *Times Literary Supplement*, March 20, 1992, p. 5.

8. Hermann Göring, "Verantwortliche Wirtschaftsführung," *Der Vierjahresplan* 1 (1937), p. 66.

9. Wirtschaftsminister Walther Funk, "Die wirtschaftspolitische Aufgabe," *Der Vierjahresplan* 2 (1938), p. 140.

10. Staatssekretär Paul Körner, "Leistung des ganzen Volkes," *Der Vierjahresplan* 3 (1939), p. 4.

11. See "Der Frontarbeiter," *Der Vierjahresplan* 4 (1940), pp. 366–369. The Frontarbeiter were militarized work crews not unlike the American "Seabees."

12. Jakob Werlin, "Der wirtschaftliche und soziale Sinn des Volkswagens," *Der Vierjahresplan* 2 (1938), pp. 472–473. Werlin visited Henry Ford in the United States in 1938 and secured the aging industrialist's blessing on the Volkswagen.

13. Hans Mommsen, "Noch einmal: Nationalsozialismus und Modernisierung," *Geschichte und Gesellschaft* 21 (1995), p. 401. See also Dominick La Capra, *History and Memory after Auschwitz* (Cornell University Press, 1998), pp. 3, 30ff.; Saul Friedländer, "The Final Solution: On the Unease in Historical Interpretation," in *Lessons and Legacies*, ed. P. Hayes (Northwestern University Press, 1991); "Introduction," in Gerald Fleming, *Hitler und die Endlösung* (Ullstein, 1987); Thomas Saunders, "Nazism and Social Revolution," in *Modern Germany Reconsidered 1870–1945*, ed. G. Martel (Routledge, 1992), p. 166.

14. See Horst Matzerath and Heinrich Volkmann, "Modernisierungstheorie und Nationalsozialismus," in *Theorien in der Praxis des Historikers*, ed. J. Kocka (Vandenhoeck & Ruprecht, 1977).

15. Falk Pingel, *Häftlinge unter SS-Herrschaft. Widerstand, Selbstbehauptung und Vernichtung im Konzentrationslager* (Hoffmann und Campe, 1978), p. 62.

16. Friedländer, "The Final Solution."

17. PS-1919, Himmler's Speech in Posen, 4/11/1943. See also La Capra, *History and Memory after Auschwitz,* pp. 25–42.

18. John Kasson, *Civilizing the Machine* (Grossman, 1976), esp. pp. 137–180; Leo Marx, *The Machine in the Garden* (Oxford University Press, 1964); David Nye, *American Technological Sublime* (MIT Press, 1994); Paul Florman, *The Existential Pleasures of Engineering* (St. Martin's Griffen, 1996).

19. Modris Eksteins, *Rites of Spring* (Anchor Books, 1989), p. 322. Eksteins (ibid., pp. 320–321) focuses on the technological means of the Holocaust as its quintessential modernity.

20. David Noble, *The Religion of Technology* (Knopf, 1997), p. 22.

21. Gustavo Corni and Horst Gies, *Brot, Butter, Kanonen* (Akademie Verlag, 1997).

22. Dr. Böhmkamp, i.A., Hauptabteilung E1, 24/2/38, "Zeitschriften-Vorgutachten 'Natur und Kultur'," Bundesarchiv Lichtefelde, R16/1272.

23. For notable exceptions, see the following: Robert van Pelt and Debórah Dwork, *Auschwitz 1270 to the Present* (Norton, 1996); Jean-Claude Pressac, *Auschwitz* (Beate Klarsfeld Foundation, 1989) and *Die Krematorien von Auschwitz* (Piper, 1993).

24. Thomas Nipperdey, "Probleme der Modernisierung in Deutschland," in *Nachdenken über die deutsche Geschichte* (Beck, 1986); Henry A. Turner, "Fascism and Modernization," in *Reappraisals of Fascism*, ed. H. Turner (New Viewpoints, 1975).

25. Mommsen, "Noch einmal."

26. Gerald Feldman, "The Weimar Republic: A Problem of Modernization?" *Archiv für Sozialgeschichte* 26 (1986), p. 1. Norbert Frei ("Wie Modern war der Nationalsozialismus?" *Geschichte und Gesellschaft* 19, 1993: 367–387) also speaks of this debate as one shrouded in fog.

27. Ian Kershaw, *The Nazi Dictatorship* (Edward Arnold, 1989), p. 418 and "Ideologe und Propagandist. Hitler im Lichte seiner Reden, Schriften und Anordnungen," *Vierteljahrshefte für Zeitgeschichte* 40 (1992): 263–271. See also Mark Walker, "Naturwissenschaftler, Techniker und der Nationalsozialismus" in *Ich diente nur der Technik*, ed. A. Gottwaldt (Nicolaische Verlagsbuchhandlung, 1995). On the organizational dimensions of modernity in industry, see Alfred Chandler, *The Visible Hand* (Harvard University Press, 1977), pp. 393–592; Chandler, *Strategy and Structure* (MIT Press, 1969); Jürgen Kocka, "Scale and Scope: A Review Colloquium," *Business History Review* 64 (1990): 711–716.

28. Mary Nolan, *Visions of Modernity* (Oxford University Press, 1994).

29. Ernst Jünger, *Der Arbeiter*, in *Werke*, Band 6: Essays II (Klett, 1964), p. 318. On Jünger, see Jeffry Herf, *Reactionary Modernism* (Cambridge University Press, 1984), pp. 70–108.

30. George Mosse, on p. vii of the 1981 edition of his book *The Crisis of German Ideology* (Schocken), withdrew his claim that Nazism was strictly an anti-modern movement, stating that "the Nazis made use of the most up-to-date technology in all fields."

31. For one of the earliest developments of this idea, see Talcott Parsons, "Some Sociological Aspects of the Fascist Movements," in Parsons, *Essays in Sociological Theory* (Free Press, 1954). See also Thomas Childers, *The Nazi Voter* (University of North Carolina Press, 1983); William Brustein, *The Logic of Evil* (Yale University Press, 1996).

32. Timothy Mason, "Zur Entstehung des Gesetzes zur Ordnung der nationalen Arbeit, vom 20. Januar 1934: Ein Versuch über das Verhältnis 'archaischer' und 'moderner' Momente in der neuesten deutschen Geschichte," in *Industrielles System und politische Entwicklung in der Weimarer Republik*, ed. H. Mommsen et al. (Droste, 1974), p. 324.

33. See Henry Ashby Turner, "Fascism and Modernization," in *Reappraisals of Fascism*, ed. Turner (Franklin Watts, 1975). Turner claimed that anti-modern ideology fired the imagination of Adolf Hitler, in whose eyes "modern industrial society was wholly and unavoidably incompatible with what they held to be the only true wellspring of social life: the folk culture" (ibid., p. 119).

34. Ralf Dahrendorf, *Society and Democracy in Germany* (Norton, 1979), p. 383.

35. David Schoenbaum, *Hitler's Social Revolution* (Norton, 1980); Jens Alber, "Nationalsozialismus und Modernisierung," *Kölner Zeitschrift für Soziologie und Sozialpsycologie* 41 (1989): 346–365; Dietmar Petzina, "Soziale Lage der deutschen Arbeiter und Probleme des Arbeitseinsatzes während des Zweiten Weltkriegs," in *Zweiter Weltkrieg und sozialer Wandel*, ed. W. Dlugoborski (Vandenhoeck & Ruprecht, 1981); Rüdiger Hachtmann, *Industrie Arbeit im Dritten Reich* (Vandenhoeck & Ruprecht, 1989), esp. pp. 54–89; Albrecht Ritschl, "Die NS-Wirtschaftsideologie—Modernisierungsprogramm oder reaktionäre Utopie?" in *Nationalsozialismus und Modernisierung*, ed. R. Zitelmann and M. Prinz (Wissenschaftliche Buchgesellschaft, 1991).

36. Hans Mommsen, "Nationalsozialismus als vorgetäuschte Modernisierung," in *Der Nationalsozialismus und die deutsche Gesellschaft* (Rowolt, 1991).

37. Wolfgang Sauer, "National Socialism: Totalitarianism or Fascism," in *Reappraisals of Fascism*, ed. H. Turner (Franklin Watts Inc., 1975), p. 275. See also Detlev Peukert, "Alltag und Barbarei. Zur Normalität des Dritten Reiches," in *Ist der Nationalsozialismus Geschichte?* ed. D. Diner (Fischer, 1987), p. 52;

38. Nipperdey, "Probleme der Modernisierung," pp. 44–59.

39. Jeffrey Herf, *Reactionary Modernism*. Herf is doubly exceptional in that he acknowledges both the Nazis' technological enthusiasm (usually avoided by self-confessing moderns) and their ferocious anti-Enlightenment bile (usually avoided by the postmoderns). Yet conflation of the Enlightenment with scientific and technological "reason" endures in his analysis, so much so that he is forced to postulate a "crisis of German ideology." German engineers supposedly had to overcome the contradiction between rejecting Enlightenment humanism and embracing modern industry, which they did through thinkers who spun a spiritual cult of romanticism around machines. Compare Ken Alder, *Engineering the Revolution* (Princeton University Press, 1997). Alder suggests that engineers at the time of the Enlightenment were more concerned with control than with Enlightened freedoms.

40. Götz Aly and Susanne Heim, "Die Ökonomie der Endlösung: Menchenvernichtung und wirtschaftliche Neuordnung," *Beiträge zur nationalsozialistischen Gesundheits- und Sozialpolitik* 5 (1987): 11–90; Götz Aly and Susanne Heim, *Vordenker der Vernichtung* (Fischer, 1993); Zygmunt Bauman, *Modenity and Ambivalence* (Cornell University Press, 1991); Zygmunt Bauman, *Modernity and the Holocaust* (Cornell University Press, 1989); Detlev Peukert, "The Genesis of the 'Final Solution' from the Spirit of Science," in *Reevaluating the Third Reich*, ed. T. Childers and J. Caplan (Holmes &Meier, 1993); Detlev Peukert, *Inside Nazi Germany* (Yale University Press, 1987); Ronald Smelser, "How 'Modern' were the Nazis? DAF Social Planning and the Modernization Question," *German Studies Review* 12 (1989); Robert Ley, *Hitler's Labor Front Leader* (Berg, 1988).

41. Detlev Peukert, "Alltag und Barbarei," pp. 51–61. See also Rainer Zitelmann, "Die totalitäre Seite der Moderne," in *Nationalzozialimus und Modernisierung*, ed. R. Zitelmann et al. (Wissenschaftliche Buchgesellschaft, 1991); Geoff Eley, "Die deutsche Geschichte und die Widersprüche der Moderne. Das Beispiel des Kaiserreichs," in *Zivilisation und Barbarei*, ed. F. Bajohr et al. (Hans Christians Verlag, 1991).

42. Norbert Frei ("Wie Modern war der Nationalsozialismus?" p. 385) claims that Hitler viewed modern technology "partly with skepticism." As evidence he cites, with implied criticism, Rainer Zitelmann, whose biography of Hitler nevertheless contains much contrary evidence drawn from private speeches and other sources.

43. Bauman, *Modernity and the Holocaust*, p. 77. See Ann Taylor Allen's, David Lindenfeld's, and Alan Beyerchen's contributions to "The Holocaust and Modernity," a special issue of *Central European History* (volume 30, 1997). See also Geoff Eley, "German History and the Contradictions of Modernity: The

Bourgeoisie, the State, and the Mastery of Reform," in *Society, Culture, and the State in Germany, 1870–1930*, ed. G. Eley (University of Michigan Press, 1996).

44. Bauman, *Modernity and the Holocaust*, p. 68.

45. Peukert, "The Genesis of the 'Final Solution,'" pp. 234–252.

46. Alan Beyerchen, "Rational Means and Irrational Ends: Thoughts on the Technology of Racism in the Third Reich," *Central European History* 30 (1997), p. 390.

47. Wilhelm Ziegler (Riechsministerium für Volksaufklärung und Propaganda), "Zusammentarbeit von Technik, Wissenschaft und Kunst," *Der Vierjahresplan* 2 (1938), p. 584.

48. Peter Wagner ("Sociological Reflections: The Technology Question during the First Crisis of Modernity," in *The Intellectual Appropriation of Technology*, ed. M. Hård and A. Jamison, MIT Press, 1998) notes the similar impetus to direct the dynamism of the industrial revolution in the Swedish "Folks Home" movement, in German National Socialism, in Stalinism, in the New Deal, and in Italian Fascism.

49. Zitelmann, "Die totalitäre Seite der Moderne."

50. Gottfried Feder, *Das Programm der NSDAP und seine weltanschaulichen Grundgedanken* (Eher, 1932), pp. 46–47.

51. Thomas Parke Hughes, *Networks of Power* (Johns Hopkins University Press, 1983), p. 6; "Modern and Post-Modern Engineering," *Sternwarte-Buch* 1 (1998): 256–275.

52. Zitelmann, *Hitler*, p. 357.

53. Nolan, *Visions of Modernity*; Richard Overy, "'Blitzkriegswirtschaft'? Finanz-politik, Lebensstandard und Arbeitseinsatz in deutschland 1939–42," *Viertel-jahreshefte für Zeitgeschichte* 36 (1988): 370–435; Overy, "Mobilization for Total War in Germany 1939–1941," *English Historical Review* 88 (1988): 613–639; Overy, *The Nazi Economic Recovery 1932–1938* (Cambridge University Press, 1982).

54. Otto Dietrich, Riechspressechef der NSDAP, "Der Führer und der deutsche Arbeiter," *Der Vierjahresplan* 1 (1937), p. 214.

55. Erich Gritzbach, "Nationale Kraft durch neue Lebensgestaltung," *Der Vier-jahresplan* 1 (1937), p. 345. See also Wilhelm Zangen, "Dienst an der Nation. Ein Rückblick über zehn Jahre nationalsozialistischer Industriearbeit," *Der Vierjahres-plan* 7 (1943), p. 4.

56. "P.," "Von künftiger deutscher Wirtschaftsgestaltung," *Der Vierjahresplan* 4 (1940), p. 805.

57. Götz Aly, *'Endlösung'* (Fischer, 1995). See also Christopher Browning, "Vernichtung und Arbeit. Zur Fraktionierung der planenden deutschen Intelligenz im besetzten Polen," in *'Vernichtungspolitik'*, ed. W. Schneider and U. Herbert (Junius, 1991); Ulrich Herbert, "Rassismus und rationales Kalkül. Zum Stellenwert utilitaristisch verbrämter Legitimationsstrategien in der nationalsozialistischen 'Weltanschauung'," in ibid.; Michael Burleigh, *Death and Deliverance* (Cambridge

University Press, 1994), pp. 98–99; Frei, "Wie Modern war der Nationalsozialismus," pp. 371–374; Avraham Barkai, *From Boycott to Annihilation* (University Press of New England, 1989), p. 114; Peter Hayes, "Big Business and Aryanization in Germany, 1933–1939," *Jahrbuch für Antisemitismusforschung* 3 (1994): 254–281. The tendency to identify Jews as a threat to a "modern" industrial economy was something that German authorities shared with the French. See Wolfgang Seibel, "Holocaust und wirtschaftliche Verfolgungsmaßnahmen—Anlaß zur Neubewertung der Strukturalismus/Intentionalismus-Debatte? Das Beispiel Frankreich, 1940–1942," delivered at Arbeitskreis Unternehmen im Nationalsozialismus der Gesellschaft für Unternehmensgeschichte, Frankfurt am Main-Höchst, January 14, 2000.

58. Hauptabteilung Planung und Boden, Stabshauptamt des RKF, *Planung und Aufbau im Osten* (Deutsche Landbuchhandlung, 1941), p. 11.

59. NO-1043, Leo Volk, undated, "Generaltreuhaenders fuer Baustofferzeugungsstaetten im Ostraum im Jahre 1940."

60. I thank Jan-Eirk Schulte for this biographical information, which he received from the Kulturamt of the Landeshauptstadt Stuttgart and the Westdeutsche Wirtschaftschronik.

61. Affidavit of Kurt Brune, Defense Document Books of Hans Hohberg. Karl Bestle; Karl Neimann, 30/6/43; Kurt May, 23/6/41, "Jahresrechnung des Vorstandes der Drucker AG, Dampfsägewerke und Holzwarenfabriken in Brünn zur Jahresrechnung 1940"; undated, unsigned, "Technische Grundlagen" and op. cit. "Protokoll der Aufsichtsratssitzung 1943," US National Archives microfilm T-976/22. Compare Hounshell on Gunnison homes: David Hounshell, *From the American System to Mass Production* (Johns Hopkins University Press, 1984), pp. 145–146, 310–314.

62. Affidavit of Kurt Brune, Defense Document Books of Hans Hohberg, International Military Tribunal, Case IV vs. Pohl et al.

63. Eno Georg, *Die wirtschaftlichen Unternehmungen der SS* (Deutsche Verlags-Anstalt, 1963), pp. 77–79.

65. Shiela Duff, *A German Protectorate* (Frank Cass, 1970), pp. 98–102.

65. Alfons Leitl, "Professor Hermann Gretsch/Eine Ausstellung seiner Arbeiten," *Die Bauwelt* 32 (Heft 26, 1941), p. 4.

66. Georg, *Die wirtschaftlichen Unternehmungen*, p. 81; Dr. Höring, Report of 16/11/41, T-976/22. Later Pohl acquired up to 94% of Drucker's stock.

67. Ibid., p. 80. The DWB recapitalizing Deutsche Meisterwerk with over 800,000 RM. Volk to Pohl, 1/9/41, "Umorganisation der Ämter," T-976/35. The chief of the SS Reich Security Head Office, Reinhard Heydrich, replaced Neurath as Protektor on Sept. 27, 1941. Heydrich opposed other SS bids for corporations, such as mineral-water companies, for motives that are not clear, but he did not oppose this furniture deal.

68. Kurt May, 23/6/41, "Jahresrechnung des Vorstandes der Drucker AG, Dampfsägewerke und Holzwarenfabriken in Brünn zur Jahresrechnung 1940, T-976/22.

69. Dipl. Ing. Hermann Gretsch, "Zeitgemäßes Wohnen," *Die Bauzeitung* 38 (1941), p. 425.

70. See Frank Trommler, "Von Bauhausstuhl zur Kulturpolitik. Die Auseinandersetzung um die moderne Produktkultur," in *Kultur*, ed. H. Brackert et al. (Suhrkamp, 1990).

71. Herman Gretsch, "Vom richtigen Wohnen," 37 (1940), p. 426.

72. Gretsch, "Zeitgemäßes Wohnen," p. 430.

73. Erich Fromm, *Arbeiter und Angestellte am Vorabend des Dritten Reiches* (Deutsche Verlags-Anstalt, 1980), pp. 142–150. Fromm's statistical samples of NSDAP party members were so small as to render his data unrepresentative. He also noted, curiously, an overrepresentation of Nazis among admirers of Neue Sachlichkeit.

74. Mechtild Rössler and Sabine Schleiermacher, "Der Generalplan Ost und die Modernität der Großraumordnung. Eine Einführung," in *Der 'Generalplan Ost'*, ed. M. Rössler and S. Schleiermacher (Akademie Verlag, 1993), p. 10.

75. Alfons Leitl, "Professor Hermann Gretsch," p. 4.

76. Gretsch, "Die Stellung des Entwerfers in der Wirtschaft," *Bauen, Siedeln, Wohnen* 19 (1939), p. 751

77. Frank Trommler, "The Avant-Garde and Technology: Toward Technological Fundamentalism in Turn-of-the-Century Europe," *Science in Context* 8 (1995): 397–416 and "The Creation of a Culture of Sachlichkeit," in Geoff Eley, *Society, Culture, and the State in Germany, 1870–1930* (University of Michigan Press, 1996), pp. 465–485. Jennifer Jenkins, "The Kitsch Collections and *The Spirit in the Furniture*. Cultural Reform and National Culture in Germany," *Social History* 21 (1996): 123–141.

78. On construction: Fritz Todt, "Regelung der Bauwirtschaft," *Der Vierjahresplan* 3 (1939): 762–764. Eugen Vögler, "Durch Betriebsrationalisierung zur Leistungssteigerung in der Bauwirtschaft," *Der Vierjahresplan* 3 (1939): 765–769. J. Jb., "Rationalisierung im Wohnungsbau," *Der Vierjahresplan* 4 (1940): 1042–1044. On craft industries: Josef Free, "Wandlung der Maschinentechnik durch die neuen Werkstoffe," *Der Vierjahresplan* 2 (1938): 530–537. Ferdinand Schramm, "Neue Ausrichtung des Handwerks," *Der Vierjahresplan* 3 (1939): 461–463. Annonymous, "Die Betriebsgrößen in kriegswirtschaftlicher Beurteilung," *Der Vierjahresplan* 4 (1940): 103–104. Heinrich Krumm, "Wert und Aufgabe der Konsumgüterindustrie im Kriege," *Der Vierjahresplan* 5 (1941): 518–521.

79. The report, by Alfred Wetzler, Walter Rosenberg, Czeslaw Mordowicz, and Arost Rosin of the Polish resistance, was published by the American War Refugee Board, Washington D.C., 1944. It has been reprinted in English and in its French versions by Jean-Claude Pressac, *Auschwitz*, p. 461. See also the similar reaction to the "largest and most efficient technical installations for the extermination of people" in *Law Reports of Trials of War Criminals* (United Nations War Crimes Commission, 1948), p. 12.

80. Memoirs of Dr. Paul Bendel, reprinted and translated from the French *Temoignages sur Auschwitz* (Amicale des Dports d'Auschwitz, 1946) in Pressac, *Auschwitz*, p. 469. Bendel worked with the Sonderkommandos.

81. Alfred Franke-Gricksch, undated report, "Umsiedlungs-Aktion der Juden," reprinted in Pressac, *Auschwitz*, p. 238.

82. Globocnik was the organizer of the Belzek, Sobibor, and Treblinka death camps.

83. Rudolf Höss, "Meine Psyche, Werden, Leben, u. Erleben," autobiography, Institut für Zeitgeschichte (IfZ) 13/4, p. 86. Franciszek Piper, "Estimating the Number of Deportees to and Victims of the Auschwitz-Birkenau Camp," *Yad Vashem Studies* 21 (1991), p. 87.

84. Pressac, *Auschwitz*, pp. 132, 184.

85. Christopher Browning, *Ordinary Men* (Harper Collins, 1992), p. xv.

86. See Dieter Pohl, *Nationalsozialistische Judenverfolgung in Ostgalizien 1941–1944. Organisation und Durchführung eines staatlichen Massenverbrechens* (Oldenbourg, 1996), pp. 241–252. For the liquidation of all remaining Jews in the Lublin district in November 1943, the concentration camp Lublin chose to resort to mass shooting even though gassing apparatus was available. See Tomasz Kranz, "Das KL Lublin—Zwischen Planung und Realisierung," in *Die National-sozialistischen Konzentrationslager, Entwicklung und Struktur*, Band 1, ed. U. Herbert et al. (Wallstein, 1998), pp. 377–378. Christopher Browning, "The Development and Production of the Nazi Gas Van," in *Fateful Months* (Holmes & Meier, 1985). Gas vans had been developed first for use against the handicapped in operation T-4.

87. Bischoff to Topf & Söhne, 11/2/43, "Krematorium III," repreinted in Pressac, *Auschwitz*, p. 360.

88. Pressac, *Auschwitz*, p. 227.

89. Bauman is the primary representative of this view. See also Langdon Winner, *Autonomous Technology* (MIT Press, 1977).

90. Testimony of Dr. Bruno Tesch, British Military Court Hamburg, 4 March 1946, pp. 18–25. Tesch's gassing specialist, Dr. Joachim Drosihn also visited Neuengamme, Sachsenhausen, and Ravensbruck and admited to having learned of the gassing of human beings in 1942. Testimony, 3 March 1946, p. 18. Testimony of Emil Sehm, 1 March 1946: 9–12 claimed that Tesch had worked with the SS to adopt gassing to killing human beings in the autumn of 1942, exactly when the SS began developments Crematorium II & III. Regarding Peters, Urteil des Schwurgerichts Frankfurt, 27/5/55, ZSL VI 439 AR-Z 18a/60: 16–17.

91. Article by Peters and E. Wustinger, "Sachenentlausung in Blausäure-Kammern," offprint from *Zeitschrift fur hygienische Zoologie und Schädlings-bekämpfung* from 1940, received by the Neubauleitung Auschwitz, 3/7/41, US Holocaust Memorial Museum microfilm RG-11.001M.03: 42 (502: 1: 332).

92. Firm H. Kori GmbH to Haumptamt CIII of the WVHA, 2/2/43, RG-11.001M.03: 42 (502: 1: 332). This organization is confirmed on much earlier drawings that are reproduced in Pressac, *Auschwitz*, pp. 31–38.

93. Hauptabteilung CIII/3, 8/6/42, "Aktennotiz," RG-11.001M.03: 28(502: 1: 138). A weekly report covering the 14th to the 19th of July, 1941 by Karl Bischoff's predecessor Schlachter, RG-11.001M.03: 34 (502: 1: 214).

94. Unsigned, 12/9/41, "Erläuterungsbericht," RG-11.001M.03: 34 (502: 1: 218). Also Karl Bischoff, 15/7/42 and 13/3/42, "Erläuterungsberichte," same film (502: 1: 220) and (502: 1: 225) respectively.

95. Chief of Amt Allgemeiner Bauaufgaben to all SS Bauinspektionen, 11/3/42, "Entlausungsanlagen," RG-11.001M.03: 43 (502: 1: 335). On Degesch chambers at Lublin, see the Polish article prepared on gas chambers with diagrams for the Eichmann trial, IfZ Eich 1427.

96. Wirths, 4/12/42, "Besprechung beim Landrat des Kreises Beilitz," RG-11.001M.03: 43 (502: 1: 332).

97. Eley, "German History and the Contradictions of Modernity," p. 103.

98. Paul Weindling, "The Uses and Abuses of Biological Technologies: Zyklon B and Gas Disinfestation between the First World War and the Holocaust," *History of Technology* 11 (1994): 291–298.

99. See Bischoff, 9/5/43, "Aktenvermerk," RG-11.001M.03: 35 (502: 1: 233) and Wirths to Hans Kammler, 10/8/44, "Bericht uber die Wirksamkeit der Kurzwellenentlausungsanlage," 11.001M.03: 43 (502: 1: 333).

100. Affidavit of Henryk Tauber, reprinted and translated in Pressac, *Auschwitz* (pp. 482–501).

101. Pressac, *Auschwitz*, pp. 171–181, 380.

102. Richard Breitman, *The Architect of Genocide* (Knopf, 1991), p. 204.

103. See for example NO-2368, affidavit of Friedrich Entress.

104. Source of quotes here and above: Lyndall Urwick, "Organization as a Technical Problem," in *Papers on the Science of Administration*, ed. L. Urwick (Institute of Public Administration, 1937), p. 49.

Technological Systems, Expertise, and Policy Making: The British Origins of Operational Research

Erik P. Rau

We live in an age of systems and statistics. In the twentieth century, their representations migrated from engineering and the sciences to popular culture. We have come to understand events in everyday life as interrelated phenomena occurring within abstract structures. Since World War II, operational research (OR) has been an important instrument by which this dual proliferation has taken place.[1] Although the intrusion of statistical and systems thinking into everyday life has recently attracted attention from historians of technology, science, and mathematics, surprisingly few historians have examined operational research.[2]

Operational research is primarily concerned with modeling the behavior of complex systems—military, regional development, health care—using statistical and probabilistic methods, with an eye toward increasing effectiveness and predictability. Since its origins in World War II, OR has spread to most nations, assisting government officials, business executives, and university and other institutional administrators in coping with complex, systems-oriented problems. It has exerted significant influence on other systems methodologies, such as systems analysis and systems engineering.

Virtually the only part of OR's past that has received any attention is its origins in World War II, primarily because many of its first practitioners were prolific scientists who either wrote autobiographies or otherwise wrote extensively about their wartime experiences. But that material fails to capture much of the significance of OR's origins. In this essay, I argue that, beyond the problems in military operations that OR helped solve, the field is significant for the system of patronage that sustained it. How did operational research become a recognized field of expertise by the end of the war? Techniques used and problems solved will not reveal the answer;

although they are certainly an important part of the story, they alone cannot explain the conferral of authority upon a rather diverse—not to say confused—array of practices. Authority was conferred and a new professional identity formed through a working relationship between scientists and military officers within a framework of patronage.

The state (including the military) as patron and promoter of technical expertise is not a particularly new subject to historians, but much of their attention has been focused on earlier periods. World War II and its aftermath remain understudied. Yet this period is crucial for understanding the modern state and recent political culture. Brian Balogh has recently observed, regarding the American case, that not until World War II did the federal government and the professions build a regular and permanent partnership, the Progressive Era and the New Deal notwithstanding.[3]

The emerging historical writing on political cultures encourages the fusion of political history's traditional focus on government and parties with social history's emphasis on race, ethnicity, class, gender, and other identities. There is no reason for professional identities to be excluded from this. Indeed, the formation of new forms of technical expertise during World War II and the Cold War sheds light on the expansion of the state, for only through patronage of new forms of expertise was this expansion possible. In some cases, these new professional identities also play a role in the recent evolution of political cultures, for political discourse is often framed by technical knowledge or findings. The Cold War also involved the transfer of technical knowledge and expertise to allied and client states. In addition, reconstruction in Japan and Europe and independence in formerly colonized states encouraged the proliferation of new forms of expertise, such as operational research.[4]

OR's origins shed light on all these processes—the growth in the state's (i.e., the military's) capabilities, the framing of strategic debates, and the sharing of technical expertise between allies. The creation of this new technical field coincided with scientists' increasing influence on and ambition in policy making; World War II caused the authority of civilian technical professions to be renegotiated. How this process took shape is the focus of this essay. Why was OR a British invention? What was its nature and how was it shaped by the constraints of practice? What led to its rapid proliferation? These questions cannot be answered in strictly technical terms. Nor are they of narrow technical interest. They ultimately structure the relationship between technique and power. I will explore these issues in three sections.

The first section will deal with the Britishness of OR's origins. OR's British-American roots owe much to the commitment of those two nations to radar technology. Other combatants during the war supported radar research, but the British first entrusted their scientists and engineers to help forge policy on how a radar system should be used as well as how it should be built. In the beginning, "operational research" referred less to a specific set of skills (at first there were no disciplinary standards) or a professional identity than to a relationship of trust established between a military command and a group of civilian radar researchers attached to it. OR, in other words, grew out of the institutional and work culture of the prewar British air defense system.

In the second section, I will explore how operational research developed its identity as a scientific field between late 1940 and late 1941. The staff of the first OR groups were primarily engineers interested in the radar system. They functioned and behaved much like engineering consultants. In contrast, Britain's elite scientists were more generally interested in controlling the integration of civilian technological expertise into the war effort, including strategic planning and policy making. Their ambitions formed the core of the Social Relations of Science movement.[5] I will argue that these elites, led by physicist Patrick Maynard Stuart Blackett, created a disciplinary identity for operational research on the foundation of the engineers' work.[6] They did so in order to realize their broader objective of influencing military decision making. Working successively for the Royal Army, the Royal Air Force, and the Admiralty, Blackett transformed an engineering activity into a field science of warfare. Historians of science have recently shown interest in the field sciences precisely because scientists cannot control the context of their work to the extent they can in a laboratory. Instead, they must negotiate the content, procedures, and priorities of their scientific work—even the nature of their scientific identities—with other social groups who share the same space.[7] This process of negotiation, I will argue, defined operational research during its early development under Blackett and others. In the absence of any professional standards or techniques, practitioners from a wide variety of technical fields fell back on the techniques of their own disciplines, adjusting as necessary to local circumstances and work cultures. Meanwhile, they negotiated with local commanders over the priority of their work and access to information and authority. The tools in this bargaining included scientific knowledge of new military technologies and (more important) of statistics and probability theory. But much also depended upon the personalities involved. Blackett devoted considerable attention to developing and institutionalizing a relationship of trust with his military patrons. Patronage, at least as much as skill with quantitative techniques, provided the basis of OR's identity during the war years.

In the third section, I will examine OR's rapid proliferation after 1941. Scientists enthusiastically promoted OR with the support of their patrons in the services and ministries.[8] Ironically, despite OR's tendency to reinforce centralized control, the scientific elite was unable (and, in some cases, unwilling) to centralize OR activities under its jurisdiction. Instead, OR groups accommodated their local military commanders or ministry officials. OR remained a localized phenomenon, defined by local circumstances. Its practice and meaning varied from command to command; differentiation among OR groups commonly included group structure, personnel, types of problems, access to information, and work procedures. A consequence of this localization was that OR groups tended to adopt the general perspective of their military patrons. This became problematic when the Western Allies retook the offensive in 1943 and military commanders debated strategic policies. Uncertainties and genuine differences over how to proceed often led to competition and even conflict among commanders, officials, and, by extension, their OR groups and scientific advisors. The development of bombing policy involved several such incidents.[9]

Instances of overlapping and competing forms of expertise not only shed light on the making and implementation of wartime strategic policy; they also foreshadow some of the problems that haunted attempts at linking knowledge to power in the later twentieth century. The development of countercultures and criticisms of technocracy in a variety of industrial states by the late 1960s indicated a more widespread discontent with the alliances among government agencies, private institutions, and the professions— alliances that seemed to preclude popular participation.[10]

In the conclusion, I suggest how historical research on operational research may shed light on postwar state building and its connection to civilian expertise.

"Born of Radar"

"Operational research was, in fact, born of radar," remembered Robert Alexander Watson-Watt after the war.[11] Though Sir Robert was hardly a disinterested observer (the Scottish engineer had helped manage Britain's radar development program), he at least correctly identified the context in which "operational research" arose. Most of the other combatants in World War II had independently and secretly begun research into the use of radio waves for detection and location of aircraft by 1939, but only Britain formed an OR program during these years. The only other nations that would follow suit during the war were English-speaking countries, and even these did so primarily in response to British encouragement.[12]

What made OR British? Most participants in OR's early years have tended to pose the question somewhat differently, regarding the lack of operational research in other nations as an abnormality needing explanation. Some authors have suggested that Germany did not spawn OR groups because science cannot develop "normally" under totalitarian regimes.[13] But historical scholarship on scientists, engineers, and physicians in Nazi Germany, the Soviet Union, and other authoritarian states reveals a range of responses in expert communities to such political climates, including accommodation and collusion.[14] Researchers' progressive self-image does not always mesh well with their historical behavior.[15] What requires explanation is the development of a new form of expertise labeled "operational research" in Britain, not its absence elsewhere. The latter approach assumes that the emergence of OR, and by extension any other form of expertise, is somehow normal. I argue quite the contrary: OR's emergence was due to peculiarities in the British radar program, in which civilian researchers enjoyed great influence in shaping air defense policy.

The construction of Britain's air defense during the 1930s was championed by civilians. Air defense was a goal that citizens across the political spectrum could embrace.[16] Within the Royal Air Force, however, few officers supported the cause, in part because the RAF's identity had been founded on the airplane's potential to wage strategic (i.e., offensive) warfare but also because few could imagine effective anti-aircraft measures against the bigger, faster, longer-range bombers then being designed.[17] In the aftermath of Japanese air attacks on the East Asian mainland in 1932,

British Prime Minister Stanley Baldwin gloomily predicted that "the bomber will always get through."[18] Nevertheless, public calls for defensive measures swelled after the Nazis' electoral victory in Germany, and they eventually found an echo within the Air Ministry. There, in late 1934, Director of Scientific Research Harry E. Wimperis and the meteorologist Albert Percival Rowe convinced Air Ministry officials to establish a Committee for the Scientific Survey of Air Defence (CSSAD) consisting of members drawn from the scientific community.[19] This committee was to survey scientific and technological developments that seemed promising for air defense. Besides Wimperis (Rowe served as committee secretary), it included Sir Henry T. Tizard (a chemist and rector of the Imperial College of Science and Technology with a long history of service to the government in general and the Air Ministry in particular), Archibald Vivian Hill (a Nobel Prize-winning physiologist who had spent World War I studying air defense problems), and Patrick Blackett (a rising star in the physics community whose previous military experience was a short wartime tour of duty in the Royal Navy that included the Battle of Jutland). The committee first convened on January 28, 1935.[20]

Winston Churchill, recently installed on the super-ministerial Committee of Imperial Defence, grumbled that an Air Ministry committee could not be objective. Yet available evidence suggests that CSSAD was dominated by its civilian membership, not by the Air Staff.[21] This set a pattern that would continue in the years before the war: civilian scientists and engineers played a leading role in shaping air defense policy by recommending what would be built and how it would be used; they shaped the discussion of policy matters. CSSAD was chaired by Tizard, and it soon became known as the Tizard Committee. Wimperis also held open the door for other civilian researchers. Shortly before the committee first convened, Tizard invited Robert Watson-Watt, then superintendent of the National Physical Laboratory's Radio Research Department, to explore the potential of using radio waves as an anti-aircraft weapon.[22] Watson-Watt's staff soon dismissed the weapons angle, but suggested that radio waves could be used to detect and locate large aircraft. The British had stumbled across the basis for radar.

The Tizard Committee immediately seized upon Watson-Watt's idea and gave its development top priority.[23] After a preliminary demonstration in

February impressed the RAF's top research and development officer, Air Marshal Sir Hugh Dowding, the Air Ministry established a development outpost, first in Orfordness and then at Bawdsey Manor (on the English coast, near the River Deben).[24] Civilian researchers thus played a critical role in launching Britain's radar development program. In practice, they also managed the research program, despite its being formally administered by the Air Ministry. Watson-Watt was appointed supervisor, and members of his staff at the National Physical Laboratory formed the initial core of the research team.

Civilians continued to guide the program in other critical ways. Possessing gear for the detection and location of aircraft would benefit Britain only if it were properly integrated into an air defense system that took full advantage of it. Civilians took early moves to make sure that radar would be. Although other radar enthusiasts besides Dowding could be found on the Air Staff, the officers had taken no steps to work on air defense procedures. Tizard seized the initiative.

Before outlining Tizard's moves, however, I should sketch out the nature of the early radar technology, for it imposed constraints on and opportunities for the overall system. Among the few who knew of its existence, radar was regarded as a device that multiplied the effectiveness of fighter aircraft. Until this point, the Air Ministry, working with limited resources and under treaty restrictions, supported a building program that emphasized bombers as a deterrent against potential enemies. The bankruptcy of this strategy became apparent as Adolf Hitler's Luftwaffe grew. It became clear that Britain could not produce the number of interceptors needed for a viable defense in the foreseeable future.[25] Radar was intended to bridge this viability gap by increasing the potency of the available fighters. This was a tall order for the relatively primitive equipment designed by the Bawdsey Manor team. The early sets produced long wavelength signals, resulting in poor resolution. The sets were also difficult to use. Instead of the round screen with a circular sweep that is familiar today, the visual output of the early radars consisted of two beams, one displaying the range of the target and the other showing displacement. Moreover, the early sets were not terribly accurate. As late as 1938, the equipment could not give reliable information about enemy altitudes. Finally, radar equipment was large and bulky. Sets small enough to be installed onboard aircraft would

not come into service until after the war started.[26] In short, the early generations of radar equipment were bulky, inconvenient, temperamental, and
extremely limited. Radar was not a technological "quick fix."

Tizard was well aware of these limitations in mid 1936, and he appreci-
ated the need to integrate radar into a system that overcame the new technologies limitations and accentuated its strengths. To do so, he turned from
technology to policy. Few military officers would have dared to intervene
in the research and development effort, yet Tizard, a senior Air Ministry
scientific advisor, began to intervene in operational matters. According to
his biographer, Tizard encountered no great opposition when he sought the
Air Staff's approval to run a series of unofficial air defense trials.[27] Intended
to last a few weeks, these trials stretched out to well over a year, during
which scientists and officers collaborated freely. The trials were held at the
RAF's station at Biggin Hill, located outside London under the major air
routes connecting the metropolitan area to Central European destinations.
The experiment consisted of a series of mock air battles in which prototype
radar instruments tracked attacking "bombers." Defensive forces
attempted to use the radar data, supplied at regular intervals to a "headquarters," to call "fighter" units into action and direct them to their prey
by radio.[28] Dowding, promoted and charged with executing Britain's air
defense, eagerly adopted the procedures that RAF officers (many of whom
were kept ignorant of the exact nature of radar technology) developed in
conjunction with Tizard and the Bawdsey team. The basic techniques and
vocabulary developed at Biggin Hill later formed the basis for air defense
operations under Dowding during the Battle of Britain.[29]

Two characteristics of the emerging air defense system deserve closer
attention since they reveal important aspects of civilian authority within
the system's organizational culture. The first is that the air defense system
was centrally controlled and was dependent upon radar equipment.
Centralized control constituted something of a departure from traditional
air defense tactics, which left much of the initiative in the hands of squadron
leaders who flew with their patrols. Radar significantly reduced the initiative of these pilots and transferred it to the senior officers on the ground.
Some pilots chafed at this loss of initiative. During the Biggin Hill trials,
Tizard felt compelled to ease the way through frequent interaction with air
crews, acting not as a scientific advisor but as a fellow pilot.[30] In the autumn

of 1940, as German bomber attacks rocked British cities, Dowding's command was itself rocked by insubordination in the "Big Wing controversy," which led to the aging commander's ouster.[31] That civilian experts strengthened centralized control against local initiative was not lost on the junior officers.

The second critical aspect of the air defense procedures was its lack of automation, despite its centralized nature. The novelty of radar and the popular appeal of fighter aircraft has blinded most observers to the fact that much of the work of the early British air defense system was done by hand, on the ground, and out of sight. Radar stations and fighter aircraft (and their pilots) have received the lion's share of historical attention, but they were only the inputs and outputs of a vast information-processing system. This system employed a great number of RAF and Women's Auxiliary Air Force (WAAF) personnel, who operated the radar and observer stations, cleaned up the information the stations produced ("filtering"), interpreted and plotted the results, and then transmitted them to the command centers ("telling"). There the aggregate information was plotted on large map tables. Senior officers reached decisions, then sent orders to launch defensive strikes back down the hierarchy to field commanders at selected air bases. Thousands all across Britain were employed in these operations; the effort expended in organizing, training, and managing them was enormous.[32] Significantly, civilian researchers enjoyed a great deal of influence over this effort, which gave rise to operational research.

The air defense system's characteristics—temperamental technology, high centralization, low automation—allowed civilian experts to invent a role for themselves within the system's emerging organizational culture. But that role required an interface that permitted civilians to operate within the hierarchical confines of a military command. Operational research provided that interface.

The term "operational research" was first coined to describe the engineering troubleshooting aimed at overcoming radar technology's shortcomings. Before Bawdsey Manor became functional as the first prototype radar station, in 1937, the Biggin Hill experiment provided the sum total of everyone's experience with a modern air defense system. No one knew what to expect as the station powered up. More detailed procedures were needed for plotting, filtering, and telling, and RAF officers needed training in operating and managing Bawdsey and other stations as they became operational. No one had more than a theoretical understanding of how the whole system would work. Not until this point could specific procedures for plotting, filtering, and telling radar data be described in detail. RAF officers also needed training in operating the radar stations. These tasks fell to Eric C. Williams, a young engineering professor leading a small research group at Bawdsey Manor. Williams and his group worked closely with Squadron Leader Raymund G. Hart, the RAF officer responsible for administering operational radar stations.[33] Neither side seems to have thought the arrangement sufficiently unusual to comment on it at the time.

This small civilian foray into military operations became more pronounced as operational issues continued to arise. When the 1937 summer air exercises revealed that the radar equipment at Bawdsey responded well but the rest of the control system performed disappointingly, the Williams team helped to identify and remedy the bottlenecks.[34] Such work may have constituted nothing more than standard engineering practice, but it nonetheless indicated a growing dependence among officers on civilian technical knowledge. The next summer, during the Munich crisis, the addition of several radar stations during the air exercises produced confusion when wildly conflicting data from the different stations overwhelmed the control system. Williams's team responded by proposing alternative methods of producing, filtering, and transmitting information that bridged over radar's weaknesses.[35] Finally, during the 1939 summer exercises, Williams's team was split into two main groups, with Williams managing a section supervising the radar operations at Fighter Command headquarters at Stanmore. A telephone engineer, G. A. Roberts, led a roaming contingent of engineers among the various radar stations.[36] A. P. Row, the erstwhile secretary of the Tizard Committee who had succeeded Watson-Watt as supervisor at Bawdsey,[37] observed that the activities of these groups had become large enough to be organized as a formal organizational function. Since Williams and Roberts focused on operations rather than on development, Rowe referred to these engineers as the "Operational Research Section." As Williams himself wryly noted after the war, the title was simply a bureaucratic convenience.[38]

Accepting the civilians' technical authority, Dowding and several members of his staff petitioned Rowe to detail a team of Bawdsey engineers to

Fighter Command. This invitation represented an attempt to tap civilian expertise for military ends without disrupting their civilian identity. Dowding did not want to induct the engineers into the military; their value lay in their links to the Bawdsey research community. Rowe obliged Dowding. For both Dowding and Rowe, the arrangement was simply for bureaucratic convenience; administrative formality was merely made to reflect and facilitate what collaborating engineers and officers had already established in practice. "Operational research" denoted no recognized skill at that time. In fact, the name was dropped when William, Roberts, and others moved to Stanmore; the group was first named the Radar Research Section, then renamed the Stanmore Research Section in February 1940. These labels illustrate that the earliest OR activities were identified primarily by technology or by patron, rather than by specific practices.

The specific practices these civilians performed seem almost banal today, consisting largely of troubleshooting the implementation of a complex technological regime at Fighter Command.[39] Can such humble beginnings alone reveal how they became the basis for what soon became a recognized and respective technical field and why this technical field developed in Britain? What the above suggests is that OR developed in Britain because it proved to be a convenient interface among the authorities of different kinds of experts, technical and military. Within a culture that celebrated civil society and in which the intermingling of state and civil society has been historically regarded as problematic, the first OR section provided an institutional solution to the problem of establishing trust between officials and civilians.[40]

"Scientists at the Operational Level"

During 1940 and 1941, operational research transcended its engineering origins and was transformed into a science, specifically a field science of warfare. The shift occurred as members of Britain's scientific elite were agitating for a more prominent role in British military policy making. OR's transformation stemmed from their ambitions. For their part, military commanders accepted civilian OR groups as a hedge against uncertainty, both technological and organizational. The first commands to adopt OR after Fighter Command—the Army Anti-Aircraft Command and the RAF

Coastal Command—were also among the commands most besieged by equipment and personnel shortages, often having to cope with the inadequate planning of military superiors or government ministries.[41] Perceived neglect encouraged officers' unorthodox maneuver of appealing to scientific outsiders for aid. These intersecting interests of officers and scientists held the potential for the innovative collaborations that later adopted the name "operational research." This was no simple or linear development from the arrangement at Fighter Command. The Fighter Command group developed from the relatively narrow concerns of getting the air defense system to work. Transformed into a science, OR became part of the scientific elite's political agenda.

The scientists' ambitions extended directly from the Social Relations of Science movement, which emerged during the tumultuous years of the 1930s and culminated during the war years.[42] Economic difficulties at home and the rise of fascism abroad had politicized Britain's scientific elite during that decade.[43] The left wing of the movement—the "scientific socialists," including the physicist John Desmond Bernal, the zoologist Lancelot Hogben, and the biologists J. B. S. Haldane and Joseph Needham—have received the lion's share of historians' attention.[44] However, the Social Relations of Science movement drew members from a wider political spectrum.[45] Further toward the political center were Tizard and Hill. Political convictions did not shape these scientists' approach to their work. Rather, the converse was true: their political convictions were shaded by hopes they shared for their profession.[46] From pamphlets and articles published during this era, it seems clear that many scientists came to politics in the 1930s out of professional frustrations—particularly over the perceived underutilization of science in the economic (and later military) emergency that Britain faced.[47]

Tizard, Hill, and Blackett were all of the SRS movement, and the radar program provided them with an outlet which they fully exploited. Tizard's influence in RAF research policy, already substantial by the mid 1930s, continued to rise during the early war years, in part because of radar. During these years his stature allowed him to help mobilize the scientific community to lobby for greater influence in the war effort. A. V. Hill was even more assertive, spearheading a successful drive to attach scientific and engineering advisory committees to Churchill's war cabinet. In 1942 he launched an attempt to form a "Scientific General Staff" that would have the kind of planning and coordinating authority over the scientific community that the military general staff enjoyed over military forces. Hill envisioned that such a body would also address the war cabinet directly. Hill's effort in this failed, but his ambition and its broad support in the scientific community signaled the transition among some scientists from a stance of apolitical objectivity to an embrace of activism.[48] Others active at the intersection of research and activism included J. D. Bernal, the zoologist Solly Zuckerman, Robert Watson-Watt, and Patrick Blackett. Not coincidentally, all these scientists became early champions of operational research. Reformulated as a scientific endeavor, OR promised scientists opportunities in military policy making. Moreover, the insights they gained at the level of military commands bolstered their confidence and authority in dealing with political and military leaders in strategic matters. In short, OR formed an integral part of scientific activists' ambition to colonize warfare.

Yet the process by which OR underwent its transformation began with surprisingly little premeditation. The program of setting up "scientists at the operational level," as Blackett put it, was not an outright attempt to somehow hijack OR from the engineers. Yet by 1942 "operational research" described scientific undertaking within military commands. Descriptions of OR became more methodologically prescriptive; more important, they often alluded to the system of patronage in which OR was inextricably embedded.

Just how was operational research claimed for science? Patrick Blackett was one of the leaders in this endeavor. A physicist trained in Ernest Rutherford's laboratory in Cambridge while a young naval officer at the end of World War I, Blackett soon developed a reputation as an experimentalist, especially in extracting information about subatomic particles from cloud chambers. His occupation required familiarity with statistical methods and probability theory, and it led him to understand modern life—warfare included—in probabilistic and statistical terms. Blackett was also a lifelong Fabian socialist, serving at one time as president of the left-leaning Association of Scientific Workers. Later in life, he served in Harold Wilson's Labour government as the scientific advisor to the Ministry of Technology. As Baron Blackett, he spoke in the House of Lords on the importance of harnessing science to alleviate social problems. Economic development in the post-colonial world was a favorite theme, as it was for many on the scientific left. Clearly, Blackett, like many others of his generation, saw social and political activism as complementary, not contradictory, to a scientific career.[49]

In 1940, Blackett's contribution to Britain's defense had reached a lull. Extremely active on the Tizard Committee through the 1930s, he had taken on a number of studies concerning radar's impact on anti-aircraft tactics.[50] The Royal Army's Anti-Aircraft Command (known as "Ack-Ack") provided close defense of metropolitan areas against the enemy bombers that eluded Fighter Command. But with the start of the war, the Tizard Committee fell dormant and Blackett turned his talents to bombsight design for the Royal Aircraft Establishment at Farnborough.[51] This was narrow work, far removed from the questions of tactics and operating conditions that he had tackled earlier. As he later wrote, "one might conclude that relatively too much scientific effort has been expended hitherto in the production of new devices and too little in the proper use of what we have got."[52]

To his friends, it seemed that Blackett was languishing at Farnborough. One such friend was Hill, who tried to engage his friend in more interesting and important work. Hill knew that Anti-Aircraft Command needed someone familiar with radar technologies. The command's top officer, Army General Sir Frederick Pile, struggled with supply and personnel shortages, for which he held rival ministry and military officials responsible.[53] Among his early difficulties was a problem involving gun-laying radar equipment. Hill met with Pile to discuss his grievances and then suggested he secure a scientific advisor. Hill arranged a meeting between Blackett and Pile in August 1940 at which Blackett agreed to serve as Pile's scientific counsel.[54]

Only scant information is available with which to analyze Pile's and Blackett's evolving relationship.[55] However, we do know roughly what transpired during their association. Blackett retained his civilian status by remaining on the Supply Ministry's payroll, yet he took his orders exclusively from Pile and recognized no other authority while in his advisory capacity. Blackett persuaded Pile to allow him to recruit a team of civilian scientists on similar terms. Within 6 months the group grew to eleven members, including representatives from the physical and biological sciences.[56] The reaction of Pile's officers is not clear, except that they understood the civilians as a group apart. Formally named the Anti-Aircraft Command Research Group, the civilians were more commonly referred to by the officers as "Blackett's Circus."

Blackett exploited the group's anomalous status to craft for it a distinct and constructive role. He interpreted the group's assignment as a purely scientific venture. In a memorandum he wrote: ". . . a considerable fraction of the staff of [the research group] should be of the very highest standing in science. . . .The atmosphere required is that of a first-class pure scientific research institution, and the caliber of the personnel should match this." As such, its members would not take on "daily routine responsibilities in relation to the staff work of the headquarters"; the team was to be kept "free for non-routine investigations and researches."[57] Their duty, in other words, was to reflect the work culture of the laboratory, not that of the engineer or that of the shop floor. Because he constructed the group's duties as qualitatively different from those of the officers, Blackett anticipated no intrinsic conflict between the two groups: ". . . very many war operations involve considerations with which scientists are specially trained . . . and in which serving officers are in general not trained."[58] For Blackett, total war in the modern era was a cooperative venture to which scientists could make independent contributions.

Of what did these contributions consist? The first problem Blackett's Circus confronted (presumably the one that had spurred Pile to seek outside help) concerned the gun-laying radars that had recently begun arriving at Ack-Ack. The gun-laying radars were intended to help aim anti-aircraft guns at German bombers during night attacks. By the late summer of 1940, British interceptors had thwarted Germany's invasion plans (the Battle of Britain) and the Luftwaffe had begun considering nighttime raids on metropolitan areas (the Blitz). At that time interceptors were not equipped with airborne radar, and thus air defense of Britain's cities fell largely to Ack-Ack. In practice, anti-aircraft guns were aimed slightly ahead of their targets, since enemy planes moved at great speed and altitude. Anticipating the future position of the target was achieved with a device called a predictor. Crews looking through a predictor's rangefinder tracked the aircraft. The predictor (actually a mechanical calculator) extrapolated the proper aiming point as its crew plotted the target's course. Night attacks, however, made visual tracking impossible. For such occasions, the gun-laying radars would replace the human operators. Unfortunately, an oversight in their design led to a fatal compatibility problem: the radar sets produced electronic output, but the predictors relied on mechanical inputs. Blackett faulted the radar designers for "a certain lack of imaginative insight into operational realities." With a technological fix not readily available, the Circus developed methods of plotting radar information manually and then feeding the data directly into the predictors by hand. They also designed some easily manufactured plotting devices to help automate the process.[59]

Yet such work was basically engineering, and it resembled what Blackett considered the "routine investigations and researches" that he wanted to avoid. Technical problem solving, he insisted, "must be left entirely to the technical branches of the Service in collaboration with the research and development establishments and to the manufacturing firms," which his scientists could assist most appropriately "by interpreting (a) the operational facts of life to the technical establishments, and (b) the technical possibilities to the operational staff."[60] Thus Blackett positioned the Circus as an intermediary between the producers and users of technology. But above all, they would be scientists. Blackett was far more interested in transferring the scientists' laboratory skills and methods—close observation, quantitative measurement, critical analysis—to the battlefield. In fact, the detail work involved in developing plotting methods for the predictors was soon handed over to a special unit that Blackett persuaded the Ministry of Supply to create: the Anti-Aircraft Wireless School in Petersham.[61]

Nevertheless, Blackett, far from being doctrinal, adopted a heterodox approach. He wrote that the Circus "must not copy in detail the technical methods of any other science, but must work out techniques of its own, suited to its own special material and problems."[62] Such "material and problems" were the very stuff of command: determining the best tactics, strategy, and operations under given constraints. Although the scientists had no experience in military matters, Blackett believed they could judge the effectiveness of a particular military operation through analysis. What led him to this conclusion was his conviction that the problems of modern warfare were susceptible to statistical and probabilistic reasoning: "It is clear that an intelligently controlled operation of war, if repeated often enough with reasonable tactical latitude allowed to the participants, will tend to a state where the yield of the operation is a maximum, or the negative yields (losses) a minimum. . . . Success in most operations of war . . . is due to the sum of a number of small victories, for each of which the chance of success in a given operation is small."[63] Not only did Blackett believe in using quantitative measures to understand the effectiveness of past or present operations, he also insisted that they could predict the likelihood of success of planned operations.[64] By developing standards of measurement, precision, observation, and analysis, Blackett began to establish a field science of warfare.

Adapting scientific practices to a martial setting required a great deal of negotiation. Blackett's vision of scientific warfare depended upon his scientists' access to battle and intelligence data. Here the scientists necessarily relied upon the military staffs they served. Access to battle scenes and to intelligence and operational reports was, of course, highly restricted. The scientists enjoyed such access only at the pleasure of the commander. That Blackett could gain access to such information indicates his skill in earning Pile's trust and/or Pile's willingness to go outside established military channels for help. More remarkable still is Blackett's expectation of freedom in criticizing officers as he saw fit: ". . . the work of [a scientific group] at a Command must inevitably involve suggestions for improvement."[65] Ensuring the trust of officers under such conditions, Blackett realized, was important to the Circus's effectiveness.

Thus, the relationship that the scientists enjoyed with officers was as crucial to success as the methods they used. In fact, interactions with military counterparts may have formed an even stronger basis for the Circus's identity than standards of measurement or precision. The quantitative techniques that the scientists employed were generic to many scientific fields; effective application of these hinged on the particular relationship the scientists enjoyed with the command. Blackett realized that the key to success was trust, which he scrupulously cultivated with Pile and his staff. He insisted that "the . . . group must be an integral Part of the Commander-in-Chief's staff and all the reports or recommendations must be to the C-in-C of the command." Otherwise, he predicted, Ack-Ack's staff officers "would rightly feel aggrieved and the intimate collaboration between operational Service staffs and . . . research workers . . . would become impossible."[66]

Through discretion, Blackett's scientists achieved a remarkable level of access to intelligence reports and were allowed to participate in planning and policy making at Anti-Aircraft Command. Blackett later summarized a small number of the group's activities during his 6-month stay at the command.[67] In one instance, scientists were called in to analyze why coastal batteries were more successful in shooting down German planes than their inland counterparts. The scientists concluded that these installations were not more successful; their "success" lay in overoptimistic "kill" claims that could not be corroborated since the reported hits occurred over water. The scientists took the opportunity to recommended statistically informed reporting techniques. In another case, Blackett was invited to suggest how limited numbers of anti-aircraft guns and radar sets could be distributed to offer the city of London maximum protection. Using geometrical modeling and probability theory, the scientists suggested a configuration that maximized the protective cover of the guns.[68]

It is difficult to evaluate the Circus's influence in concrete terms, for too little information is available to link their policy recommendations to successful outcomes. The evidence that does exist has tended to be anecdotal, the postwar reminiscences of participants. Some are counterfactual or inconclusive. For instance, last-minute deliveries of guns and radars made Blackett's advice on the placement of London's anti-aircraft gun installations moot. Debating whether or not they would have worked is strictly an exercise in conjecture, in which Blackett indulged after the war. Perhaps the best indication of the civilians' value that we currently have is the Circus's long-term popularity in the army. The Circus eventually grew into an OR organization for the entire army.[69] And when Blackett was lured away to another command, Pile complained: "They have stolen my magician."[70]

In March 1941, with the German aerial onslaught beginning to ebb, Blackett seized the opportunity to transfer his analytical methods and organizational strategies to another military unit. Like the army's Anti-Aircraft Command, the RAF's Coastal Command suffered supply and personnel problems due to its relatively low status within its own service. Although first assigned reconnaissance duty, it soon shouldered the job of U-boat hunter, for which it was woefully prepared. Its crews had no operational doctrine, little equipment, and practically no training to counter the U-boat threat.[71] Since the beginning of the war, its effect on German submariners had been primarily psychological; its "kill rate" stood at only 2 to 3 percent.[72] Thus, its commander, Air Marshal Sir Frederick Bowhill, had as much incentive as Pile to turn to outside help.

Nevertheless, Blackett's arrival at Coastal Command seems to have been orchestrated from above. Once again, Hill figured in the chain of events. In early 1941, he pushed the War Cabinet's Scientific Advisory Committee (which he had helped create) to recommend the assignment of a scientific advisor to Bowhill. The committee discussed the matter with the Secretary of State for Air, Archibald Sinclair, who backed the idea. The Air Staff, on the other hand, hesitated, in part because Blackett's appointment to Pile was known to have been a source of embarrassment to the army general staff. Air Staff resistance finally folded in the face of pressure from Sinclair. The Secretary of State for Air further dictated the selection of Blackett as director of the effort.[73]

Sinclair's actions may have been influenced by an idiosyncratic Air Staff officer, Air Marshal Sir Philip Joubert de la Ferté. Joubert felt little loyalty to Air Staff's sacred cows or prejudices. He had been, along with Dowding, an early supporter of radar, especially its application to the anti-submarine problem. Early in the war, Joubert was named Assistant Chief of the Air Staff for Radio, and thus the Air Staff's radar champion. He promptly pushed for the creation of a Development Unit for Coastal Command to look into anti-submarine applications.[74] In the autumn of 1940, at Joubert's request, A. P. Rowe assigned John C. Kendrew, a leading chemist, to lead the unit, bestowing upon him the title Radio Operational Research Officer. Kendrew's tasks coalesced around educating Coastal Command staff officers in the intricacies of radar.[75] It was a limited role, and both Sinclair and Joubert wanted to eliminate it in favor of something akin to Blackett's group at Ack-Ack. It was Joubert who, after Blackett had been lured away, broke the news to Pile, with the consolation that Blackett would be returned to the army officer within 6 weeks.[76] But Blackett never returned, and within 3 months Joubert joined Blackett at Coastal Command headquarters in Northwood after relieving the aging Bowhill. Blackett could hardly have chosen a more receptive patron.

Blackett's arrival at Coastal Command completed the transformation of operational research into a science of warfare. Until that point, Blackett had not bothered to give his activities a name—probably because he regarded them not as the basis for a new discipline but as a vehicle that permitted scientists to participate more fully in the war effort.[77] With Blackett's arrival, however, Kendrew's position as Radio Operational Research Officer was abolished and the title assigned to the new group of scientists Blackett recruited: the Operational Research Section. From that point forward, the name became inextricably linked to the general study of military operations by scientists at the command level.

With Joubert at the helm at Coastal Command, the OR Section grew rapidly, eventually boasting a staff of more than 60. Moreover, the civilian scientists began to enjoy the right to review most tactics and operations; later, they would even intervene in strategic matters. Their specific tasks have been commented on elsewhere.[78] The important point is that OR's identity (if not its practice) remained rooted not to the local peculiarities of technique but to the attempt to institutionalize alliances between local commanders and civilian scientists.

Blackett tried to set the terms of such alliances in a brief he wrote for the Admiralty staff in the autumn of 1941. This brief, intended to persuade the Admiralty to adopt operational research, prescribed how OR groups should be organized and attached to military commands. Blackett spent far less energy prescribing the content of the work, which he considered negotiable. Blackett's brief also appears to have been intended for a broader audience than just the Admiralty. Despite his later modesty about the memorandum, its broad prescriptions are indicative of a general blueprint rather than a specific pitch to a limited audience.[79] In fact, within 3 months of Blackett's presentation, copies of it were circulating not only among Britain's top officers but also among American military and scientific personnel. Most likely, he intended it to serve as a primer for a wider audience of military and government officials.[80]

"Not as Questioning and Independent as They Could Have Been"

The rapid promotion and proliferation of operational research during 1942 should be seen as part of the more general effort among the scientific elite to influence policy making in the war effort. That year also saw A. V. Hill make his bid to establish the "scientific general staff" in the War Cabinet.[81] Scientific advisors to the Admiralty successfully lobbied for Blackett's appointment as the Navy's OR expert.[82] Charles G. Darwin, grandson of the famous Cambridge naturalist, was appointed to oversee OR's expansion in the army.[83] And in the RAF, nearly every major combat command adopted an OR section during 1942 or shortly thereafter.[84] The scientific leaders also promoted OR overseas, encouraging their counterparts in allied nations to follow suit. An OR service was soon established in the Royal Canadian Air Force, followed soon thereafter by one each in the Mediterranean, Indian, and Far Eastern theaters.[85] Tizard, Watson-Watt, and others urged American scientists to take up the cause. Vannevar Bush, the engineer responsible for mobilizing much of the American research community for the war, was told by Darwin that OR was "held here [in Britain] to be far and away the most important application of science to military affairs, being so much more direct than the technical developments which have hitherto been the main contribution."[86] By 1942, OR enthusiasts had concluded that scientific expertise could make much broader contributions to the war effort than simply aiding research and development projects, and that new weapons might not even be the wisest expenditure of Allied scientific capacities.

The campaign to proliferate operational research proved generally effective, for it spread among the English-speaking combatants during the rest of the war. This proliferation also caused some unexpected difficulties. Ironically, despite its wide adoption, OR remained a localized phenomenon. Blackett himself had encouraged this when he had advised OR scientists to develop methods suited to local circumstances.[87] Their success is reflected in OR's rapid adoption. Commanders were not compelled to adopt an OR section, and few officers would have bothered taking on the scientists if they had not derived value from the association.

OR's malleability reflected the scientists' tendency to adopt the general terms and goals of their military patrons. To be sure, the scientists did not uncritically adopt the opinions of their commanders. But almost all those who served in OR groups and documented their experience after the war recount episodes of having to win over officers' acceptance. Moreover, their memoirs often exaggerate the level of independent action or thinking. David Edgerton has recently suggested that British scientists' postwar image as loyal patriots but independent and principled opponents to military excesses does not stand up to historical scrutiny.[88] Freeman Dyson, for instance, has railed against the excesses of the strategic bombing campaign while acknowledging his participation in RAF Bomber Command's OR Section, which rationalized portions of that campaign.[89] The convergence between OR groups and their commanders is a consequence of the work environment in which OR practice was carried on. Even things as basic to a scientific undertaking as data collection were predicated on the system of patronage that supported OR. OR data was by definition military intelligence, and thus restricted. Without a basic agreement about the mission of a command, no trust could develop between a command and its OR section; without trust, no access to data; without data, no meaningful operational research.

With local circumstances affecting both the content and the practice of operational research, variation among OR sections in organization, in personnel, in practices, and in research agenda became all but inevitable. It is of no small symbolic weight that members of some OR groups, particularly those in overseas commands, relinquished their civilian status, took commissions, and donned uniforms. Such variation held potential difficulties. Although none of the scientists advocated standardization of practice, most recognized that the patronage system under which OR scientists worked might compromise their control over the field. Starting about 1942, Tizard and others advocated techniques such as rotating personnel between OR groups and development programs. Besides acquainting OR practitioners with the latest weaponry developments, rotation increased scientists' control over OR. This and other schemes failed to materialize, however, leaving prerogative in the hands of officers.

The provincial outlook of most OR sections held the additional threat of encouraging conflicts among scientists working for different OR groups. The potential for such conflicts grew as the Western Allies gradually reclaimed the initiative from Germany during 1943. The increasing priority placed on coalition warfare led to competition among commanders, ministers, and cabinet officials over how the war should be fought and which commands and missions would receive priority. The establishment of a Supreme Headquarters for Allied Expeditionary Forces, with US Army General Dwight D. Eisenhower presiding as chief, added a fillip to this trend. The information produced by OR groups often provided grist for strategic debates, and occasionally the OR sections themselves participated in them. In an atmosphere of uncertainty, with access to intelligence limited by local circumstances, OR sections often became advocates for their patrons or partisans for their own strategic views. In any case, the ideal of objectivity was difficult if not impossible to achieve. Uncertainty created the possibility of disagreement among scientific advisory bodies, including OR sections.

The scientists themselves, however, never considered such tensions to be products of the patronage system under which they worked. Preferring to see themselves as independent partners of and equals to the officers they served, they were hard pressed to reconcile the image of objective science with emerging differences among scientists over strategic policy. Some went as far as to accuse colleagues of having political motivations or of being bad scientists. As one scientist put it, "not all scientists . . . were as questioning and independent in their judgments as they could have been."[90] Ironically, the very patronage system that gave OR its identity precluded such independence.

The Western Allies' strategic bombing campaign opens a window onto this situation. The strategic bombing program was planned and carried out amid great uncertainty surrounding its effectiveness to the Allied cause. The theory of strategic bombing, to which the RAF largely owed its status as an independent branch of the British military, promised that air power could destroy the enemy's capacity to wage war. Just what constituted legitimate and effective targets (military positions, industrial areas, worker housing, national morale), how those targets could best be attacked (by daylight "precision bombing" or more indiscriminate "area bombing"), and with what weapons (high explosives, incendiaries, either with fast or slow fuses) remained under debate for most of the war.[91] Military planners and government officials hoped that operational research groups, together with other scientific bodies—particularly the Ministry of Home Security's Department of Research and Experiments—would generate sufficient empirical data on the effects of bombing to dispel some of the uncertainty. In view of the variety and magnitude of the stakes invested in the bombing program, however, the potential for clashes remained high.

As it happened, Britain's first empirical data on the effects of bombing was extracted not from German ruins but from its own urban rubble, the result of Germany's bombing attacks. Stricken cities were visited by civilian experts from the Ministry of Home Security's Department of Research and Experiments (better known by the name of its headquarters, Princes Risborough). The Princes Risborough group included architects, engineers, physicists, physiologists, insurance actuaries, fire fighters, and others. From mountains of debris, these men painstakingly catalogued the bombs and incendiaries in the German arsenal and assessed their effects—structural damage, production stops, injuries, subdued morale.[92] In the early war years, the physicist J. D. Bernal and the zoologist Solly Zuckerman compiled reports on the attacks on Hull and Birmingham, helping to establish the Princes Risborough group as an authority on bombing effects. These reports, and the assumptions about industrial economies that underlay them, formed the basis from which the Western Allies planned their bombing campaigns against German-occupied Europe. Not until the postwar strategic bombing surveys were conducted did it become apparent that the data and the assumptions were often incomplete and misleading.[93]

In an effort to conserve bomber aircraft, Britain did not launch a sustained bombing program until the summer of 1941. That August, Churchill asked his scientific advisor, Frederick Lindemann, to undertake a study of the progress to date. Lindemann reviewed more than 650 strike photographs taken during June and July, and his conclusions disturbed the War Cabinet. He found that, of the 6103 sorties flown in the previous 2 months, only 4065 (roughly two-thirds) actually resulted in an attack on the designated target; in the remaining sorties, crews never found the target. Of the bombers that attacked their designated target, only one-third dropped their load within 5 miles of it.[94] At subsequent War Cabinet meetings, officials and officers debated the merits of the strategic bombing offensive in light of Lindemann's evidence. Put on the spot, the Bomber Command's leader, Air Chief Marshal Richard Peirse, immediately established his own Operational Research Section in order to investigate his outfit's poor performance. A. P. Rowe's radar researchers again formed the early core of the group, which focused primarily on navigation. D. G. Dickins, who had helped Tizard run the 1936 experiments at Biggin Hill, assumed leadership of the section. Peirse's conversion to OR, however, could not save him. The Air Staff relieved him of his responsibilities on February 22, 1942, preferring the no-nonsense Air Chief Marshal Arthur Travers ("Bomber") Harris.[95] Harris adopted the now-orphaned OR group, first assigning it tasks related to navigation and then broadening its assignments to cover radar operations, bombing accuracy, defensive gunnery, and other aspects of bombing missions.[96] It would seem that by 1942 patronizing an OR group held political appeal as well as technical utility.

The shakeup at Bomber Command did not end War Cabinet scrutiny, however. The following February, Lindemann continued his investigations into bombing effects. In the absence of solid German data, Lindemann decided to use that of the Princes Risborough team as a proxy. Zuckerman had not yet worked out conclusions from the collected mass of information, but under pressure from Lindemann he handed over the data. Lindemann then drew his own conclusions.[97] Most important of these was that bombing missions at night, although too inaccurate to cause significant damage to Germany's military or industrial infrastructure, might still be effective in reducing German morale. This reasoning served as a moral justification for what was later euphemistically called "area bombing," the indiscriminate bombing of built-up urban areas under cover of darkness. Churchill's subsequent support of area bombing stemmed from several sources, but Lindemann's report was undoubtedly influential.[98] Originally intended as a temporary measure until precision bombing became feasible, area bombing became the theoretical backbone of Harris's strategic operations for the rest of the war.

In April, when Zuckerman finished his report, his conclusions opposed Lindemann's. Bernal and Zuckerman had found that residents of Hull and Birmingham suffered no long-term decline in morale from the destruction visited upon them, and this undermined area bombing's raison d'être.[99] The Air Staff simply shrugged off these conclusions. But because the shift to area bombing held strategic repercussions for Britain's other two military service branches, Churchill turned over the scientist's conclusions to the army staff and the Admiralty for comment. At the Admiralty, the Director of Naval Intelligence asked Blackett, newly ensconced as the chief of naval operational research, to comment on Lindemann's argument, while the Air Staff referred their copy to Tizard. Both these men reached separate, but almost identical, estimates of the number of Germans killed by the area-bombing campaign; these estimates were far below Lindemann's. Blackett concluded that traffic accidents killed more

Germans per month than the bombing campaign.[100] Faced with this sting-
ing criticism, Lindemann revised his report, this time making housing stock
rather than German lives the objective. Again, Blackett and Tizard found
his estimates far too optimistic. Regardless, the War Cabinet continued its
support of the area-bombing program.[101] Blackett and Tizard both believed
that British bombers could be more effectively used by Coastal Command
against submarines than by Bomber Command against the German heart-
land. Tizard felt increasingly marginalized by the Air Staff after his criticism of Lindemann was ignored; he eventually withdrew from his advisory
role and returned to academic life as president of Magdalen College at
Oxford.[102] The subsequent feud that developed between Tizard and
Lindemann divided the British scientific community far into the postwar
years.[103]

Blackett would soon suffer a fate similar to Tizard's. Early in 1943, he
was given another chance to present his opinion to the War Cabinet. The
imminent delivery of highly accurate American-made microwave radar sets
had precipitated a debate over their distribution.[104] Not surprisingly, the
Air Staff and strategic bombing advocates on the War Cabinet wanted them
assigned to Bomber Command, while the Admiralty urged their installation on Coastal Command's aircraft. Perhaps Joubert, a high-ranking RAF
officer, could have strengthened the case for the Admiralty, but Joubert's
days at Coastal Command were numbered. His ardent support of anti-submarine operations had run afoul of the Air Staff, and he was replaced in
November 1942 by the more conventional and conservative Air Marshal
John Slessor, a stalwart supporter of the bomber offensive.[105] In late March
1943, Blackett drafted the Admiralty's position on the allocation of
microwave radar sets. Based on kill rates predicted by his former colleagues
at Coastal Command, he argued that a total force of 260 heavy bombers,
rigged as submarine hunters and outfitted with microwave radar, would
strike a mortal blow to the U-boats. This implied not only that the majority of the new radar should be sent to Coastal Command, but also that 190
of Bomber Command's precious aircraft should be transferred to Coastal
Command.[106] The Admiralty forwarded Blackett's discussion of the issue
directly to the War Cabinet.

In military circles, Blackett's report caused a stir. Slessor was furious
with Blackett for not having vetted the report with him before submitting

it. Describing Blackett's report as "slide rule strategy of the worst kind," Slessor aimed at discrediting the scientist. Reflecting from his retirement, Slessor carefully praised the use of scientists in the war, but added that "they must stick to their lasts."[107] Blackett may have thought it his prerogative to deny Slessor a first reading, but the air marshal steadfastly believed that the scientist had overstepped his authority and competence. In the event, the War Cabinet reached a compromise, transferring 72 of the heavy bombers to Coastal Command, supplemented by two American squadrons of B-24 Liberators. These extra planes, along with the combined military and intelligence forces of Britain and the United States, effectively defeated the U-boat threat in the North Atlantic later that autumn.[108] Like Tizard, Blackett found himself shut out from the inner circle of strategic decision making. He felt that the military elite had not only conspired against him and Tizard but had also sullied their reputations as disinterested advisors.[109] Interestingly, Blackett leveled his strongest accusation against his colleague at Bomber Command. Dickins, Blackett charged, was guilty of complicity with the air-power conspiracy through his failure to undermine area bombing. Had "a strong and trusted operational research section . . . been established at Bomber Command," Blackett later asserted, then the pro-bombing stalwarts would never have been able to shrug off Blackett's and Tizard's criticisms. Dickins, Blackett believed, had failed to stand up to his patron's mistaken assumptions.[110]

Dickins became a favorite whipping boy for those scientists who, like Blackett, questioned the wisdom of the area-bombing campaign. Another detractor was Zuckerman, who had in the meantime become the scientific advisor of Air Marshal Arthur William Tedder, the Allies' commanding air officer in North Africa and the Mediterranean.[111] Because of his experience in analyzing the effects of bombing in Britain, Zuckerman was recruited by Tedder to examine and critique Allied bombing operations in the Mediterranean. The scholarly Tedder, somewhat of an outsider among the Air Staff, got on well with the urbane zoologist. Zuckerman's studies of the bombing operations against Pantelleria and Sicily convinced Tedder that air power could be used to strategic effect if applied against railroad marshaling yards and other transportation targets, which would serve to harm both the enemy's long-term industrial capacity and his military logistics.[112] At the end of 1943, when Tedder was appointed to coordinate

Allied air plans for the impending invasion of Europe (Operation Overlord), he insisted that Zuckerman return to Britain with him and join in the planning sessions.

Nothing Zuckerman had experienced before then prepared him for the raucous debates among British and American airmen and their learned advisors as they vied for Eisenhower's favor. Despite the mountain of data that had been collected by 1944, enough uncertainty persisted to spark heated battles over the deployment of air support for Overlord. Zuckerman, whose "Transportation Plan" was far from flawless, had to defend his proposals against other advisors, who argued with some justification that the situation in the Mediterranean did not apply to Central Europe. Zuckerman later remembered: "I was visited by emissaries [of other advisory groups] . . . each of whom tried to persuade me that I was misguided in relying on the results of my analysis of the Mediterranean bomber offensive."[113] For Dickens, still at Bomber Command and supportive of continuing the area-bombing campaign, Zuckerman reserved special abuse. It was he who Zuckerman accused of not being "as questioning and independent" in his judgment as he might have been. In the end, Zuckerman's "Transportation Plan" was adopted in a modified form that accommodated certain US and RAF interests.[114]

Accusations such as Zuckerman's and Blackett's grossly oversimplified the complex web in which operational researchers and other scientific advisory groups worked. To be sure, obsequiousness was possible within the system of patronage that sustained OR, but no firm evidence has ever suggested that any scientist flagrantly abrogated professional conduct. Scientists' opprobrium against one another indicates primarily their own insecurity about their role in policy making. Patronage fostered interdependence, not independence. Expecting conclusions to be consistent across OR groups denied the variation in working conditions and the lack of standardized methodology that were facts of life for OR practitioners. In fact, standardizing their methodology might well have prevented scientists' access to the important technical policy issues they hoped to address through operational research. Instead, they interpreted OR's success as a matter of personal integrity and competency, a conclusion that underestimated the effects of local conditions and patronage.

Conclusion

The British scientific community, particularly its liberal and leftist elites, developed operational research as an interface between the research community and military policy makers. Through this interface, they hoped to control the deployment of wartime technical expertise and to influence military policy. Lacking a standardized methodology, OR scientists adapted to the peculiar conditions and personalities within local military commands. Military patronage, therefore, became the firmament of OR's early identity. The enthusiastic response by military commanders and government ministers encouraged OR's proliferation, creating the possibility that OR scientists and other scientific advisors might come into conflict once the defensive concern with tactics shifted toward an offensive preoccupation with strategy. This possibility became reality in the debates over strategic bombing.

Few scientists let their spirits be dampened by these squabbles, which they remembered as a few distasteful instances in an otherwise exciting period. Unsurprisingly, many members of the wartime elites—Blackett, Zuckerman, Bernal, and a host of others—continued to promote operational research after the war.[115] These postwar developments are beyond the scope of this essay, yet some observations suggest how research on OR's past provides historical insights into the postwar events concerning technology, science, and political culture.

In Britain, OR promoters built on their postwar success by suggesting civilian applications for the field. The new Labour government's nationalization of several major industries provided an opportunity, which several scientists (particularly Bernal) exploited. OR's early entry in civilian government, however, was rather checkered, and its adoption in industry was gradual.[116] For this reason, the center of the OR world soon shifted from Britain to the United States, where the bonds of the Cold War's iron triangle—military, industry, and the academy—continued to provide an abundant source of patronage.

Building on this support, American wartime participants promoted operational research from within military services, universities, and private nonprofit "think tanks" such as the Rand Corporation. Within 10 years these promoters constructed major conduits for OR's distribution: OR courses at major research universities, the Operations Research Society of America (supported by the National Research Council and the military OR groups), and OR's adoption by top-flight business consulting agencies.[117] Eager markets for OR emerged in government agencies, large corporations, and service organizations, particularly hospitals and philanthropies. By the early 1960s, the governments of the United States and Britain supported attempts to export OR to other nations—first to Japan and to members of the North Atlantic Treaty Organization, then to the developing world. Yet this impressive growth was not without its frictions. Operational research met with resistance, especially from social scientists (who claimed priority on some innovations) and from competing fields such as systems analysis and systems engineering.

OR's postwar proliferation and fate raises a number of issues that tie the history of technology and science to the study of social and political history, a connection Thomas Hughes has invited and fostered throughout his career. OR's spread suggests that, as the twentieth century progressed, powerful interests in Great Britain, the United States, and other industrialized nations found it attractive to represent the world in terms of interrelated technological systems whose states and processes, not to mention human users and makers, could be understood as abstract statistical and probabilistic phenomena. For some physical scientists and engineers, OR seems to have become a new metaphysics for the systems age, providing the definitive synthetic and technical overview of modern existence. Such a view naturally privileges the role of technical expertise and implies an important role for experts in policy making. Less speculatively, one may notice, as has Theodore Porter, that the tools, methods, and procedures of science and those of administration converged in the twentieth century, emphasizing greater quantification and precision.[118] Some historians of policy, politics, and political culture, notably Brian Balogh, have suggested links among the proliferation and specialization of expertise, the sustained growth of government agencies, and the fate of American liberalism in the twentieth century.[119] Undoubtedly, OR's rapid development provides some clues to this process. Since operational research did not long remain under Anglo-American stewardship, having quickly spread to societies with strong, central states (such as France), and even authoritarian regimes (such as the

Soviet Union), it affords us comparative insights into how technical expertise and other kinds of authority were reconciled and mutually reinforced within political cultures of the late twentieth century.

Notes

1. "Operational research" is the British term. In the US, the field is known as "operations research." Because this chapter treats the British origins of this field, "operational research" is used throughout.

2. On the spread of systems thinking, see *Systems, Experts, and Computers*, ed. A. Hughes and T. Hughes (MIT Press, 2000). For recent work on the spread of statistical and probabilistic thinking, see Gerd Gigerenzer et al., *The Empire of Chance* (Cambridge University Press, 1989) and Theodore M. Porter, *Trust in Numbers* (Princeton University Press, 1995). Most of the literature on OR's past has been written by practitioners or their sponsors. The British wartime experience is outlined in the following works: Air Ministry, *The Origins and Development of Operational Research in the Royal Air Force* (Her Majesty's Stationery Office, 1963); P. M. S. Blackett, "Operational Research: Recollections of Problems Studied, 1940–45," in *Brassey's Annual* (Macmillan, 1953); Freeman Dyson, *Disturbing the Universe* (Harper & Row, 1979); Harold Larnder, "The Origin of Operational Research," in *Operational Research '78*, ed. K. Haley (North-Holland, 1979); Jonathan Rosenhead, "Operational Research at the Crossroads: Cecil Gordon and the Development of Post-War OR," *Journal of the Operational Research Society* 40 (1989): 3–28; C. H. Waddington, *O.R. in World War 2* (Elek Science, 1973); Sir Robert Watson-Watt, *Three Steps to Victory* (Odhams, 1957); Solly Zuckerman, *From Apes to Warlords* (Harper & Row, 1978). American wartime founders recount their experiences in the following works: John Burchard, *QED* (Wiley/Technology Press, 1948); Charles W. McArthur, *Operations Analysis in the US Army Eighth Air Force in World War II* (American Mathematical Society, 1990); Joseph F. McCloskey, "US Operations Research in World War II," *Operations Research* 35 (1987): 910–925; Philip M. Morse, *In at the Beginnings* (MIT Press, 1977); Jaycinto Steinhardt, "Operations Research in the Navy," *US Naval Institute Proceedings* 72 (1946): 649–655; Lincoln R. Thiesmeyer and John E. Burchard, *Combat Scientists*, ed. A. Waterman (Little, Brown, 1947). Historical work on the subject is largely of recent vintage. For historians' perspective on the British story, see M. Kirby and R. Capey, "The Air Defense of Great Britain, 1920–1940: An Operational Research Perspective," *Journal of the Operational Research Society* 48 (1997): 555–568; M. Kirby and R. Capey, "The Area Bombing of Germany in World War II: An Operational Research Perspective," *Journal of the Operational Research Society* 48 (1997): 661–677; M. Fortun and S. S. Schweber, "Scientists and the Legacy of World War II: The Case of Operational Research (OR)," *Social Studies of Science* 23 (1993): 595–642. For the American story: Fortun and Schweber, "Scientists and the Legacy of World War II"; P. Mirowski, "Cyborg Agonistes: Economics Meets Operations Research in Mid-Century," *Social Studies*

of Science 29 (1999): 685–718; Andy Pickering, "Cyborg History and the World War II Regime," *Perspectives on Science* 3 (1995): 1–48; Erik P. Rau, Combat Scientists: The Emergence of Operations Research in the United States during World War II, PhD thesis, University of Pennsylvania, 1999; Robin E. Rider, "Operations Research and Game Theory: Early Connections," *History of Political Economy* 24 suppl. (1992): S225–S239; "Operational Research," in *Companion Encyclopedia of the History and Philosophy of the Mathematical Sciences*, ed. I. Grattan-Guinness (Routledge, 1994); Stephen P. Waring, *Taylorism Transformed* (University of North Carolina Press, 1991); Waring, "Cold Calculus: The Cold War and Operations Research," *Radical History Review* 63 (1995): 28–51.

3. I have found Brian Balogh's writing particularly helpful in thinking about political patronage and expertise since 1940. I believe his style of analysis transcends its American subject matter. See "Reorganizing the Organizational Synthesis: Federal-Professional Relations in Modern America," *Studies in American Political Development* 5 (1991): 119–172, *Chain Reaction* (Cambridge University Press, 1991), and the introduction to the *Journal of Policy History*'s special issue on social policy (8, 1996: 1–33).

4. For a particularly lucid account of the exportation of American expertise to Japan after World War II, see pp. 159–182 of Olivier Zunz, *Why the American Century?* (University of Chicago Press, 1998). International exchange of social policy expertise and opinion has recently become a field of interest among political and social historians. An influential work is Daniel T. Rodgers, *Atlantic Crossings* (Belknap, 1998). Most of this literature, however, focuses on the period before World War II.

5. For more on the Social Relations of Science movement, see William McGucken, *Scientists, Society, and the State* (Ohio State University Press, 1984).

6. For more on Blackett, see Sir Bernard Lovell, "Patrick Maynard Stuart Blackett," *Biographical Memoirs of Fellows of the Royal Society* 21 (1975): 1–115.

7. See Henrika Kuklick and Robert E. Kohler, eds., *Science in the Field* (University of Chicago Press, 1996).

8. See Rau, Combat Scientists, especially pp. 77–78 and chapter 2.

9. For more on the uncertainties and disputes concerning the allied strategic bombing program, see Alfred C. Mierzejewski, *The Collapse of the German War Economy, 1944–1945* (University of North Carolina Press, 1988), pp. 61–85. For an excellent overview of the Anglo-American development of the strategic bombing doctrine, see Michael S. Sherry, *The Rise of American Air Power* (Yale University Press, 1987).

10. Balogh, "Reorganizing the Organizational Synthesis," pp. 169–170; Balogh, *Chain Reaction*, pp. 221–301.

11. Watson-Watt, *Three Steps to Victory*, p. 203. Watson-Watt published a second, more expansive autobiography 2 years later: *The Pulse of Radar* (Dial, 1959). For a biographical sketch of this idiosyncratic engineer, see J. A. Ratcliffe, "Robert Alexander Watson-Watt," *Biographical Memoirs of Fellows of the Royal Society* 21 (1975), 549–568.

12. See Rau, Combat Scientists.

13. Kirby and Capey have used this argument in their accounts.

14. On science, medicine and engineering in the Nazi case, start with the following: Monika Renneburg and Mark Walker, eds., *Science, Technology, and National Socialism* (Cambridge University Press, 1994); Robert N. Proctor, *The Nazi War on Cancer* (Princeton University Press, 1999); Franz W. Seidler, *Fritz Todt* (Herbig, 1986).

15. For a recent exploration on how scientists' behavior can diverge from their highly touted communitarian ideals, see Daniel J. Kevles, *The Baltimore Case* (Norton, 1998).

16. Support for air defense could be found in such politically divergent camps as the leftist Cambridge Scientists' Anti-War Group and the conservative nationalists.

17. For a discussion of the Air Staff's strategic thinking and budgetary realities between the wars, see Neville Jones, *The Beginnings of Strategic Air Power* (Frank Cass, 1987); Barry D. Powers, *Strategy without Slide-Rule* (Croom Helm, 1976); Malcolm Smith, *British Air Strategy between the Wars* (Oxford University Press, 1984); John Terraine, *The Right of the Line* (Hodder and Stoughton, 1985). The official history of the RAF's rearmament can be found in Sir Charles Webster and Noble Frankland, *The Strategic Air Offensive against Germany, 1939–1945*, volume 1, part 2 (HMSO, 1961). During the early 1930s, the main responses of most British leaders to air attack was to advocate disarmament or deterrence. Defense was not considered a feasible option until after the development of radar. See Kirby and Capey, "The Air Defence of Great Britain," p. 557.

18. Baldwin's comment was delivered to Parliament on November 10, 1932.

19. Ronald W. Clark, *Tizard* (MIT Press, 1965), pp. 105–113.

20. For more on Tizard, Blackett, and Hill, see Clark, *Tizard*; Lovell, "Blackett"; Sir Bernard Katz, "Archibald Vivian Hill," *Biographical Memoirs of Fellows of the Royal Society* 24 (1978): 71–149. Clark provides an adequate overview of the CSSAD's early activities.

21. Clark, *Tizard*, pp. 111–112, 120–148; Thomas Wilson, *Churchill and the Prof* (Cassell, 1995), pp. 34–43. Churchill was assuaged only when his friend and scientific advisor Sir Frederick Lindemann (later Lord Cherwell) was installed on the CSSAD. This set the stage for numerous disagreements between Lindemann and the others, which eventually embroiled the entire British scientific community. For an overview and an analysis of the controversy, see D. E. H. Edgerton, "British Scientific Intellectuals and the Relations of Science, Technology, and War," in *National Military Establishments and the Advancement of Science and Technology*, ed. P. Forman and J. Sánchez-Ron (Kluwer, 1996), pp. 20–22. McGucken (*Scientists, Society, and the State*, pp. 220–225) recognizes that the bitterness stemmed not from petty jurisdictional jealousies but rather from very different ambitions regarding the organization of science for war. Lindemann was later dropped from the committee, and the ill will generated by this episode created controversy far into the postwar years.

22. Robert Buderi, *The Invention That Changed the World* (Simon and Schuster, 1996), pp. 54–59 and 64–76.

23. Ibid.

24. Ibid.

25. Kirby and Capey, "The Air Defence of Great Britain," pp. 557–558.

26. Buderi, *The Invention that Changed the World*, pp. 78–82.

27. Clark, *Tizard*, pp. 149–163.

28. Ibid.

29. Ibid.

30. Ibid. (Tizard had learned to fly during his military service in World War I)

31. Terraine, *The Right of the Line*, pp. 194–205.

32. For a description of the "Dowding System," see ibid., pp. 173–180.

33. E. C. Williams, "The Origin of the Term 'Operational Research' and the Early Development of the Military Work," *Operational Research Quarterly* 19 (1968): 111–113; Larnder, "The Origin of Operational Research," pp. 3–12; Air Ministry, *Origins and Development*, pp. 1–7.

34. Williams, "The Origin of the Term," pp. 111–113.

35. Ibid.

36. Air Ministry, *Origins and Development* , pp. 6–7.

37. Watson-Watt was promoted as the Air Ministry's Director of Communication Development, thus Rowe's boss.

38. Williams, "The Origin of the Term," pp. 111–113.

39. For a detailed description of the kind of tasks the Operational Research Section undertook, see Air Ministry, *Origins and Development* , pp. 10–32. This account is not necessarily chronological and does not provide a sense of how the nature of this work may have changed over time.

40. Such trust helped to make the (quantitative) findings of this and future OR groups acceptable and valid. For a discussion on the relationship between quantitative data, their social validity, and the creation of social institutions, see Porter, *Trust in Numbers*, pp. 33–48.

41. For more on the RAF Anti-Aircraft Command, see General Sir Frederick Pile, *Ack-Ack* (George G. Harrap, 1949). Pile served as Ack-Ack's commander. For more on the RAF Coastal Command, see John Buckley, *The RAF and Trade Defence, 1919–1945* (Keele University Press, 1995).

42. For more on the Social Relations of Science movement, see McGucken, *Scientists, Society, and the State.*

43. Ibid., p. 4.

44. The standard work on the "scientific socialists" is Gary Werskey, *The Visible College* (Holt, Rinehart and Winston, 1978).

45. McGucken, *Scientists, Society, and the State*, pp. 357 and 363.

46. McGucken (ibid., pp. 355–356, 361–362) points out that, after the Social Relations of Science movement had ensured that scientists would have some access to government decision makers, many scientists supported a second movement, Freedom in Science, in order to prevent government planning of science.

47. Pamphlets with titles such as *The Frustration of Science* (Allen & Unwin, 1935) speak for themselves. Others include *Scientific Research and Social Needs*, ed. J. Huxley (Watts, 1934); *What Is Ahead of Us?* (Macmillan, 1937); and J. G. Crowther, *The Social Relations of Science* (Macmillan, 1942). Blackett, the zoologist Solly Zuckerman, and the leftist science writer J. G. Crowther collaborated on several of these.

48. McGucken, *Scientists, Society, and the State*, pp. 225–230, 248–249.

49. Lovell, "Blackett," pp. 75–85.

50. These studies are catalogued in the finding aid to Blackett's papers deposited at the Royal Society in London. See Jeannine Alton, *Report on Correspondence and Papers of Patrick Maynard Stuart Blackett, Baron Blackett (1897–1974), nuclear physicist, 1920–1974* (Royal Commission on Historical Manuscripts, 1979). Blackett's papers are not generally open to the public.

51. Lovell, "Blackett," pp. 53–56.

52. P. M. S. Blackett, "Scientists at the Operational Level" (1941), reproduced in his *Studies of War* (Hill and Wang, 1962).

53. Pile, *Ack-Ack*, pp. 69–104.

54. Ibid., p. 114; Lovell, "Blackett," p. 56.

55. Blackett's papers are not yet open to the general public.

56. Joseph F. McCloskey, "The Beginnings of Operations Research: 1934–1941," *Operations Research* 35 (1987), p. 148.

57. Blackett, "Recollections," p. 93.

58. Blackett makes these arguments in "Scientists at the Operational Level." This memorandum, written for the Admiralty, represents Blackett's mature thinking on OR. It is widely regarded as a founding document within the OR community.

59. Blackett, "Recollections," pp. 89–90.

60. Ibid., p. 93; "Scientists at the Operational Level," pp. 171–176.

61. J. G. Crowther and R. Whiddington, *Science at War* (Philosophical Library, 1948), pp. 94–97. Crowther suggests this was set up by physicist J. D. Cockcroft. Blackett may have nudged the Ministry of Supply to contact Cockcroft.

62. Blackett, "A Note on Certain Aspects," p. 177.

63. Ibid., pp. 178–186.

64. Ibid.

65. Blackett, "Recollections," p. 93.

66. Ibid., p. 93.

67. Ibid., pp. 88–92.

68. Ibid., pp. 90–91.

69. See O. H. Wansbrough-Jones's contribution to "Operational Research in War and Peace," *Advancement of Science* 4 (1948), pp. 321–322. This is part of a post-mortem on British scientists' war experiences hosted by the British Association for the Advancement of Science held in Dundee in the summer of 1947.

70. Pile, *Ack-Ack*, p. 114.

71. Buckley, *The RAF and Trade Defence*, p. 165. On the war against the U-boat, compare the official history, Captain S. W. Roskill, *The War at Sea, 1939–1945* (HMSO, 1954), with the recent revisionist account by Clay Blair, *Hitler's U-Boat War* (Random House, 1996 and 1998).

72. Waddington, *O.R. in World War 2*, pp. 213–214.

73. Buckley, *The RAF and Trade Defence*, pp. 165–173.

74. Ibid.; Air Ministry, *Origins and Development*, p. 75.

75. Air Ministry, *Origins and Development*, p. 74

76. Buckley, *The RAF and Trade Defence*, p. 172.

77. I have found no evidence suggesting that Blackett bothered to give a name to his activities. In any event, he appears to have been reluctant look upon OR as a new discipline. Although he supported the founding of the Operational Research club and its successor after the war, he regarded OR as an interdisciplinary activity and only reluctantly supported the trend toward professionalization. For early evidence of his attitudes, see P. M. S. Blackett, "Operational Research," *Operational Research Quarterly* 1 (1950): 3–6.

78. For the present purposes, these tasks are addressed adequately by Waddington.

79. Blackett, *Studies of War*, pp. 169–170.

80. On the spread of operations research in the US, see Rau, Combat Scientists.

81. See Lovell, "Blackett," pp. 75–85.

82. Ibid., p. 60.

83. Rau, Combat Scientists, pp. 101–102. See also Joseph F. McCloskey, "British Operational Research in World War II," *Operations Research* 35 (1987), p. 463.

84. This is made clear in Air Ministry, *Origins and Development*.

85. Air Ministry, *Origins and Development* , pp. 125–126, 155–177.

86. Quoted in Rau, Combat Scientists, p. 101.

87. See P. M. S. Blackett, "A Note on Certain Aspects of the Methodology of Operational Research" (1943), reproduced in *Studies of War*, pp. 176–198.

88. See Edgerton, "British Scientific Intellectuals."

89. Dyson, *Disturbing the Universe*, pp. 19–32.

90. Zuckerman, *From Apes to Warlords*, p. 234.

91. See Mierzejewski, *The Collapse of the German War Economy*; Sherry, *The Rise of American Air Power*.

92. For more on some of the activities at Princes Risborogh, see Zuckerman, *From Apes to Warlords*, pp. 141, 242. For an American perspective on the doings at Princes Risborogh, as well as an overview of corresponding American activities, see J. Burchard, ed., *Rockets, Guns and Target* (Little, Brown, 1948).

93. Mierzejewski (*The Collapse of the German War Economy*, pp. 61–85), better than most writers, conveys the enormous complexity confronting the allied commanders as they struggled to reach a consensus on the air war.

94. Terraine, *Right of the Line*, pp. 292–298, 459–460; Sir Charles Webster and Noble Frankland, *The Strategic Air Offensive against Germany, 1939–1945* (HMSO, 1961), volume 1, pp. 178–188.

95. Terraine, *Right of the Line*, p. 460; Air Ministry, *Origins and Development* , pp. 43–46. The Air Ministry history details the tasks the Bomber Command OR Section undertook and confirms that its origin in August (when the Butt report was submitted) coincides with the command's heightened interest in navigational aids.

96. For brief summaries of Bomber Command ORS activities, see Air Ministry, *Origins and Development* , pp. 46–73.

97. Wilson, *Churchill and the Prof*, pp. 73–75.

98. Ibid.

99. Zuckerman, *From Apes to Warlords*, p. 144.

100. Blackett, "Recollections," pp. 100–101; *Studies of War*, pp. 109–110.

101. Zuckerman, *From Apes to Warlords*, pp. 141–144.

102. Clark, *Tizard*, pp. 311–316.

103. For an analysis of this debate's conflation of morality and "good" vs. "bad" science, see Edgerton, "British Scientific Intellectuals."

104. Terraine, *Right of the Line*, pp. 436–438. For more on the American development of radar, see Buderi, *The Invention that Changed the World*. The official history is Henry E. Guerlac, *Radar in World War II* (Tomash, 1987).

105. Philip Joubert de la Ferté, *The Fated Sky* (Hutchinson, 1952), pp. 238–239; Buckley, *The RAF and Trade Defence*, pp. 125–126. Slessor's autobiography, covering this period, is Sir John Slessor, *The Central Blue* (Cassell, 1956).

106. Lovell, "Blackett," pp. 62–63.

107. Slessor, *The Central Blue*, p. 525.

108. Terraine, *Right of the Line*, pp. 440–452.

109. Blackett, *Recollections*, p. 102. In this brief essay of his wartime OR activities, Blackett recalls this episode with a note of bitterness.

110. Blackett, *Studies of War*, p. 111.

111. For more on Tedder, see Terraine, *Right of the Line*, p. 339. For evidence of Tedder's admiration for Zuckerman in his own words, see Lord Tedder, *With Prejudice* (Little, Brown, 1966). For Zuckerman's account of their relationship see Zuckerman, *From Apes to Warlords*, pp. 166–215.

112. Zuckerman, *From Apes to Warlords*, pp. 210–211. For a retrospective on the relative strengths and weaknesses of Zuckerman's analysis, see Mierzejewski, *The Collapse of the German War Economy*, p. 180.

113. Zuckerman, *From Apes to Warlords*, p. 227.

114. Mierzejewski, *The Collapse of the German War Economy*, pp. 101–102.

115. The Operational Research Club was begun immediately after the war, eventually morphing into a full-fledged professional society, the Operational Research Society. Blackett, who presented a paper at the society's inauguration, expressed unease with operational research as a discipline. For him and others among the first generation of practitioners, OR served primarily to rationalize policy making and planning. Blackett and his colleagues continued to call for the social engagement of scientists. As noted above, Blackett served the Wilson government and, having earned a peerage for his wartime activities, spoke out on the issues of international development and science policy. See Lovell, *Tizard*, pp. 105–106.

116. Rosenhead, "Operational Research at the Crossroads," pp. 24–26.

117. Operational research courses, aimed at graduate students in established sciences, were first taught at MIT, at Johns Hopkins, and at the Case Institute of Technology. The first two examples grew out of the administrative support that MIT and Johns Hopkins offered the navy and the army, respectively, which manifested itself in the Operations Evaluation Group (Navy) and the Operations Research Office (Army). Not to be outdone, the Air Force formed the Rand Corporation in addition to its in-house Operations Analysis Division even before the service had gained independence from the army. In the hope of coordinating military expenditure on research and development, the Secretary of Defense decided that the Joint Chiefs of Staff should have their own OR group, the Weapons Systems Evaluation Group. The same scientists who encouraged these developments coaxed the National Research Council into forming a Committee on Operations Research, which culminated in the creation of the Operations Research Society of America. They also encouraged the adoption of OR at consulting firms such as Arthur D. Little, MacKenzie and Company, and Booz, Allen, Hamilton.

118. See Porter, *Trust in Numbers*; *The Values of Precision*, ed. M. Wise (Princeton University Press, 1995).

119. Balogh, "Reorganizing the Organizational Synthesis," pp. 70–172; Balogh, introduction to *Journal of Policy History* 8 (1996), pp. 23–28.

Technology, Politics, and National Identity in France

Gabrielle Hecht

What is national identity? How does it form and reform? Why does it matter? For some time now, such questions have preoccupied cultural theorists—especially historians of modern France, who have examined how "the nation" was invoked in the French provinces in the pursuit of local economic and political interests, shown how diverse groups throughout the country imagined themselves as national citizens, and explored the role of national identity talk in debates over issues such as Americanization, modernization, immigration, and colonization.[1] By and large, however, French historians have not considered how the discourses and practices of technological change intervened in conceptions, meanings, and uses of French national identity.

This absence is particularly glaring—and curious—for the post-World War II period. France witnessed profound technological changes after the war: the construction of massive hydroelectric and nuclear power plants, the development of supersonic jets, the growth of the telephone and railway networks, the spread of domestic appliances, and much more. These changes were accompanied, and often fueled, by widespread debates over the meaning of Frenchness. The changes and the debates came at a time when concern about France's greatness—a concern driven by decolonization and fears of Americanization—penetrated all manner of political and cultural debate. Yet most of the scholarship that addresses French national identity remains focused on how pre-twentieth-century symbols and events become reconstituted in French memory.[2] Through its choice of subject matter, the very scholarship that seeks to demonstrate the continually contested and renegotiated nature of Frenchness has paradoxically ended up locating the essence of French culture in châteaux, monuments, literature, and revolutionary events.

This inadvertent essentializing, I submit, is but one manifestation of broader divisions in the humanities and the social sciences that go well beyond French history: namely, divisions between material and cultural approaches to historical and social change. This essay proposes one way of transcending the material/cultural divide by demonstrating how intersections between technology and national identity are crucial to understanding postwar France. Technologies served as important symbols for national identity after the war. This much cultural historians have conceded, if not seriously investigated. More important, though, the relationship between technological change and the meanings of national identity went both ways. Technologies did not merely have symbolic importance; they were themselves the outcomes of cultural processes in which ideas about national identity played an important role. Understanding the metaphorical and physical (re)building of the French nation requires simultaneous attention to cultural discourse and technological practices.

Whereas cultural historians have been guilty of ignoring technology, historians of technology have been guilty of putting culture in a black box. In practice, this has meant using culture as an *explanation* for technological change rather than a *process* that itself requires explanation. Historians of technology have tended to treat national identity as an unproblematic, second-order phenomenon. In such treatments, expressions of national identity might justify large-scale technological development, but they do not shape that development. They are hood ornaments, not engines. The connection between national identity and technological development thus appears to be "merely" symbolic. Ironically enough, such a conclusion does not differ markedly from that reached by cultural historians

Interdisciplinary dialogue thus has much to contribute to theoretical frameworks in both the history of technology and cultural history. In this essay I will demonstrate, through a series of three vignettes, how these two disciplines, broadly construed, might contribute to each other. Following the principle of close empirical investigation that lies at the core of this volume, I will focus on the mutual construction of nuclear power and national identity in France. Through the deployment of a number of theoretical concepts, however, I also propose strategies of analyzing cultural and technological processes *simultaneously*—strategies which I hope will benefit those who pursue other topics. Before proceeding to my empirical material, therefore, let me first discuss the concepts that constitute my analytic tools.

National Identity

My conception of "national identity" is inspired by a broad range of scholarship on nationhood and nationalism. I ground my treatment in Benedict Anderson's classic formulation of the nation as an "imagined community." At the most basic level, this means that nations are not autochthonous social units but rather communities whose coherence is imagined through political and cultural practices. The content and the function of these imaginings vary according to time and place. However stable a sense of nationhood may appear, national identity is in fact continually subject to negotiation and contestation. Ideas about national identity do not grow by themselves. They must be actively cultivated in order to persist. Further, articulating and rehearsing these ideas often reformulates them.[3] So, I argue, does grounding these ideas in the material realities of technological systems.

Discussions of national identity typically refer back to the past. But ultimately, national-identity discourse is not about the past per se or even about the present. It is about the future. National-identity discourse constructs a bridge between a mythologized past and a coveted future.[4] Nations and their supposedly essential characteristics are imagined through a telos, in which the future appears as the inevitable fulfillment of a historically legitimated destiny. This process naturalizes change; it makes proposed novelties appear to be logical outgrowths of past achievements. Proponents of large-scale technological systems might thus justify modernization by placing their systems in a direct historical lineage with past national achievements, thereby giving their technologies political and/or cultural legitimacy. Invocations of national identity are thus not gratuitous acts: this is one reason why historians of technology must take them seriously. Deliberately or not, people usually invoke the nation to perform political, cultural, and sometimes even technological work.

Historicizing the Relationship between Technology and Politics

A common site for the invocation and negotiation of national identity is in the discursive formulation of the relationship between technology and politics. My next analytic concept, therefore, is best formulated as a question: How do the historical actors whom we study themselves conceptualize this relationship?

Historians have put great effort into examining the ontology of the relationships between technology and politics. Sociologists have argued that we cannot decide ahead of time what counts as technology and what counts as society, insofar as these categories emerge from, rather than precede, the construction of an artifact or system.[5] But these scholarly efforts—and debates over technological determinism more generally—can overlook an important dimension of the story they seek to tell. Historical actors themselves *do* have a priori ideas about the nature and relationship of technology and politics (or society, or culture). Their beliefs—be these beliefs in technological determinism or more complex ideas about how technology and politics relate—shape their actions and decisions. We must therefore ask how engineers (and workers, though I do not discuss them in this essay[6]) themselves conceptualized such relationships, and we must explore what is at stake in those conceptualizations. Thus, for example, French state technologists did not conceive of technology as something radically separate from politics (or culture, for that matter). Quite the contrary: many of them saw technology as a thoroughly political entity. Comprehending the reasons behind and the manifestations of this view is absolutely crucial for understanding the shape of the nuclear program and the political behavior of the technologists who built it.

I do not mean to deny that we should seek our own understanding of these relationships. Of course we must. But in doing so, we cannot simply dismiss the conceptualizations of historical actors.

Technopolitics

The above considerations lead to the next theoretical concept, a notion that I have termed "technopolitics." I use this term to refer to the strategic practice of designing or using technology to constitute, embody, or enact political goals. Here I define technology broadly to include artifacts as well as nonphysical, systematic means of making or doing things. Two examples of technopolitics are nuclear reactors designed with the express goal of creating and implementing military atomic policy and optimization studies aimed at shaping industrial policy. From the very beginning, engineers and administrators consciously conceived of these reactors and these optimization studies as hybrids of technology and politics. Many of the criteria that shaped their technical choices were deliberately political. Calling these hybrids "politically constructed technologies" is correct; however, it is not sufficient, because technologists intended them as tools in political negotiations. At the same time, these technologies are not, in and of themselves, technopolitics. Rather, the *practice* of using them in political processes and/or toward political aims constitutes technopolitics.

Why not just call that practice "politics"? The answer lies in the material reality of the technologies. These technologies cannot be *reduced* to politics. The effectiveness of technologies as objects designed to accomplish real material purposes *matters*—among many other reasons, because the material effectiveness of technologies can affect their political effectiveness. It is precisely the technological aspect of these hybrids that shapes the *kind* of political voice that technologists have—that shapes, in other words, the nature of expert authority. (Other factors shape that voice too, of course—especially educational background, institutional provenance, and sociopolitical hierarchies.) Technologists did not participate in French political life as members of a party, or thanks to their clever way with words. They participated because they engaged in, or supervised, or organized the design of material artifacts. Their skills differentiated them from ordinary politicians and contributed greatly to their authority and their influence. For all these reasons, "politics" does not fully capture either the nature or the power of the strategic practices in which experts engaged.

Technopolitical Regimes

The final theoretical concept I develop here is that of technopolitical regimes. These regimes are grounded in institutions, and they consist of linked sets of people, engineering and industrial practices, technological artifacts, political programs, and institutional ideologies which act together to govern technological development and pursue technopolitics. This concept is anchored in Thomas Hughes's notions of technological system and technological style. A technological system is a linked network of artifacts, knowledges, and institutions operating in a coordinated fashion toward a series of specified material goals.[7] Thus the French nuclear program is a technological system whose components include state agencies, private companies, reactors, laboratories, uranium mines, university curricula, factories, and portions of the electricity distribution network. The technopolitical regimes that I examine operate *within* this system. They emanate from different institutions, and they have distinct (though sometimes overlapping) goals and ideologies. For convenience, I have labeled the regime based in the Commissariat à l'Énergie Atomique the *nationalist* regime and the one based in Électricité de France the *nationalized* regime. (These labels are associated with institutional stereotypes, and I will try to characterize the regimes more subtly in my analyses.) Both regimes sought to shape the French nuclear system.

I have chosen the "regime" metaphor for three reasons. The first reason relates to the use of "regime" in political parlance to refer at once to the people who govern, to their ideologies, and to the various means through which they exert power. By analogy, "technopolitical regime" is a good shorthand for the institutions, the people who run them, their guiding myths and ideologies, the artifacts they produce, and the technopolitics they pursue. The term aims both to evoke the similarity with political regimes and to convey the difference that technology makes. Second, "regime" conveys the idea of *regimen* or *prescription*. The regimes I examine aim, through the pursuit of technopolitics, to prescribe not just policies and practices but also broader visions of the sociopolitical order.[8] Third, "regime" captures the contested nature of power. The two technopolitical regimes I examine aimed at governing nuclear development at a national level and at governing specific technological practices at an institutional level. But these regimes were not uncontested. Just as national political regimes (democratic or otherwise) must grapple with opposition, these technopolitical regimes had to contend with varying forms of dissent or resistance, both from outside and from within the institutions they governed. These regimes were neither static nor permanent: a technopolitical regime is easier to topple than the technological system within which it operates.

I will now turn to my empirical material in order to suggest ways in which these concepts and analytic tools can be used to understand the simultaneous construction of technology, politics, and national identity. The first and briefest vignette shows how elite state experts or "technologists" conceptualized the relationship between technology and politics. The second vignette describes the two technopolitical regimes of the nuclear program and shows how each regime built reactors as forms of technopolitics and manifestations of national identity. The final and longest vignette explores the "war of the systems," a prolonged battle over the future of France's nuclear program, in order to demonstrate how technopolitics both complicates and elucidates the construction of national identity.

Technologists and the Politics of National Identity

World War II led many French elites to rethink the role of the state in directing the economy (in general) and industrial, scientific, and technological development (in particular). Investing in and guiding the modernization of French industry would help the state revive the economy and restore the country to greatness. Several state institutions were created to accomplish these goals, including the Commissariat à l'Énergie Atomique (CEA) and the nationalized electric utility company Électricité de France (EDF). A growing class of state experts took charge of these institutions. From their positions of leadership, these engineers, economists, and professional administrators vigorously forged their visions of a French technological identity.

Fundamentally, these processes involved an effort to define the relationship between technology and politics. This effort had long historical roots, from the French revolution to the technocracy movements in the 1920s and the 1930s.[9] In the postwar period, this effort constituted one of the mechanisms through which experts sought to participate in state politics and in debates on the restitution of French greatness. On one front, technologists tried to articulate an identity for themselves that opposed the one they constructed for politicians. More significantly, technologists sought to represent their creations as essentially French, a representation they pursued by seeking to efface boundaries between technology and politics. Let me briefly outline the first strategy, then examine the second one in more detail.

An essential component of the technologists' identity discourse involved characterizing politicians as corrupt, dishonest, and ineffective. Consider this typical remark, made by an alumnus of the École Polytechnique in a speech titled "The Cardinal Virtues of the High-Class Engineer":

Do we want the best of our engineers to participate, through their acts, their pens, and their words, in making our economy healthy again? Well then! We must immunize them, from their very first steps, against the disease known as political thinking of which Machiavelli was the champion. . . . The Florentine did not hesitate to claim that the individual should sacrifice to the State not just his fortune and his life, which is just fine, but also his honesty, which is detestable. . . . In the twentieth century, . . . a democracy cannot accommodate long-standing deception.[10]

As had been the case since at least as far back as the French Revolution, "politicians" who practiced corrupt machinations were thus the "other" in technologists' identity discourse.

But at a deeper level lay the implication that, because of their values and knowledge, technologists were ultimately better equipped to pursue political activities than politicians were. At this level, politics had to be stripped of connotations of ideology and corruption and had to take on a broader meaning that could encompass some of what technologists did and should do. In this redefinition, technologists sought to efface what they claimed was an outdated boundary between technology and politics.

One important site for the effacement of the alleged boundary between technology and politics was the discourse about the role of technology in the future of France. This discourse attempted both to describe what a future technological France would look like and to define a specifically French form of technological and industrial development. Invoking national identity was thus essential to blurring the boundary between technology and politics.

The fundamental premise of discussions about a future technological France was that in the postwar world technological achievements defined geopolitical power. But this did not mean that technological pursuits all over the world were identical. Indeed, one of the great dangers of adopting this standard was the loss of cultural specificity. France's leading technologists worried about this, declaring in 1965: "From now on our presence in the world depends on our ability to imprint our mark on this civilization by means of significant contributions from French technology and French science."[11] Such language fit into a whole discourse about *le rayonnement de la France* (the radiance of France). "Radiance" was synonymous with the grandeur of France; the notion referred back to glorious days past, invoking Louis XIV, Napoleon, and the heyday of French imperialism. French "radiance" was precisely what postwar geopolitics and decolonization threatened, exactly what President Charles de Gaulle sought to reestablish throughout the 1960s. Even in colonial relations, technologists could achieve what eluded politicians. Writing just a year after the Algerian crisis that had brought de Gaulle to power, one civil servant remarked: "In 1958, destiny knocked on the door. . . . In came the technocrats who would build the Franco-African industrial community. In them, the science of engineers is united with the will of captains. The new French strategy, French peace, will be brought to the world."[12]

It might be tempting to take proclamations about French "radiance" as empty promotional or nationalist rhetoric, but this discourse achieved much more than that. For technologists, it formed part of a very deliberate strategy to colonize national politics. By repeatedly associating their technological achievements with the "radiance" of France, technologists sought to position themselves as legitimate actors in a political space. Their ability to define and represent French culture could potentially justify other excursions into political life.

This strategy entailed identifying the Frenchness of French technology. In doing this, technologists often appealed to a general sense of tradition and history. Demonstrating a historical continuity between past and present provided reassurance that technological culture need not imply the loss of a French essence. For example, one group of technologists wrote that "the beautiful" was "a traditional export of France." Cultivating the esthetic dimension of industrial projects would not only define their Frenchness but also enhance national prestige:

> . . . beauty . . . brings prestige because it represents a considerable attraction. . . . The CEA's installations, which receive numerous foreign visitors, would certainly gain nothing by being hideous, and the esthetic of certain nuclear reactors, whose cost is insignificant with respect to [the cost of the] equipment, does more for the radiance of France than would ten times as many millions spent on propaganda."[13]

History—incarnated by great men and greater achievements—was another important tool. A 1957 promotional film[14] about French industrial achievements opened with shots of Versailles and the words of Jean-Baptiste Colbert to Louis XIV: "Sire, the grandeur of a State rests on its arts and manufactures." A 40-minute overview of technological France followed. The film closed with a view of two nuclear reactors under construction. The voice-over spoke of "the latest great accomplishment of the century of the atom, the future's answer"—at this point the film switched to a shot of the Eiffel Tower—"this other great symbol of French industrial grandeur outlined in the sky of Paris." The relationship between modernity and history exemplified in such attempts to create a new cultural iconography worked both ways. According to one prominent engineer, "all that subsists in France in the way of tradition . . . will only have real value, will only be able to shine, if the nation as a whole is solidly of our time." Hence, just as history and tradition were necessary in order to make French technology truly French, modernity and technology were necessary in order to make France truly France.

Engineers and planners thus deliberately sought to represent their objects as political and cultural artifacts. They did not argue that their technological achievements were universal, inevitable products of worldwide industrial logic. On the contrary, they aimed at contributing to the national debate about the future of France and the preservation and modernization of French culture. Ignoring this debate would have marginalized them. By contributing something new and demonstrably "French," they carved out a central position in that debate. In this manner, defining a relationship between technology and national identity formed the basis of a political strategy for technologists.

This strategy was not confined to the level of discourse. Technologists also invoked national identity in the process of building large-scale technological systems. The next vignette discusses the conflation of technology and politics in the origins of the French nuclear power program.

Technopolitical Regimes and Technopolitics

In the 1950s and the 1960s, the French nuclear power program was jointly run by the CEA and EDF. The CEA directed nuclear research and development; EDF aimed to develop a national and unified power generation system. Institutional goals thus overlapped when it came to developing nuclear power plants.

In some respects the two institutions had much in common. Both had been created immediately after the war for the purpose of modernizing France's scientific and technological infrastructure. Both were state institutions, directed by elite state technologists, with a mission to uphold the national interest. Both were committed to state-directed technological development, and more specifically to articulating a vision of France's future and national identity that relied on technological prowess. In both, engineers and managers sought to direct national policy through their technological work.

Despite these similarities, however, technologists in each institution had different visions of what France's future should look like and of how that future should be attained. In order to characterize these differences, I argue that distinct technopolitical regimes evolved within each institution.

The CEA was governed by what I call a nationalist technopolitical regime.[15] Created in 1945, the CEA was originally headed by a Communist physicist and a Gaullist administrator; its leaders reported directly to the prime minister. Communists and Gaullists had both been instrumental in the Resistance, and that gave them considerable moral legitimacy after the war. But the Cold War ensured that this shaky partnership did not last long: by the early 1950s, many Communists had been purged from the CEA. The new leaders of the CEA were Gaullist nationalists who emphasized national independence through military might.

This was a tricky position to maintain in the 1950s, a decade of considerable political instability in France. Concerned with rebuilding the nation's economy, successive governments repeatedly maintained that France was interested only in peaceful applications of nuclear power. The new leaders of the CEA, however, believed that France should develop military nuclear capability, and that their job, as guardians of the French national interest, was to implement this goal. These convictions shaped their decision in the early 1950s to pursue gas-graphite reactors rather than another type.

In choosing which type of reactor to develop in the early 1950s, the CEA's leaders first had to choose a reactor fuel: natural uranium or enriched uranium. They chose natural uranium for two reasons. First, France and its African colonies had ample natural uranium supplies. Building an enrichment plant was beyond their financial means; buying enriched uranium from the United States would have contradicted the goal of national independence. Second, fission reactions in natural uranium produce, as a byproduct, plutonium that can be used in atomic bombs. Having chosen natural uranium, the CEA's leaders selected graphite as the moderator and carbon dioxide gas as the coolant.

The decision to develop gas-graphite reactors was immediately followed by the appointment of Pierre Guillaumat (a Gaullist and a graduate of the École des Mines) to the CEA's top administrative post. Together, these events signaled the emergence of a Gaullist nationalist technopolitical regime in the CEA. The CEA's new regime upheld a vision of nationalism that excluded Communists and valued national grandeur and military technological prowess. It also valued institutional autonomy and nuclear expertise.

In subtle contrast, EDF developed a nationalized technopolitical regime. EDF emerged in 1946 as the product of a consensus between state engineers, labor unions, and representatives of a wide range of political parties. These men agreed that the new France should have a standardized electrical network run by one public utility. After heated negotiations, they agreed to create a nationalized institution in which labor representatives would participate in management.[16]

Despite the existence of diverging ideological factions within EDF, most could agree on a few basic issues. Foremost among these was EDF's mission: to make France energy independent by producing and distributing the most electricity at the least cost. More generally, the electric distribution network united France symbolically as well as technologically: complete electrification would enable all French citizens to participate in the modernization of their nation.[17] Finally, although the implementation of worker participation provoked some disagreement, few denied its value in principle. Nationalization thus made room for everyone at EDF to embrace an ethos of public service. Working for EDF meant apprehending and serving the entire nation through the production and distribution of electricity. These values, and the system of producing and distributing electricity in which they were embedded, drove EDF's technopolitical regime within the nuclear program.

Thus, two technopolitical regimes guided the development of the nuclear program. These regimes were closely tied to broader political visions of how France should modernize—visions that both came from and guided other arenas of state-directed development. In principle, both the CEA and EDF reported to higher state authorities, and indeed the ministries responsible for overseeing these institutions had to approve all major policy decisions. In practice, however, the institutions also had a great deal of autonomy. The technopolitics produced by each institution constituted this autonomy and determined the degree to which each institution shaped national policy, rather than the other way around.

Though distinct, the two technopolitical regimes in question cannot be understood as radically opposed: although each regime developed its own reactor site, limited resources forced them to collaborate on both sites. This collaboration was fraught with conflict: each regime sought to mold its reactors into components of its technopolitics. Reactor design and industrial contracting reveals two dimensions of these technopolitics. The first involved French industrial policy, the second national nuclear policy. I will briefly address each in turn by comparing two reactor projects: the G2 reactor, developed by the CEA at its Marcoule site beginning in 1953, and the EDF1 reactor, developed by EDF at its Chinon site starting in 1955.

Both regimes considered the question of the relationship between themselves (as embodiments of the state) and private industry to be of paramount importance. Engineers in each institution produced the overall design of their reactors, but construction had to be contracted to private manufacturers. How should these contracting relationships be structured? Leaders in both regimes believed these relationships had prescriptive potential: the prestige of nuclear development provided the opportunity to set an example for other large-scale industrial endeavors, and thus to sketch out the nation's industrial future. More was at stake, therefore, than the reactor projects.

The CEA's leaders argued that French companies should not waste time or resources competing against one another. In order to stand up to increasingly large foreign corporations, French industry needed to consolidate its resources and develop its strengths. Accordingly, they espoused what they called a "policy of champions."[18] This involved hand-picking a single company to design each major reactor component, without issuing a request for bids. Out of these industrial "champions," one company would be selected to act as the general contractor for the project. Building G2 wouldn't make the "champions" rich, but it would give them know-how, confidence, and prestige, which would help later in obtaining foreign contracts. This policy would enhance France's industrial base in the short term and its economy in the long term. The "policy of champions" hence guided the contracting of G2's construction and constituted one prescriptive dimension of the CEA's technopolitical regime.

When it came time to collaborate on the EDF1 reactor, the CEA's leaders wanted EDF to pursue the same "policy of champions." But the engineers who headed the EDF1 team had different ideas. They espoused the anti-capitalist sentiment that had spawned nationalization. By building EDF1, they were providing a public service. They should therefore optimize the cost and efficiency of the reactor themselves, rather than seek to promote specific private companies. The best way to keep costs down was to divide the reactor into components and request bids for each one. EDF engineers would direct the basic design of each component; private companies would function as competing suppliers, rather than as conceptualizers. In EDF's technopolitical regime, the utility would direct nuclear power development in the best interests of the state; private industry would merely follow orders. Overriding the objections of the CEA's furious leaders, EDF engineers pursued this method in issuing contracts for their EDF1 reactor.[19]

What kind of reactor did each regime produce, and how were these technopolitics?

Figure 1 schematizes the CEA's G2 reactor at Marcoule. The core, contained in a large cylinder, was made up of a stack of graphite bars and uranium fuel rods. When enough rods were loaded into the reactor, the core went critical, setting off a self-sustaining fission reaction. This fission liberated a great deal of heat. Carbon dioxide gas flowing through the core absorbed this heat and then traveled to the "energy recuperation installation," where the heat was converted into electricity.

When talking to prime ministers and other government officials, the CEA's leaders emphasized the flexibility of G2's design. They acknowledged that the reactor could produce plutonium, but they emphasized that it was a prototype for an electricity-generating reactor. France, after all, was not supposed to be developing military nuclear capability. A quick look at two aspects of G2's design, however, reveals how its electricity-generating function was relegated to secondary status.

Consider first the "energy recuperation installation." The heat exchangers stood outside the building that housed the reactor. In order to reach the exchangers, the coolant had to travel through many meters of heat-dissipating pipes, thereby losing a significant amount of the energy generated by the fission reaction. Clearly, this was not an optimal way to generate electricity from nuclear power.

Consider next the reactor's fuel loading mechanism (figure 2). Recall that plutonium is a by-product of fission in natural uranium. In order to get *weapons-grade* plutonium, though, operators had to remove the fuel rods

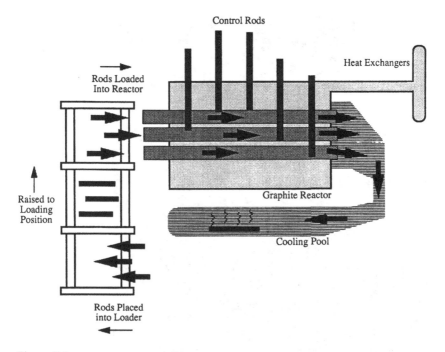

Figure 7.1
Schematic diagram of G2 (not to scale). Source: *Bulletin d'Informations Scientifiques et Techniques du CEA,* no. 20 (1958). (drawing by Carlos Martín)

at the right time. If they waited too long, the fission reaction would "poison" the plutonium, making it unfit for use in a bomb. To ensure that the plutonium in the core would be of weapons grade, and to get enough plutonium for a bomb within the next few years, they needed a mechanism to remove the fuel rods rapidly. Had G2 been primarily a prototype for an electricity-producing reactor, engineers would have sought to leave the fuel in the core as long as possible, in order to extract as much heat as possible. But such was not the CEA's goal. Engineers therefore chose a continuous loading system, which could remove and replace fuel rods while the fission reaction was taking place and thereby expedite the production of weapons-grade plutonium.[20] Thus, although G2 may have appeared to be a prototype for an electricity-generating reactor, it was in fact optimized for producing weapons-grade plutonium.

In contrast, EDF's engineers optimized EDF1 (schematized in figure 3) for electricity generation. EDF didn't have the expertise to design a different

Figure 7.2
The loading machine for the G2 reactor. (CEA/MAH/Jahan)

kind of reactor, so EDF1 had some basic features in common with G2: it ran on natural uranium, was moderated by graphite, and was cooled by carbon dioxide gas. But EDF1's design looked quite different from G2's. Most obviously, the heat exchangers were right next to the pressure vessel that contained the core (rather than many meters of energy-losing pipes away), and they were inside the reactor building (rather than outside). Further, fuel loading took place while the reactor was stopped. Unlike the CEA, EDF wanted to burn up the fuel rods as much as possible in order to extract the maximum

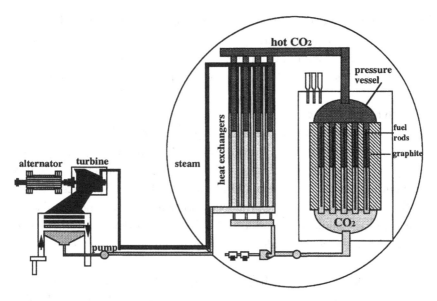

Figure 7.3
A schematic diagram of EDF1 (not to scale). (drawing by Jay Slagle; source: EDF,
Rapport de sûreté Chinon A1, 1980)

amount of heat. They did not want to incur the extra expense for a device
that could move rods through the reactor quickly. They also hoped that a
loading device that could be used only when the reactor was off line would
limit how much weapons-grade plutonium the CEA might later demand
from EDF1.[21] EDF engineers thus produced a design they felt would make
most efficient use of both fuel rods and investments and would be sufficiently
simple to provide a good basis for future reactors.[22] They had designed a
reactor whose performance and capabilities reflected their regime. They
hoped that in future collaborations the CEA would have to work with their
parameters.

For both regimes, the nuclear program provided a means of recasting
French national identity in technological form. The language with which
engineers described their accomplishments, both to engineers in other industrial sectors and to the general public, makes this clear. Marcoule's director,
for example, repeatedly noted that "the Arc de Triomphe . . . would easily
fit in the vast metallic structure that shelters . . . G2."[23] The prestige that
Marcoule derived from being associated with historical monuments was

transitive: as a great and uniquely French achievement, Marcoule would embody and strengthen French greatness. EDF engineers used similar language to discuss Chinon, emphasizing that building nuclear reactors fulfilled their mission of public service to the French state and the French people.[24]

Both technopolitical regimes thus aimed at tightening the links between technological (nuclear) prowess and national identity. Both regimes proposed visions of France's political and industrial future and, through this future, France's identity. They did so not just with their language but also by building reactors that were hybrids of technology and politics. G2 and EDF1 were neither inevitable products of some inherent technological logic nor infinitely malleable products of political negotiation. Rather, each resulted from a seamless blend of political and technological goals and practices.

The two reactors can thus be understood as technopolitics. CEA and EDF technologists deliberately—even proudly—sought to make their technologies instruments and embodiments of politics. Politics and policy making gave the reactor projects significance, both within each regime and in the interactions each had with its surroundings. Thus, for example, EDF1 was important not because it would actually produce economically viable electricity (EDF's first truly commercially viable reactor, EDF4, went on line in 1969) but because it represented the first step in a nationalized nuclear program that would enact and strengthen the utility's ideology and its industrial contracting practices. At the same time, the technological form of their politics mattered. French military nuclear policy in the 1950s did not take the form of a ministerial decree; it took the shape of G2 (and other related technologies). This meant, for example, that in 1954 the assurances of the CEA's leaders that Marcoule reactors were both electricity-generating prototypes and plutonium producers enabled one prime minister to abstain from deciding about a bomb. Official French state policy declared a purely peaceful interest in atomic energy, while the CEA's actual technopolitics performed a military nuclear policy.

The CEA's nationalist technopolitical regime found form in its Marcoule reactors and in its "policy of champions," which ultimately sought to promote the consolidation of French industry into a single company per industrial sector. EDF's nationalized technopolitical regime found form in its

Chinon reactor and in its efforts to micro-manage industrial contracting, which (somewhat ironically for a nationalized company) sought to preserve competition between private companies as a means of ensuring cheap electricity. Both regimes sought to develop prescriptions for governing nuclear development within their institutions and for directing nuclear and industrial policy on the national stage. Embedding these prescriptions in artifacts and practices constituted a strategic move in which technology and politics were deliberately conflated.

The reactors at Marcoule (figure 4) and Chinon (figure 5) thus functioned as technopolitics in several interconnected ways. Through these reactors, the two regimes aimed to control both the technological and the political dimensions of nuclear development. Both regimes claimed the knowledge and authority to decide what was right for the nation, and to hereby define aspects of national identity. These claims, in turn, had two purposes: to validate some technopolitical choices with respect to others, and to validate technopolitics as a decision-making process with respect to the state and the rest of the nation.

The tight connections that each regime drew between its technopolitics and its vision of France's national future would have lasting effects on the development of the nuclear program. The effects of these connections

Figure 7.4
G2 and G3 in 1960. (CEA/MAH/Jahan)

Figure 7.5
An aerial view of the Chinon site in 1966. (EDF Photothèque)

became especially visible in the late 1960s, a time of renewed crisis both in the nuclear program and in the state's consideration of how industrial change would express and guide the nation's future and identity. In order to demonstrate this point, let me now turn to the third and final vignette.

The War of the Systems

After the reactors described above, EDF went on to build five more gas-graphite reactors in France. The CEA, meanwhile, erected three reactors at Marcoule, built several atomic bombs (military development became officially sanctioned by the government in 1958), and worked hard on other aspects of the fuel cycle (mining, manufacturing fuel rods, and processing spent fuel).

Over the course of the 1960s, however, the leadership of EDF's technopolitical regime gradually began to shift from engineers to economists,[25] and this led to changes in the regime itself. The new leadership had different ideas about the meaning of nationalization and public service for the regime. They also grew increasingly impatient with the CEA's role in the development of nuclear power plants. Meanwhile, private companies had been complaining about EDF's industrial contracting policy, which retained design control within EDF rather than distributing it to the contractors. EDF's new managers were considerably more sympathetic to the demands of private industry than the utility's engineers.

The tensions caused by these rising conflicts over nuclear development came to a head in 1966–67, when US companies began selling light-water reactors on the international market at prices below cost. The American reactors appeared cheaper than the French gas-graphite reactors, and so they attracted the interest and attention of EDF's new managers. Purchasing an American reactor license, they mused, might be a good way to keep French private companies (who would then build reactors under the license) happy, and might provide an escape from the CEA's purview.[26]

But EDF's new economist-managers faced major obstacles to implementation of this plan. Gas-graphite reactors were so tightly associated with French national identity that purchasing an American license could be tantamount to an act of treason. Within the nuclear program, engineers from both the CEA and EDF who had invested their entire professional careers in the gas-graphite design strenuously objected to a change in course. At the highest level of government, meanwhile, de Gaulle and some of his advisors firmly believed that the gas-graphite program was absolutely crucial to the maintenance of France's independence and identity. The biggest problem now facing EDF's new managers, therefore, was precisely the conflation of technology and politics that had made the program so successful in the first place. Gas-graphite reactors had become unassailable in large part because they were technopolitical (not just technical) artifacts. In order to buy an American license, EDF's managers would have to dissolve the particular blend of technology and politics that had crystallized around the question of national identity. This dissolution was no small task: it took the form of a three-year debate conducted in a wide variety of locations that came to be known as "the war of the systems."

A series of technical mishaps at Chinon's reactors triggered the start of this war.[27] I will skip over the details of these mishaps here,[28] but I will note the bad publicity they attracted in the national press. In *Le Monde*, Nicolas Vichney blamed both the private builders and EDF. French technology,

Vichney wrote, could not meet the high standards demanded by nuclear plants. Furthermore, EDF had flawed industrial contracting practices.[29] The utility had tried to build something too complicated too fast, and the technical abilities of its personnel could not rise to the occasion. The satirical weekly Le Canard Enchaîné interpreted events more bluntly: "In short, our home-grown nuclear equipment doesn't hold up." Le Canard gleefully noted de Gaulle's displeasure at the incidents: "Heads will roll, citizens!"[30]

Within the nuclear program, the technical mishaps and the publicity they generated gave EDF's new managers the leverage they needed to propose redefining the meaning of "public service" within the utility's technopolitical regime. They wanted to serve the public not simply by making electricity but also by creating a climate congenial to the development of private industry—which meant reshaping the utility's industrial contracting policy, and perhaps ultimately changing reactor systems. This outlook worried labor militants, who wanted to preserve a regime in which EDF would control industrial development. The proposed change in orientation would betray EDF's original mission as the model for a new society because it would rob employees of decision-making powers.[31]

Would nationalization continue to entail the contractual and technological subordination of private industry? Or would it acquire a new, more ambiguous meaning in which EDF would make national energy policy by supporting rather than dominating private industry? The war of the systems provided the terrain on which these questions played themselves out.

The war of the systems also became a proving ground for a larger debate about France's future that had been raging in the upper echelons of the French state. De Gaulle wanted France to conduct independent industrial and technological development. His prime minister, Georges Pompidou, wanted French industry to become economically "competitive" in the international market: for him, the nation had to forge a distinctive identity primarily through its economic activity. This meant interdependence with other nations, not independence from them. The war of the systems mirrored these two poles. The technopolitical importance of the nuclear program meant that the process and the outcome of this war would carry great weight in broader debates about France's industrial future and identity.

The first battle in the war was waged in a joint CEA-EDF committee appointed to study the various reactor systems operating in Europe and America. Headed by Jules Horowitz of the CEA and Jean Cabanius of EDF,[32] the committee presented its results in late January 1967. It immediately became obvious that the effort to smooth over differences by appointing a joint committee had failed. Though Horowitz and Cabanius agreed on some basic numbers (such as the capital costs for various reactors), they differed on the technopolitical meaning of these numbers—so much so that they produced two separate reports. For my purposes here, the four most important points of contention between the two reports were (1) the meaning of public service, particularly with respect to private industry, (2) the proper criteria for evaluating the performance of the gas-graphite system, (3) uncertainty in the data, and how to handle it, and (4) the "political" dimensions of the issue, and who had the responsibility to analyze these dimensions. The CEA's report maintained tight links between technology and politics discussed above in order to justify the gas-graphite system.[33] EDF's report severed these links in order to advocate the switch to the light-water design. In his introduction, Cabanius wrote: "Political considerations, in particular those relating to the acquisition or manufacturing . . . of enriched uranium are up to the public authorities and will not be raised in this study. The rapporteur has strictly limited himself to the industrial, technical, and economic side [of the issue]."[34]

Cabanius went on to characterize the state of French industry, and in so doing he expressed a desire to redefine public service within EDF's technopolitical regime. EDF, he wrote, played the dual role of customer and supplier to French industry. As a supplier, it had to offer its clients—namely, the manufacturers (privileged here over the general public)—the cheapest possible electricity in order to help them compete with foreign companies. As a customer, it had to help French companies reorganize themselves into large consortia capable of taking on the massive investments required by reactor manufacturing. Encouraging these consortia to work under a foreign license would further help French industry because the dynamism and success of the licensers would provide important financial and technical support for the licensees.[35] Ostensibly, then, EDF's first priority should be to strengthen French industry.

Horowitz focused his definition of public service on the gas-graphite system. He proudly noted that the system had already exceeded expectations in several ways: the price of natural uranium fuel had dropped faster than anticipated, the fuel rods had proved technically reliable, and the reactor cores had performed well. As for system costs, Horowitz argued that these would have been much lower without the many problems that plagued the construction and startup of Chinon's reactors. He blamed EDF's technical incompetence and inconsistent attitude toward private contractors for this poor performance, rather than the system itself. Marcoule, after all, had proved that French reactors could run smoothly and efficiently.[36]

Cabanius, in contrast, blamed the mishaps not on EDF but on the technology. The complexity of the gas-graphite system posed particularly delicate construction problems. Yes, Marcoule had performed well, but its reactors were smaller and less complex than EDF's, and the difficulties of building gas-graphite reactors increased dramatically with the scale of the reactor. "The natural uranium-gas-graphite system," wrote Cabanius, "therefore contains a source of incidents which could have serious consequences. . . ."[37] The gas-graphite system was inherently flawed. The light-water system, however, was not. Its capital costs were low. Furthermore, with so many reactors already on order from the United States, manufacturing could be standardized (which would lead to even lower costs and greater reliability than the gas-graphite system had).[38] Cabanius portrayed the spread of light-water reactors as inexorable, thereby sowing the seeds of technological determinism.

Horowitz admitted no such determinism. Yes, American utilities had ordered a remarkable number of light-water megawatts in the last two years. Even more amazing, he noted snidely, this enthusiasm was based on the actual performance of just two 200-megawatt reactors! True, the American program would probably succeed, thanks to its technical rigor and the vast resources of its manufacturers. But this did not mean that the same program would succeed in France. National context mattered deeply. Insufficient data made good predictions nearly impossible.[39] Whereas Cabanius treated the American numbers as reasonably accurate characteristics of the *technology*, Horowitz treated them as rough estimates based on the *context*. In Cabanius's analysis, the technology, abstracted from its context, was the most important variable. For Horowitz, what worked for the United States would not necessarily work for France. Technology did not develop along a single path, irrespective of context.

Despite his skepticism toward the American numbers, Horowitz used them in his calculations (they were, after all, the only ones available). The two men thus ended up with similar estimates for the cost of conventional and nuclear kilowatt-hours.[40] But they handled the uncertainty in the data differently. Cabanius used quantitative methods, calculating the variation in the cost of the light-water and gas-graphite kilowatt-hours for different values in parameters such as fuel burnup rate, capacity factor, and capital costs. Plotting uncertainty offered a sense of control, suggesting that because it was quantifiable, it was manageable. Naming and describing uncertainty, in other words, eliminated the need for qualitative assessment. For Horowitz, however, uncertainty required qualitative judgment. In the absence of hard "facts," political acumen had to guide choices. This came through most clearly in Horowitz's discussion of enriched uranium supplies. Cabanius had dismissed this topic in a single short paragraph: though light-water plants would initially rely on foreign supplies, eventually France or Europe would build enrichment plants. Further discussion fell into the realm of "politics," and therefore beyond Cabanius's self-defined mandate. Horowitz also saw this as a political issue, but he felt that it fell squarely within his mandate. Whereas Cabanius had written of foreign "supplies," Horowitz wrote of foreign "dependence." "Political reasons" (which he left unspecified) would make a European enrichment plant impossible, and France could never afford one on its own. Furthermore, the enriched uranium produced in France or Europe would cost considerably more than American uranium, thereby negating the cost advantage of the light-water system.

There were many other differences between the two reports, but these suffice to demonstrate how each man constructed his argument in terms of France's national interest. For Horowitz, the gas-graphite system equaled French independence in the world. Cabanius sidestepped this claim and redefined French national interest in economic terms, arguing that EDF had to reshape its regime in a way that would help French industry compete in world markets. In this regime, pursuing light-water technologies under a foreign license made national sense. Horowitz did not repudiate the goal of helping French companies compete internationally; indeed, as we saw in the previous section, this had been a goal of the CEA's "policy of champions." For the CEA, however, industrial "champions" had always meant French companies building French technology. Abandoning this pursuit now was folly. In arguing for different systems, the two men promoted different versions of France. Choosing a system was also about shaping the national future. Technology and politics were thus inseparably entwined for both men—even if only one of them admitted it.

Although EDF's economist-managers sought to exclude politics from their analysis, they could not exclude politics from their world. Indeed, their arguments in favor of economic criteria held little sway with President de Gaulle, who ultimately had the final say on this matter. The president held fast in his commitment to French independence and glory, and his close advisors assured him that these were synonymous with the gas-graphite system. But advocates of light-water reactors did not give up the fight. And by the end of 1967 it had become increasingly clear that positions did not fall out along neat institutional lines. EDF's top management, private manufacturers, and a few of CEA's top officials advocated light-water reactors. Some CEA employees adopted a middle ground: if it became necessary or expedient to build light-water reactors, France should develop these itself rather than buy an American license. Advocates of the gas-graphite system included labor unions and rank-and-file employees in both institutions who had devoted their careers to this technology. The impasse was referred to the PEON commission.

Founded in 1955, PEON (Commission pour la Production d'Electricité d'Origine Nucléaire) was a government-appointed commission composed of top EDF and CEA leaders, ministerial officials, and a few industrialists. Its ostensible purpose was to give the government nuclear advice. It was not, however, a decision-making body. At least until the late 1960s, programmatic decisions were negotiated within and between the CEA and EDF. PEON did little more than discuss and bless such agreements.

The commission's role grew subtler and more complex during the war of the systems. Amid the contentiousness fueled by technopolitical uncertainty, PEON meetings provided a place for constructing notions of objective arbitration. Discussions and reports provided a stage on which members could play hybrid roles: although they were there to represent specific institutions, their membership in PEON symbolically separated them from their institution and gave them a larger constituency—the nation. This hybridity conflated the self-proclaimed disinterestedness of state technologists with that of the nation. At least in principle, any PEON conclusion or document represented an arbitrated negotiation for the greater good of the nation among otherwise competing interests. When reporting back to their home institutions, PEON's members carefully nurtured the commission's status as objective arbiter. The same policy conclusion would carry more weight all around if reached by PEON.

PEON inherited the Horowitz-Cabanius mission to investigate the ramifications of each reactor system and provide a rationale for future programmatic choices. Many of the reports produced by its members sought not to define and describe the artifacts directly but rather to define and describe the context in which they would operate.[41] For industrial leaders, this context was the Common Market. This context demanded the pursuit of nuclear technology regardless of the current price of fuel. One industrialist noted slyly: "Just imagine the position of French industry in a Common Market in which, nuclear power having succeeded, German industry dominated this sector."[42] Since light-water reactors dominated the world markets, he continued, the light-water system was the only plausible choice for this context.[43] Meanwhile, CEA's representatives to PEON sought to limit the context to national frontiers. Here they met with stubborn resistance from the industrialists, who apparently refused to discuss matters in these terms.[44] PEON's April 1968 report masked these disagreements, instead seeking to redefine the French context: "It is pointless to hope for total independence. . . . The potential for economic independence can be defined as the capacity to maintain economic competitiveness over the long term and on the international front. . . ."[45] France should immediately build one American-style reactor. And, pending a reevaluation in 1970, no new gas-graphite reactors should be ordered in the next two years. One industry periodical disingenuously proclaimed these conclusions the result of a "profound unanimity" stemming from the separation of technology from politics: "men in good faith . . . were eventually bound to agree over the analysis of such a complex question from the moment that this [question] was entirely depoliticized and subjected to the objective analysis of the real problems involved."[46]

Ultimately, PEON's 1968 report did not resolve anything. Its main achievement, rather, was to legitimate two of the main strategies of the supporters of light-water reactors: the separation of technology from politics, and the redefinition of the context of nuclear development as Common Market economics. Still, de Gaulle continued to favor the French system. And within both EDF and the CEA employees remained split. Not everyone agreed that technology and politics should be separated, or that the right context for the nuclear program should be primarily economic in nature.

Meanwhile, a new source of consensus was beginning to emerge: the breeder reactor. As a technology that existed primarily on paper (only one prototype existed: the CEA's Rapsodie), the breeder was still flexible enough to fulfill a variety of technopolitical scenarios. Fearing the demise of their program, gas-graphite enthusiasts began to transfer onto breeders the burden of France's technological glory. The proponents of light-water reactors, meanwhile, used the vision of a breeder-reactor future to build a stronger constituency for the American solution.

Some EDF engineers argued that breeders provided the best reason for maintaining the gas-graphite system, which could supply both the plutonium and the experience required for the development of breeder reactors. Such a course would ultimately allow France to surpass the United States, which had no breeder-reactor experience. Horowitz of the CEA enthusiastically endorsed this viewpoint. So did EDF board member and labor militant Claude Tourgeron, who envisioned a socialist future of breeders. Tourgeron juxtaposed his argument for breeders with an argument for the "formation of nationalized companies that would free this industry from the joint pressure of large capitalist monopolies and military management."[47] These nationalized companies would provide the basis for a true socialist democracy, which could only lead to national economic growth. Breeder technology would take some time to mature, though, so France had to pursue an intermediate system in order to maintain its nuclear knowledge. Only a system based on natural uranium would allow France both to escape the clutches of American imperialism and to produce plutonium for the future breeders. And cost calculations that disadvantaged the gas-graphite system were due to nefarious capitalist practices. In particular, the latest gas-graphite estimates had been inflated by capitalist monopolies in their thirst for profit and their desire to tip the balance in favor of the American design. Thus, gas-graphite's success, breeders, and a socialist order were mutually dependent.[48] The technopolitical circle was complete.

Proponents of the American system seized on the emerging consensus to propose a different path to the breeder-reactor future. In February 1969, EDF's director-general, Marcel Boiteux, sent a memo to the prime minister in February 1969 in which he argued that France should make every effort to research and develop breeders, "the system of the future." But he contended that the main road to that future went through the American system. Using an American license would allow France to recover from the disappointment of the gas-graphite experience and to "catch its breath while waiting for a new breakthrough—that of the breeders—to which it will devote all its research and development efforts."[49] Not even the CEA's experience in designing a light-water reactor for submarines would go to waste. Instead, this experience would help French teams "mix French intelligence with American experience to build a 'Frenchified' reactor."[50] Thus, they too transferred the burden of French grandeur over to the breeders. Further, the nebulous "Frenchifying" of American reactors would preserve French nuclear know-how (and, presumably, pride). In April 1969, Boiteux and Hirsch pushed through a "plan of action" that essentially reframed the proposals and arguments of the PEON report by placing them in the context of a breeder-reactor future.[51] Meanwhile, EDF's managers had already begun to prepare for the first light-water reactor.[52]

In the mid 1950s, the CEA capitalized on the ambiguity of the gas-graphite design to move forward with the military atom. In 1969, advocates of light-water reactors capitalized on a variety of ambiguities to move forward with plans to buy an American license. Each successive step tightened the case for light-water reactors, using a combination of strategies that involved managing uncertainty, renegotiating the meaning of national independence, and defining a place for light-water reactors in the construction of French identity. With each refinement, American light-water reactors appeared increasingly necessary for the future of France.

As 1969 wore on, opposition to light-water reactors became increasingly difficult to orchestrate. French advocates of the American system—who occupied the top administrative positions in the CEA, in EDF, and in private industry—had developed irreproachable goals: to give France cheap energy and to make breeders the new symbols of French identity and independence. And they had not actually proposed terminating the gas-graphite program. The case was not closed: the government had not yet made a decision on the choice of a system. But things looked bad for gas-graphite. It was clear that EDF management, private industry, and top CEA officials were poised to buy American. It was equally clear that buying American would come at the expense of the French system. Furthermore, by then the one man seen as a guarantee against the purchase of a foreign license—Charles de Gaulle—had resigned from office. He had been replaced by Georges Pompidou, who was distinctly sympathetic to the American system.[53]

By October 1969, rumors had begun to circulate that the CEA's programs would be cut back and that layoffs would ensue. The CEA's five main labor unions joined forces to protest the layoffs, the introduction of American light-water reactors, the implied slurs on their technical competence, and the incoherence of national nuclear research policy. Unions avidly defended the performance of the French nuclear program.[54] The price of the gas-graphite kilowatt-hour was already 30–40 percent lower than the most optimistic estimates of several years earlier. The unions demanded a coherent nuclear program whose main criteria of success would be continuity, independence, and the development of a national electromechanical industry. This policy "must first and foremost be translated into the development of the gas-graphite system."[55] It would be "stupid" to abandon this and other national technologies. In a separate statement, the Communist Confédération Générale du Travail called for the publication of reports that would "reestablish the truth that is indispensable to the defense of French atomic energy. . . . The CGT's engineering and white-collar worker section will not hold back in its efforts to ensure that France remains independent in the energy sector."[56]

The worst fears of the unions were confirmed in mid October. At a press conference held for the inauguration of the Saint-Laurent 1 reactor (the pride and joy of the gas-graphite program), Marcel Boiteux congratulated the site's teams on their success. Too bad, he added, that the gas-graphite system was not commercially viable. From then on, he said, EDF would be building light-water reactors under an American license.[57] This announcement sent a shock wave throughout the nuclear program, the government, and the press. Everyone knew that this was the direction in which the program was headed, but no one—not even Pompidou—realized that a decision had been reached.

Indeed, Boiteux himself emphatically denied reports in the press that he had announced an actual decision. He had merely stated that because the economic success of Saint-Laurent was less certain than its technical success, the future of the gas-graphite system remained uncertain. It was "regrettable that his words were given the political meaning that they were."[58] Journalists had misinterpreted his responses to their questions. He had merely indicated that EDF had a *preference* for a light-water system.[59] The press had neglected to mention that he had referred all final decisions to the government.

Disclaimers notwithstanding, Boiteux's statement was widely understood to signal the end of the gas-graphite program. The layoff of a group of sub-contracted workers at the CEA seemed to confirm the death knell. Seven hundred CEA employees launched a series of strikes that would last more than a month. On October 31, news leaked that the administration planned to announce another 2000 layoffs. The unions responded by broadening their demands and intensifying their strike actions. Though they continued to express outrage on behalf of subcontracted workers, protests now focused primarily on the issue of nuclear policy. The strikes continued through the end of November and spread to all the CEA's research and production centers.[60]

Echoing earlier arguments, strikers denounced the termination of the gas-graphite program and the threat of an American industrial takeover.[61] They contested the assertion that gas-graphite reactors were not competitive and argued, furthermore, that "profitability [was] not the only important criterion."[62] National independence had to count too—especially independence from the United States. Never had the threat of American capitalism loomed larger. "We are," one tract claimed, "in the process of losing our national independence; we are on the path to under-development and colonization."[63] The problem lay in the fact that the government had not handled either industrial or research policy properly. "Such an important decision . . . should be preceded by consultations with employee representatives, not announced on the fly by a bureaucrat [Boiteux], no matter how highly placed he might be."[64] Only the unions had the nation's welfare firmly in sight: "Our goals are clear. We are in favor . . . of funding research which will ensure the intellectual, economic, and social future of an entire people and guarantee its independence."[65]

On November 14, Pompidou formally announced the termination of the gas-graphite program for the foreseeable future. In response, the CEA strikes continued to intensify. On November 17, between 4000 and 6000 protesters descended on the Place des Invalides in Paris and marched past the Eiffel Tower (figure 6).[66]

Doubtless realizing that purely political tactics would have little effect in a debate whose terms were defined by its dominant participants as economic and apolitical, a group of union engineers, scientists, and technicians prepared a counter-report on the relative merits of the competing nuclear systems. One major difference between the union report and those written by Cabanius and PEON lay in how the reports posited the relationship between technology and politics. Advocates of the American system, as we have seen, sought to *remove* what they derisively called "political" considerations from the decision-making process. Building on Horowitz's earlier arguments, union advocates of the French system sought to retain such considerations—all the more so because their extended analysis of the financial calculations conducted by the nuclear industries in both France and the United States led them to conclude that Pompidou's decision followed not the logic of an abstract economic rationality (by their calculations, the gas-graphite system was slightly cheaper for France) but the logic of politics.[67]

Indeed, the unions saw the decision to terminate the gas-graphite program as a capitulation to capitalism—American capitalism in particular. "Everyone is aware of the concerted offensive launched by American industrial consortia to get hold of the French electromechanical [industry]."[68] Pompidou's announcement merely confirmed the "Americanization of the French electronuclear [program]."[69] But the report did not argue that politics should have been left out of the decision. Instead, it argued, the *wrong* politics had guided policy makers. When the uncertainty of the economic data was taken fully into account, the resulting estimates were "sufficiently close for other criteria of choice (currency flow, capitalizing on existing investments, national independence, full employment) to be considered on the same plane."[70] In other words, rather than base a decision purely on the politics of capitalist development, the government should have also taken the politics of social relations into account. And it should have weighted other political elements (such as national independence) differently. In

Figure 7.6
CEA protesters marching past the Eiffel Tower. Photograph by Philippe Mousseau, Lumifilms. (courtesy of CFDT archives)

conclusion, the report called for the creation of a new commission—composed of ministerial officials, EDF and CEA management, and labor unions—to reexamine the case.

The government never took this report seriously, and in the end the light-water system triumphed over the gas-graphite one. Though the final door to this triumph was opened by de Gaulle's resignation and Pompidou's ascendance, the triumph itself was the process of long and complex technopolitical negotiations. State experts, not politicians, determined the terms of these negotiations.

Between 1970 and 1973, EDF broke ground on four Westinghouse-licensed reactors—a "modest" number, as prescribed by the 1970 PEON report. Any impulse to remain modest disappeared during the 1973 oil crisis. In March 1974, Prime Minister Pierre Messmer announced a new energy plan calling for the launch of thirteen 1000-megawatt light-water reactors in two years. By the time I began my research in 1989, France was obtaining more than 70 percent of its electricity from pressurized-water reactors, and engineers were eager to tell me how the light-water system had become *francisé*—Frenchified.

Conclusion

Advocates of light-water reactors had understood that attacking a system which continued to incarnate the French nation would lead nowhere. The only way to break this powerful association was by rhetorically separating technology from politics. Such a separation undermined the links between gas-graphite technology and the nation. Equally important, excluding politics from technological choice privileged economic selection criteria. Admittedly, the data that constituted these criteria were uncertain. But advocates of light-water reactors subjected this uncertainty to quantitative analysis in order to claim control over it. Advocates of gas-graphite reactors subjected the uncertainty to qualitative—specifically, political—analysis in order to do the same. As long as de Gaulle remained in power, this qualitative reasoning held. Once he stepped down, quantitative reasoning took over. The triumph of the light-water system meant that it came to be defined as the "economic" system, while gas-graphite became the "political" system. Eventually the discourse of national identity crept back into the light-water program, as builders and developers began discussing the "Frenchification" of the system. The emphasis had merely shifted from making a French technology to making a technology French. Nonetheless, the triumph of the light-water system signaled that, henceforth, economic criteria could legitimately participate in the articulation of French national identity.

Thus, the ways in which technologists discursively and materially constructed relationships between technology and politics—and the role of national identity in those constructions—had real consequences for French industrial strategy. Different kinds of relationships made possible different industrial strategies, and different constructions of national identity.

The analysis presented in this essay has, I believe, two implications for cultural historians—one historical, one theoretical. The historical implication is that in the late twentieth century technological development constituted an important site for negotiating French national identity. (Indeed, I suspect that detailed studies would yield a similar conclusion for other times and places—it is already clear, for example, that technological development plays a big symbolic role in the construction of American national identity.) In the case of French cultural history, this is not a trivial point. Indeed, it provides a way of understanding France's experience of modernity in the second half of the twentieth century.

Theoretically, this essay suggests one way to transcend sterile dichotomies between cultural and material analysis. As some historians have already observed, cultural analysis need not "imply an anti-materialist position." Instead, it can show how the material world both derives meaning from, and performs, culture.[71] As I have shown here (and other historians and sociologists of technology have shown elsewhere), technological artifacts and practices (the supposed epitomes of the material world) are deeply cultural and political. Locating the construction and the performance of cultural forms such as national identity in the practices of technological change shows how these forms are grounded in the material world. This, in turn, demonstrates not just the political power of cultural forms but also their material power. Ultimately, then, the practices of technopolitics reveal that the political power of culture cannot be separated from its material power. Only by simultaneously unpacking culture, technology, and politics can historians make this point effectively.

Notes

1. See, e.g., Peter Sahlins, *Boundaries* (University of California Press, 1989; Caroline Ford, *Creating the Nation in Provincial France* (Princeton University Press, 1993); Richard Kuisel, *Seducing the French* (University of California Press, 1993); Kristin Ross, *Fast Cars, Clean Bodies* (MIT Press, 1995); Gérard Noiriel, "French and Foreigners," in *Realms of Memory, I*, ed. P. Nora (Columbia University Press, 1996); Gérard Noiriel, *The French Melting Pot* (University of Minnesota Press, 1996); Herman Lebovics, *True France* (Cornell University Press, 1992). For an overview of debates about how American scholars have treated the concept of French national identity, see *French Historical Studies* 19, no. 2 (1995), especially the following papers: Richard F. Kuisel, "American Historians in Search of France: Perceptions and Misperceptions" (307–319); Michèle Lamont, "National Identity and National Boundary Patterns in France and in the United States" (349–365); Eric Fassin, "Fearful Symmetry: Culturalism and Cultural Comparison after Tocqueville" (451–460).

2. See most of the essays in volume 1 of *Realms of Memory*.

3. For an overview of how narrative has become a historiographical category, see Sarah Maza, "Stories in History: Cultural Narratives in Recent Works in European History," *American Historical Review* 101 (1996) 5: 1493–1515.

4. For example, in *True France* Herman Lebovics argues that "searches for France embody beliefs, values, projets, in short, ideologies of what France should be." In a different type of scholarly endeavor, Stuart Hall makes a similar point: "Though they seem to invoke an origin in a historical past with which they continue to correspond, actually identities are about questions of using the resources of history, language and culture in the process of becoming rather than being: not 'who we are' or 'where we came from,' so much as what we might become, how we have been represented and how that bears on how we might represent ourselves. Identities . . . relate to the invention of tradition as much as to tradition itself. . . ." (*Questions of Cultural Identity*, ed. S. Hall and P. du Gay, Sage, 1996, p. 4) On mythologizing the past, see *Realms of Memory*; see also Robert Gildea, *The Past in French History* (Yale University Press, 1994).

5. For examples see Wiebe Bijker, Thomas P. Hughes, and Trevor Pinch, eds., *The Social Construction of Technological Systems* (MIT Press, 1987); Bruno Latour, *Science in Action* (Harvard University Press, 1987).

6. On nuclear workers, see Hecht, *The Radiance of France* (MIT Press, 1998), chapters 4, 5, and 8.

7. Thomas P. Hughes, *Networks of Power* (Johns Hopkins University Press, 1983).

8. This becomes especially clear in an examination of reactor operation. Through artifacts and work practices, the workplaces in these regimes performed distinct visions of the sociopolitical order. See Hecht, *The Radiance of France*, chapter 5.

9. Ken Alder, *Engineering the Revolution* (Princeton University Press, 1997); Eda Kranakis, *Constructing a Bridge* (MIT Press, 1997); Richard F. Kuisel, *Capitalism and the State in Modern France* (Cambridge University Press, 1981).

10. André Léauté, "Les Vertus Cardinales de l'Ingénieur de Grande Classe," *La Jaune et la Rouge* 120 (1958), p. 41.

11. Groupe 1985, *Refléxions pour 1985* (La Documentation Française, 1964), pp. 13–14.

12. J. L. Cottier, *La Technocratie, Nouveau Pouvoir* (Edition du Cerf, 1959), p. 41.

13. Groupe 1985, *Refléxions pour 1985*, p. 88.

14. *Le Grand Oeuvre: Panorama de l'Industrie Française*, commissioned by the Ministère des Affaires Étrangeres. The first phrase in the title, literally as "the great work," also refers to "the philosopher's stone." Two copies were ordered the following year by the Ministère de l'Equipement. Today the film can be viewed at the vidéothèque of the Ministère de l'Equipement.

15. The metaphor of "regime" also evokes the dynamics and contestation of power. Although I cannot examine this dimension of "regime" in a short essay, it is an important aspect of the story. In the case of the CEA, for example, not everyone adhered to the nationalist ideologies espoused by agency leaders. Here the point is that nationalist ideologies *governed* the CEA, not that they were *ubiquitous* therein. I explore contestation to technopolitical regimes in *The Radiance of France*.

16. As Robert Frost argues in *Alternating Currents* (Cornell University Press, 1991), this ideal often did not reflect reality. It did, however, reflect official institutional and labor union ideology, which is my point here.

17. One example among many can be found in "Les prévision du IVe plan: l'avenir," *Contacts* 43 (1963): 12–25.

18. Source: interviews conducted by the author. See also Georges Lamiral, *Chronique de Trente Années d'Equipement Nucléaire à Electricité de France* (Association pour l'histoire de l'électricité en France, 1988).

19. Etude Préliminaire d'une Installation de Récupération sur EDF1, 1956; Georges Lamiral, letter (March 4, 1958) to CEA's Direction Industrielle accompanying report Installations de Récupération d'Energie G2-G3 (personal papers of Claude Bienvenu); Programme de centrales nucléaires EDF, DPP 55/505, 29 June 1955 (EU archives, JG 27/6).

20. A. Ertaud and G. Derome, "Chargement et Déchargement," *Bulletin d'informations scientifiques et techniques* 20 (1958): 69–88; P. Passérieux and R. Scalliet, "Installations de récupération d'énergie," *Bulletin d'informations scientifiques et techniques* 20 (1958): 99–114; J. Kieffer, "La centrale de Marcoule: expérience, résultats et enseignements dans le domaine de la production d'électricté," *Energie nucléaire* 5 (1963), June.

21. Memo RETN 1, July 25, 1957; Etude des Réacteurs Enégétiques EDF, Projet d'Organisation dans le cas d'un réacteur du type Uranium Naturel - Graphite - CO_2, March 8, 1957 (Bienvenu papers).

22. From my interviews it is quite clear that simplicity of design was a major goal for EDF engineers. It should be noted here that, while EDF engineers strove for low cost in designing EDF1, they did not attain that goal (in part because of an unforeseen event: the spherical steel containment vessel cracked, and repairing the crack

added tremendous costs and delays to the project.). But EDF1 represented a first step toward optimizing reactor costs, and the general goal of minimizing costs held throughout EDF's reactor program.

23. M. de Rouville, "Le Centre de production de plutonium de Marcoule: sa place dans la chaine industrielle de l'énergie nucléaire," *Revue de l'Industrie Nucléaire*, 40 (1958), June, p. 486.

24. See, e.g., Yvan Teste, "Les Installations de Production d'Energie de Marcoule et la Centrale Nucléaire de Chinon," *Mémoires de la Société des Ingénieurs Civils de France*, tome 110, fascicule II (mars-avril 1957), p. 75.

25. Frost, *Alternating Currents*.

26. The idea that France might pursue other reactor technologies did not appear out of the blue. The research and military branches of the CEA had been investigating other designs for some time. These included a small light-water submarine reactor as well as heavy-water, high-temperature, and breeder prototypes. EDF helped the CEA with some of these prototype efforts, especially on the heavy-water and breeder reactors. EDF's nuclear division also sought reactor projects not tied to the CEA. In the late 1950s, the United States concluded an agreement with Euratom that favored the importing of American reactor designs; reluctantly, the French government allowed EDF to cooperate with Belgium in the construction of the first such reactor in 1960. These efforts notwithstanding, support for the gas-graphite system held fast in both technopolitical regimes through the mid 1960s. In 1964, for example, the PEON commission—a government advisory group composed of officials from EDF, the CEA, and private industry—issued a report speculating that light-water reactors might have lower investment costs than gas-graphite reactors; it emphasized, however that pursuing the light-water option would mean either depending on America to supply enriched uranium fuel or building an expensive enrichment plant that would negate the still uncertain cost advantage of the light-water system. France, therefore, should stay on the gas-graphite track (Philippe Simmonot, *Les nucléocrates*, Presses universitaires de Grenoble, 1973, pp. 237–245). In forwarding this recommendation to Prime Minister Georges Pompidou, Gaston Palewski (the minister in charge of atomic affairs) urged him to make a rapid decision to engage in "massive" development of gas-graphite reactors. The consequences of the choice, Palewski said, mattered not only for France but also for the rest of Europe. Were France to give up its native reactors, "the ensuing technical lag would be felt in our economy and in our policy" (G. Palewski to Georges Pompidou, 4 juillet 64, CEA archives, box F3-24-25). Pompidou approved the PEON plan.

27. EDF-Conseil d'Administration, no. 244, November 25, 1966; no. 249, March 10, 1967 (EDF archives)

28. For a more complete account of these incidents, see Hecht, *The Radiance of France*.

29. Nicholas Vichney, "La centrale nucléaire EDF3 de Chinon est arrêtée pour six mois," *Le Monde*, December 2,1966; Vichney, "Les incidents survenus à la centrale de Chinon amènent à poser le problème des rapports entre l'Electricité de France et le Commissariat atomique," *Le Monde*, January 24, 1967.

30. "Le crêpage de Chinon," *Le Canard Enchaîné*, December 7, 1966.

31. CFDT, Federation EGF, Direction de l'Equipement: La Situation de l'Equipement Nucléaire, 5 Mai 67 (Bienvenu papers).

32. Groupe de Travail Commun CEA-EDF sur les Filières à Uranium Enrichi, 4 mai 66; note from A. Decelle to R. Hirsch, 4 mai 66 (both in Bienvenu papers).

33. J. Horowitz, Examen des filières electro-nucléaires dans le contexte français actuel (1 Février 1967), p. 1.

34. Jean Cabanius, Rapport du Groupe de Travail Placé sous la responsabilité de Monsieur Cabanius (EDF) (25 Janvier 1967), p. 2.

35. Ibid., pp. 4–5.

36. Horowitz, Examen des filières electro-nucléaires, passim.

37. Cabanius, Rapport du Groupe de Travail, p. 9.

38. Ibid., p. 10.

39. Horowitz, Examen des filières electro-nucléaires, p. 10.

40. They gave identical figures for the cost of a conventional kilowatt-hour (3.35–3.95 centimes, depending on the plant's capacity factor) and the cost of the pressurized water kilowatt-hour (2.67 centimes). They differed only on the cost of a gas-graphite kilowatt-hour: Cabanius priced it at 3.14 centimes, Horowitz at 3.04 centimes. The difference occurred because Horowitz was optimistic about experimental data which suggested that the CEA's new fuel rods would yield lower costs, whereas Cabanius refused to rely on the same data. Horowitz thus emphasized the paucity of data on actual operating light-water reactors while expressing great confidence in the CEA's equally unconfirmed estimates. Cabanius took the opposite approach, expressing confidence in the American estimates and skepticism in the CEA's. They based their calculations of the cost of a pressurized-water kilowatt-hour on a meeting they had with Framatome, the company that had managed the construction of the Chooz reactor under license to Westinghouse. They had also tried to get figures for the cost of building a boiling-water reactor, but no French company or consortium had a license with General Electric yet. Source: Conditions de construction de centrales à eau ordinaire par l'industrie française. Réunions du 7 juillet avec Framatome-Westinghouse, du 13 juillet avec Alsthom-GECO, memorandum from Jules Horowitz to Administrator-General, July 27, 1966 (CEA archives, F 6-13-20).

41. Commission Consultative pour la Production d'Electricité d'Origine Nucléaire, Groupe de Travail Général, Rapport de Conjoncture (no date); Prix des fuels a moyen et long terme (dated "9/11/67"); "Hypotheses de travail" (dated "27.9.67"); Hypotheses de travail concernant le développement nucléaire (dated "8.11.67"); réunion du 15 sept 1967, Compte rendu de l'activité du groupe de travail général (rédigé par Jacques Gaussens) (all in CEA archives, F6-13-20). For a full analysis, see Hecht, *The Radiance of France*.

42. Letter from Ambroise Roux to Jean Couture, 29 Mars 1968 (also signed by Baumgartner, Blancard, de Calan, Gaspard, Glasser, Jouven, Malcor), accompanying Note pour la Commission PEON, 28 Mars 1968 (CEA archives, DEDR DIV 219, DPA).

43. Industrialists added weight to their argument in favor of the light-water technology by stipulating that they would offer warranties for it but not for the gas-graphite technology. The mere act of offering warranties transformed the light-water system into a more reliable technology than the gas-graphite system without any technological work per se. Reported in Observations sur le Projet de Rapport soumis à la Commission le 29 février, memorandum, Pierre Tanguy to M. le Directeur des Piles Atomiques (Jules Horowitz), 28.2.68 (CEA archives, DEDR DIV 219, DPA).

44. Ibid.

45. Quoted in Simmonot, *Les nucléocrates*, p. 48.

46. "Le Rapport Couture" (editorial), *Revue Française de l'Energie* 201 (1968), Mai: 435.

47. Claude Tourgeron, "La production d'électricité d'origine nucléaire en France," *économie et politique*, janvier 1969, p. 2.

48. Ibid., p. 13.

49. MB/CH, Politique des Réacteurs Nucléaires (Bienvenu papers), p. 6.

50. Ibid., p. 7.

51. Memo, CEA, EDF, Programme d'action dans le domaine des centrales électronucléaires, 21 April 1969; approved by both Boiteux and Hirsch with identical letters to each other, dated 24 and 25 April respectively (Bienvenu papers).

52. Memo, Cabanius to Directeur de la REN1, Chef du SEPTEN, Chef du SEGN, 29 nov 68; memo, Direction de l'Equipement to Directeur de la REN1, Chef du SEPTEN, Chef du SEGN, no date (probably late 1968); SETPEN memo, DG/MGo, Centrale Nucléaire a Eau Légère, Organisation du travail SEPTEN-REN1, 13 Dec. 1968 (Bienvenu papers).

53. De Gaulle resigned after losing a referendum vote in April 1969; the loss is widely interpreted as an aftershock of the 1968 strikes. For more on French government politics in this period, see Robert Gildea, *France Since 1945* (Oxford University Press, 1996); Serge Berstein, *La France de l'expansion* (Seuil, 1989); Jacques Chapsal, *La vie politique sous la Ve République* (Presses Universitaires de France, 1981, 1993).

54. CGC, CFDT, CGT, CGT-FO, SPAEN, Pour un Politique Française de l'Energie Nucléaire (Déclaration des Organisations Syndicales du Commissariat à l'Energie Atomique), 8 Oct. 1969 (flyer, in personal papers of Bernard Laponche).

55. Ibid.

56. "Un passé qui est un exemple," *Le Compagnon d'Energies Nouvelles (Journal des Ingénieurs et Cadres CGT du CEA)* 7 (1969), suppl. 143: 1.

57. A memorandum from Hirsch to the Ministre du Développement Industriel et Scientifique (15 oct 1969, CEA archives F3-24-25) provides further confirmation that Hirsch and Boiteux worked out the American light-water plant together. The memo outlines their recommendation for how the decision should be shaped and worded.

58. EDF Conseil d'Administration, 278, 24 October 1969 (EDF archives)

59. Ibid.

60. "L'Action dans les Centres," *Energies Nouvelles (CGT, FSM, Journal du syndicat national des travailleurs de l'énergie atomique)*, Dec. 1969: 3.

61. CGT CFDT, CGT-FO, Hier à Palaiseau, 1200 Grèvistes manifestaient . . . , 31 Oct. 1969; CGT CFDT, CGT-FO, Pourquoi la grève du 6 novembre, 5 Nov. 1969 (tracts, in Laponche papers).

62. CGC CFDT, CGT, CGT-FO, SPAEN, Appel aux cadres du CEA, 17 Nov. 1969 (Laponche papers).

63. CFDT CGT, CGT/FO, Pourquoi l'Énergie Atomique en Grève, 6 Nov. 1969 (tract, in Laponche papers).

64. Force-Ouvrière - CFDT, Conference de Presse, Hotel de Ville de Massy (Essonne), 30 Oct. 1969 (Laponche papers).

65. CGT CFDT, CGT-FO, Pourquoi une grève à l'échelon national au Commissariat à l'Energie Atomique? Pourquoi cinq agents du CEA en sont aujourd'hui à leur onzième jour de grève de la faim? 1969 (Laponche papers).

66. Strikes continued at CEA sites in the provinces too. See "L'Action dans les Centres," *Energies Nouvelles (CGT, FSM, Journal du syndicat national des travailleurs de l'énergie atomique)*, Dec. 1969: 3.

67. I discuss their economic analysis at greater length in chapter 8 of *The Radiance of France*.

68. CGC, CFDT, CGT, CGT-FO, SPAEN, Comparaisons Économiques et Politique Industrielle dans le Domaine Électronucléaire, p. 1.

69. Ibid.

70. Ibid., p. 10.

71. William H. Sewell, "Toward a Post-Materialist Rhetoric for Labor History," in *Rethinking Labor History*, ed. L. Berlanstein (University of Illinois Press, 1993). For an overview of these debates, see also the other essays in the same volume. The following works take this "post-materialist" stance: Laura Lee Downs, *Manufacturing Inequality* (Cornell University Press, 1995); Richard Biernacki, *The Fabrication of Labor* (University of California Press, 1995).

The Neutrality Flagpole: Swedish Neutrality Policy and Technological Alliances, 1945–1970

Hans Weinberger

In his 1999 presidential address to the Society for Historians of American Foreign Relations, the diplomatic historian Walter LaFeber acknowledged surprise at the fact that historians of US foreign relations had dealt so little with technology while publishing pathbreaking works in other fields. "Some of these fresh approaches touch on technology," he noted, "but aside from the enormously important work on the atomic and hydrogen bombs and the resulting diplomacy, and a few other revealing monographic studies, the field has only begun to understand this force that has so largely shaped the past 150 years."[1] LaFeber then urged diplomatic historians to follow the insights of Thomas Parke Hughes and others in the field of the history of technology who argue for a contextual understanding of technology, not as artifacts, but as an open system seamlessly integrated with culture. "By discussing technology," he observed, "we are also discussing political, economic, and social choices and possible alternatives."[2]

In this essay I will be following the advice of LaFeber and using the history of technology to analyze Sweden's neutrality during the Cold War. The use of such a perspective will expose a different facet of the development and substance of neutrality. If one looks at the interconnectedness of military technology across national borders and the way in which Sweden's political leaders and high-level military officers have used technology to give concrete substance to the country's neutrality, one's interpretation of this neutrality is likely to change.

Sweden's neutrality has had a long tradition, going back to the beginning of the nineteenth century and the decline of Sweden as a great power in Northern Europe. The first declaration of neutrality came in 1834, when Sweden anticipated a war between England and Russia—a war that never took place. And since then Sweden has prospered from peace. The long period of peace seemed to indicate a rather successful political line and a strong national tradition. A security policy formed during the early decades of the nineteenth century had evolved into a military, political, and cultural foundation of some success and gravity. Much of Sweden's understanding of its role in the global arena during the Cold War was based on this historical experience. Neutrality seemed to have served Sweden well during a long period, including two world wars, and it was regarded as being firmly embraced by a vast majority of Swedes. In addition, neutrality as a national security policy could be portrayed as a logical outcome of the cultural self-image of the modern Swede as a peaceful, moderate creature. As many national and international historians have pointed out though, the country had, on numerous occasions, avoided being involved in war only by a combination of pragmatic "realpolitik" and historical destiny due to its geopolitical position. This observation is vividly illustrated by the events of World War II, when Sweden made important concessions to Nazi Germany after Germany's occupation of Denmark and Norway but then cooperated with Allied forces during the latter part of the war. Sweden also showed a remarkable ability to withstand German pressure for more concessions. With the Cold War, Sweden faced the question of continued neutrality or alignment with the West. As will become evident, the choices made were far from clear-cut, even to the degree of being partly unknown to most Swedes.

For obvious reasons, the ideas and concepts of neutrality as well as its history have mainly attracted scholars from those countries concerned. Studies dealing with the United States and neutrality during the Cold War are few in comparison, and fewer still if one discount books on neutralism in the Third World. In the indexes of books that count as having influenced US foreign policy, "neutrality" occurs seldom indeed. It is entirely absent from the large historiography *America in the World*.[3] One important book that looks at the United States is Jörg Martin Gabriel's *The American Concept of Neutrality after 1941*. Gabriel claims that his book is not a history of US policy toward neutral countries but rather a history of the American conception of neutrality.[4] Still, since the conception of neutrality is to be found in the practical implementation of foreign policy, it does contain substantial historical accounts of the United States' relations with

Sweden and Switzerland—the two most outspoken neutrals in Europe. Gabriel writes: "Scholars seem to have lost interest in knowing what the United States thought about neutrality" and concludes that one reason for this has been a mental spillover from a Cold War mindset of American culture into the academic world.[5]

The lack of references to neutrality in overall accounts of the Cold War might be well deserved in the grander scheme of things, since neutrality in this period was tied to smaller European countries not at the heart of the conflict. But the lack of the concept in general is somewhat more surprising since the idea of neutrality played such an important role as an alternative, for instance, in the case of Germany (the country at the heart of the conflict), as Marc Trachtenberg argues in *A Constructed Peace*.[6] On both sides of the Iron Curtain, but at different times, the idea of a united, demilitarized, neutral Germany played a role in internal policy discussions. And the European countries adhering to neutrality in 1956 included Austria, Finland, Ireland, Sweden, Switzerland, and Yugoslavia. This was not a large share of the overall number of countries, to be sure, but four of them represented geopolitical wedges separating the two power blocs of the Cold War.

In 1985 a leading Swedish historian characterized the matter of Sweden and the Cold War as having been neglected by researchers. It is still fair to ask for more studies of both domestic and foreign policy during the Cold War.[7] While the number of studies has increased, in Sweden political scientists and political historians have focused most on high politics, intentionality, and foreign policy declarations. Historians dealing with the Cold War in Sweden usually adopt one of three perspectives: a power perspective, which emphasizes the role of credibility and tradeoffs between neutrality and other issues of foreign relations such as free trade and participation in international organizations; a ideological perspective, which mainly links foreign policy to the ideological dimensions of social democracy in the domestic sphere (the Social Democratic Party having been continuously in power from 1933 to 1976); and a democratic perspective, which concentrates on foreign policy formulation as a domain of expertise and on the paradigm of consensus within Swedish politics as a way of showing strength and unity in the international arena.[8] A few additional studies have dealt directly with military aspects of neutrality and the reliability of defense in order to uphold neutrality in war.[9] Fewer still have dealt with the military cooperation between Sweden and the Western countries during the Cold War. None have focused on technology or analyzed cooperation from a perspective emphasizing the impact of technological choices, design, and standards on the meaning and conduct of neutrality.[10]

Technology and the Cold War

Although it is not difficult to agree with Walter LaFeber's overall judgment, the issue (or even problem) of technology, science, and diplomatic history is far from one of simple ignorance or exclusion. Diplomatic history has been dominated by a focus on high politics, ideology, and the intentionality of actors, but it has also, during recent decades, undergone large changes in methods and become substantially more heterogeneous, incorporating new perspectives. Saying that studies that directly address technology and science are rare is not the same as saying that technology and science do not play a vital role as an implicit factor in many accounts of the Cold War. The concepts of military power, balance of power, and national security are used by scholars to describe and explain the Cold War. All those concepts are, of course, intimately connected to technology and science. In the same obvious way, studies linking economy to policy formulation and conduct address issues of the military-industrial complex and its influences on the Cold War.[11]

It would be hard to imagine a bipolar world without nuclear weapons and other military technology. As David Holloway writes, "it was nuclear weapons above all that made the Soviet Union a superpower."[12] In regard to LaFeber's observation of the lack of interest in technology among diplomatic historians, the situation for the history of technology is in certain ways a mirror-image twin when it comes to the Cold War. Much as science and technology figure only implicitly in diplomatic history, histories of technology (with some exceptions) usually relegate the Cold War to a status of silent implication. Some have addressed the Cold War directly, but Walter McDougall, Stuart Leslie, Paul Forman, Michael Dennis, and David Hounshell all describe how the Cold War shaped the content of science and technology rather than vice versa.[13] One reason for this might be the focus of the discipline on innovation, on new technologies, and on the methodologically sound understanding of technology as a social construct. Arguing for studies of the political, social, and cultural effects of technology too easily invokes the notion of technological determinism.

But there are exceptions. Paul Edwards shows in *The Closed World* how the conditions of the Cold War shaped computers, and how they, seen as metaphors, helped shape Cold War discourse.[14] Another exception is *Inventing Accuracy*, Donald MacKenzie's study of missile guidance, which clearly illustrates both how technology was socially constructed by the Cold War and how nuclear strategy was shaped by technology. MacKenzie even argues that close scrutiny of the development of inertial guidance makes categories such as politics and technology obsolete.[15]

MacKenzie is well aware of the limitations of traditional accounts of the Cold War, especially those of the realist school's anthropomorphism in seeing the state as a homogenous decision maker. He pertinently calls this political determinism, noting the appearance of a direct causality between political decisions, policy, and practice.[16] He identifies at least three different levels of policy making, or decision making, and also stresses that these levels need not be consistent. The first level is the production of "stated posture"—a declared and official rational description of policy, usually produced at the highest level of formal political and military power. The stated posture is usually the outcome of a number of considerations concerning domestic politics and international relations. The second level is the operational planning within the military establishment, usually protected from public scrutiny, not least because of the need for secrecy. This operational planning is usually carried out in a context of competition among different branches of the armed services. The third level is the process of deciding design criteria for specific weapon systems. At this level there is an interplay between specialized agencies within the military sector and corporations. These levels interact to produce "bureaucratic politics." In essence, then, as MacKenzie points out, "'the state,' intends nothing, decides nothing, does nothing. Policy is not 'decision' . . . but 'outcome.'"[17]

My argument then would be that we need to pursue the path suggested by MacKenzie in a broader and more general way. We must be careful not to travel the old road of technological determinism, but we should not be afraid of raising the issue of science and technology as part of the Cold War, and we must try to see how engineers and technicians within the military sphere have formed and transformed foreign policy.

In the case of Sweden's neutrality, the social construction of technology has shaped the content of the policy to such a degree that the actual material world of the defense systems in fact belied the official policy of Sweden. The maxim of nonalignment in peace and neutrality in war was effectively upset by Sweden's technological orientation toward the West.

It was military engineers who turned "neutrality" into something quite different. This was done secretly, in order to maintain the official picture of Sweden's neutrality, but also because of the way any military establishment needs to function within a democratic framework, i.e. the conflict between democratic transparency and the necessary secrecy of military plans and preparations. In part the secrecy enrolled politicians and the government, in part it excluded them. A small number within the Swedish government, among them the prime minister and the minister of defense, knew more about the cooperation with the West than other members, among them the foreign minister. But the exclusion also had to do with what information actually was given to the Swedish government from the defense establishment.

Crucial to this "political" aspect of engineering is the question of ontology as presented by the actors involved in shaping Sweden's neutrality. It is both trivial and necessary to point out that the understanding of technological decisions differs among various groups of actors, as the theory of social constructivism tells us. Any particular technology means different things to different groups, and different groups perceive different consequences in any decision to implement or deploy a certain technology. There is ample evidence that members of the Swedish government did not see specific technological decisions as especially important, and that instead they saw them as mundane and neutral. Those at the political decision-making locus of Sweden's neutrality policy turned something of a blind eye to technology. To them technology was not unimportant, but it was merely instrumental. On the other hand, the military establishment did not neglect the importance of technology, but it saw technology as a means of bringing Sweden's defense system into closer affinity with the West. It is also clear, however, that the dividing line of interpretation did not follow any simple social grouping. There was a distinct difference between the ontological positions of Prime Minister Tage Erlander and those of Minister of Foreign Affairs Östen Undén in this respect, just as there were distinct differences within the military establishment as to how closely Sweden should be associated with the West.

Another methodological consideration of importance for the study of Sweden's neutrality is the systemic approach of Thomas P. Hughes. It has had a remarkable impact on the way that historians of technology view technological development and understand the social fabric of technology. In his most impressive work, which bears the consciously ambiguous title *Networks of Power*, Hughes has not only shown how large technological systems develop over time and gain momentum, but also how they are essentially sociotechnical, built as a web of hardly discernible threads of technological, political, economical, social, and cultural forces.[18] A number of very different actors—inventors, entrepreneurs, financiers, legislators, bureaucrats, scientists, agencies, schools, corporations—together build and maintain a system. But even though Hughes has mainly studied large technological systems, it can be argued that the systemic nature of technology exceeds those kinds of technologies conventionally thought of as constituting systems, precisely because systems are not well defined. They range from strongly integrated and hierarchical to loose and decentralized. There need not be one single system builder or one centralized organization at a system's center. The historian can use the concept of system to describe technologies, interconnections, and structures in order to interpret an array of actions and make them coherent. Standardization and the path dependence of technologies, for example, are social artifacts. The systematic nature of technology can be distributed across borders and diffused into different cultural and political structures.

I take a systems approach to the history of the Swedish defense system during the Cold War. By pointing to specific technological characteristics of the Swedish defense system, I want to explore the argument that Sweden was strongly connected to the military systems of the West. Focusing on the material world of mundane artifacts, one can see a systemic pattern: in essence, Western military technology and strategy enveloped Sweden's neutrality. This is not just another perspective on Sweden's neutrality. As documents become available, they make possible a closer and more detailed study of the political processes underlying policy declarations, but a traditional focus will not grasp the fundamental integration that took place at the military and the technological level, beyond rhetoric and discourse alone. In this respect, the story of Sweden's neutrality cannot be understood apart from technological practices.

My argument will of necessity be exploratory. I will focus on air-warfare technology and strategy, though I could almost equally well have focused on naval technology. A significant source and starting point has been an official governmental investigation and its report *Om kriget kommit . . .* [Had There Been a War . . .].[19] That report, issued in 1994, was based on classified interviews and documents, which were gradually released afterward. Some material is still classified. The report was also based on material in other countries. In the United States, documents from the early period are accessible, but many from the mid 1950s and after are still classified. This also holds for British, Danish, and NATO material. The example nevertheless amply illustrates that a history of technology perspective can greatly benefit studies of the Cold War. I will try to show that the history of Sweden's neutrality during the Cold War can not be understood without attention to both the systemic technological aspects of modern warfare and the social carriers of military technology. An ever-growing cooperation can be found in the "technological" interaction between Sweden and the outside world.

"A Paradise for Wishful Thinkers"

The term "Cold War" is both apt and paradoxical. It, in fact, represents a period of peace (at least in Europe) as a period of war, creating a war discourse that structured most thinking and analysis of international relations.[20] In the Swedish case, this transformed the use of the concept of neutrality, which in theory is only possible during war, into a peacetime term. In the never-never land of the Cold War, neutrality had to be interpreted anew. It was exactly the blurred boundary between the states of peace and war that elevated the neutrality issue into such a complicated problem for Sweden's political and military leaders.

For small countries, the end of World War II meant the prospect of a new world order based on mutually assured security. At Yalta, in February 1945, Churchill, Roosevelt, and Stalin had discussed the international security organization of the postwar period.[21] The idea that all countries would unite against any country that threatened peace never really stood a chance against the opposite model, according to which the great powers were to manage the world. The United Nations Organization was to function as an "elite club for managing the globe."[22] For a short time, however, it seemed that a new world organization might perhaps be able to succeed in a way that the older League of Nations had not. When the negotiations ended in June 1945, the UN Charter gave China, Great Britain, the United States, France, and the Soviet Union permanent places on the Security Council, with the right to veto any sanctions against themselves or their associates. This construction deeply concerned smaller countries.

For four years before 1949, when the North Atlantic Treaty Organization (NATO) was created, international tension increased. Even before the end of World War II, distrust began to replace whatever hope there had been to continue the cooperation that had defeated Hitler. As the Iron Curtain fell around Eastern Europe, the United States slowly proceeded toward a policy of containment. As Stalin tightened his grip over Eastern Europe, the Western powers responded with growing anxiety and pessimistic visions of the future. With the creation of the Marshall Plan, the formulation of the Truman Doctrine, and the birth of NATO, the overall structure of the Cold War soon became fixed.

During the period 1945–1949, the United States' attitude toward Scandinavia went from a rather detached and absent-minded one to one that increasingly emphasized the need for the Scandinavian countries to choose a side. US policy had long been the province of experts within the Department of State, but it was slowly evolving to include the highest echelons of power. The Scandinavian countries were seen as deeply democratic, with their heart in the right place, but perhaps as having a somewhat confusing and naive foreign policy. Although this was tolerated for a while, from 1947 on the United States put considerable pressure on Denmark, Norway, and Sweden to fully declare their loyalty to the Western democratic nations in the struggle against growing Soviet aggression.[23]

Sweden's foreign policy was heavily influenced by the foreign minister, Östen Undén, a professor of civil law who had been active in the Social Democratic Party since 1909. He held the office of Minister of Foreign Affairs from 1945 until 1962. The thinking of Undén is best characterized by the tension between his belief in neutrality as an effective foreign policy for Sweden and his conviction that international peace and collective security could be gained only by an international legal system.[24] Toward the end of World War II, he had proclaimed that if a new world order of security based on an effective, solidarian international organization with executive power to reinforce its statutes were to develop, Sweden should give up its neutrality. He thus clearly saw that neutrality was not a priori compatible with international solidarity, and he always gave solidarity a higher priority. Undén envisioned a legal system in which all nations would adhere to accepted rules of international conduct. With East-West tensions increasingly suffocating the negotiations preceding the formation of the UN and then the workings of the new organization, he saw no reason to give up Sweden's neutrality.

Undén firmly believed that neutral countries like Sweden contributed to a more peaceful world, that by their mere existence they would point to the possibility of staying aloof in a bipolar conflict. At the same time, the rhetoric surrounding Undén's foreign policy created a strong impression that, in reality, a neutral country could play a role as an intermediary. This belief and this impression resulted in a rather ambivalent strategy. On the one hand, Sweden's neutrality fostered an autarchic military position, manifested in a policy of nonalignment and in the development of specific Swedish technological solutions; on the other hand, the bridge-building element suggested that Sweden should try to seek a middle ground in the Cold War, both strategically and ideologically.

H. Freeman Matthews, a career diplomat, served as the United States' ambassador to Sweden during the years 1947 through 1950—three of the most formative years for Sweden's foreign relations in the Cold War.[25] While in Stockholm, Matthews labored to change Sweden's traditional policy of neutrality in order to make Sweden realize its dependence on the West. Matthews's perspective on Sweden's neutrality naturally differed from Undén's. Matthews felt that Sweden had shown during the war that it could not be trusted to enforce its neutrality, and that under pressure from any strong belligerent state Sweden would make concessions just to avoid military conflict. Thus, under the concept of strict neutrality, Sweden in reality conducted a not-so-glorious pragmatic policy as long as it benefited the country. The risk seemed great that Sweden would not be able to handle a serious threat from the Soviet Union. The US Department of State did not question Sweden's Western orientation or its democratic capacity to withstand communist ideological propaganda; it was Sweden's incapacity to adequately judge its own strategic position and to foresee the probable course of an outright war that troubled US diplomats.

Much to Matthews's irritation, Sweden's military and some of Sweden's leading politicians believed that their country would receive assistance from the West in case of Soviet aggression, without any prior planning or cooperation. Sweden's war planning did in fact presume help from the outside in the case of aggression extended over a longer period of time, and most of the scenarios explicated a strategic defense aimed at buying time.[26] Matthews described this dual position as "a specious tenet in the creed of Swedish self-deception."[27] Calling Stockholm "the paradise of wishful thinkers," he produced a list of what he called twelve basic fallacies of Sweden's neutrality thinking and sent it to US Secretary of State George Marshall.[28]

The foremost fallacy cited by Matthews was that Sweden thought it would be able to stay out of a third world war, and that somehow the great powers would find Sweden's neutrality advantageous. Though it was not official policy, one Swedish argument against alignment with the West rested on the assumption that Sweden was able to provide a middle ground —a third way, in a political and ideological sense, between the East and the West—and that Sweden's neutrality was an important factor in the Soviet Union's attitude toward Finland. A neutral Sweden created an ideological and strategic spectrum across the Scandinavian peninsula that lessened tensions and made the Finish solution possible; a closer and formal association with the West would only change the USSR's attitudes toward Finland in a negative direction, creating pressure for closer Finnish-Soviet military cooperation. The balance also worked in the other direction, according to Swedish thinking. If the USSR intensified pressure on Finland, Sweden would be forced to join NATO, and that would make the Soviet position in the Baltic more problematic.[29] Matthews saw nothing relevant in this thinking. In another letter to Marshall, he argued that Sweden's "sensitivity to what happens in Finland [gave] the Kremlin an opportunity for political blackmail against Sweden."[30]

Matthews found some allies among the Swedes. One of them was the Chief of the Swedish Air Force, General Bengt Nordenskiöld, who in

Matthews's eyes estimated the balance of power realistically and understood that Sweden needed to cooperate in advance to receive any Western assistance. Nordenskiöld was one of many leading military officers favorably disposed toward a Swedish alliance with the Western countries.[31] Matthews sometimes thought (correctly) that he discerned fractions even within the Swedish government—the Minister of Defense and friendly Swedish ambassadors versus the Minister of Foreign Affairs and leading Social Democratic politicians. The major newspaper *Dagens Nyheter*, under charismatic editor-in-chief Herbert Tingsten, consistently criticized the official neutrality policy.[32] Tingsten shared Matthews's morally founded critique of Undén's inability to see the cold international climate as anything other than a struggle between two power blocs. They both thought that Undén fled into formalism, trying to place Sweden on the middle ground to create credibility for its neutrality policy. In numerous editorials, Tingsten argued for joining any upcoming Western alliance.

A coherent theme in the reports from the American Embassy in Stockholm to Washington was a supposition that the sentiments of the Swedish public on the neutrality issue would change. Change never came. Matthews explained his ambition of "getting the Swedes off the neutrality flagpole."[33] And the American ambassador chose a harsh tone to get them "stirred up" and force them to realize that the price of neutrality would be both a "failure to hold neutrality" and "the loss of potential help from the West." Once the majority of the Swedish people and their politicians realized that there was no Western help in sight with the chosen policy, the "last-ditch neutrality boys like Undén [would] be overruled" at last.

In order to put more than words behind his tactics, Matthews recommended isolation when it came to military matters. "The cure for isolationism is isolation," Matthews argued, and he recommended strict restraint when it came to military promises of assistance or Sweden's procurement of US military technology.[34] Since Sweden was able to buy some surplus military equipment from the United States during this period, Matthews's recommendations were not always followed in the State Department, but there was a period of constraint, for instance, in matters of radar technology, which Sweden needed badly. In addition, Britain sold Sweden a substantial volume of military technology at the same time, a fact that made Matthews despair in his ambition to get the Swedes off the "neutrality flagpole."

Unusual among his diplomatic peers, Matthews seems to have thought systematically about military technology and about Sweden's dependence on the United Kingdom and the United States. His argument to counter isolationism with isolation surely implies this. He knew that Swedish war planning assumed Western military assistance, and he knew that Sweden would not be able to become self-sufficient in military technology (an insight shared by some of his Swedish friends and one of their common grounds for understanding). Matthews's strategy was to undermine Sweden's ability to build up its military forces by denying it access, not to new military technology generically, but to new Western technology, thus barring Sweden's integration into Western defense strategies aimed at the Soviet Union. This would make the "last ditch neutrality boys" understand how untenable being neutral actually was. They would not be able to make that neutrality credible without a strong defense, which could be had only by integrating into Western defense in general.

A Scandinavian Defense Pact

In 1950, Walton Butterworth replaced Matthews, also a career diplomat with the Department of State. At the same time, a change in the United States' attitude toward Sweden was noticeable. All along, Matthews and Foggy Bottom had not seemed to have had entirely corresponding views of how to handle Sweden. Matthews's diplomatic dispatches were often drafted as arguments against Sweden's neutrality, but their tone was often heated and hortatory, as if Matthews were actually trying to convince the State Department, not the Swedes, of the perils of neutrality. Even within the State Department there were different ideas about how to handle Sweden's neutrality and reform Nordic thinking.[35]

The main reason for the change of attitude has often been attributed to the unsuccessful negotiations of Denmark, Norway, and Sweden concerning a Scandinavian Defense Pact (SDP) in 1949, and to the ensuing decisions of Denmark and Norway to join NATO. The question of a defense pact rose to prominence in the spring of 1948, when Sweden made a proposal to Denmark and Norway that led to formal, partly secret, unsuccessful negotiations that ended at the start of 1949. But according to dispatches from the American legation in Stockholm, the idea of a Nordic bloc had been aired at least as far back as 1946.[36] Undén had been a strong opponent of a defense pact, but sentiments among the Swedish public made the Social Democrats change their attitude. A speech by Norway's Minister of Foreign Affairs, Halvard Lange, also indicated for the first time that Norway might be interested in the formation of Scandinavian military cooperation. In the autumn of 1948 the three countries created a committee to investigate the possibilities of a Scandinavian defense pact. For many Swedish critics of the neutrality policy, those negotiations indicated that the government was at last willing to reconsider its policy.

The Swedish historian Krister Wahlbäck has described the Swedish initiative and the subsequent events as a Swedish-American tug-of-war over Norway that the United States eventually won. All the Scandinavian countries were willing to create an SDP, but they differed on the global status of the pact. Both Norway and Denmark argued for association with Western powers, whereas Sweden wanted the pact to be an extension of Sweden's neutrality. All three nations saw the benefits of military cooperation and realized that none of them could withstand a Soviet attack for long without British and American assistance. The basic dividing line was to what extent such assistance would have to be prepared in advance. Denmark and Norway argued that the scale of preparations, as well as the Scandinavian need for procurement of British and American technology, required some sort of formal agreement with the West—something the Swedes saw as impossible because of their neutrality policy.

During the whole period of negotiation, US diplomats argued forcefully against an SDP, at least in its Swedish version. Ironically, the success of US diplomacy to shepherd Denmark and Norway into NATO strongly undermined Matthews's tactics. After the talks broke down, Sweden returned to an outspoken nonalignment policy, and thus Matthews's goal became even more distant than it had been. But now that the United States had secured access to Greenland and Spitsbergen through Denmark and Norway, Sweden's neutrality was recognized as much less problematic. The State Department reconsidered its relations with Sweden, the strongest military nation among the Nordic countries, and now regarded a strong Swedish defense capacity as a first line of defense in the North.[37] Rather than try to cure isolationism with isolation, the United States now accepted Sweden's neutrality and tried to make the best of the situation.

During the SDP negotiations, much effort was made to analyze various military preconditions for a common defense of the three Scandinavian countries. It is surely fair to say that a coherent picture as well as a silent agreement evolved among military experts from all three countries as to how the Scandinavian forces could best be coordinated if a Soviet attack were to come. The negotiations focused on two different options: first and foremost, a formal defense pact with mutual responsibilities; if that proved impossible, a "small solution," with more informal cooperation. The report implicitly concluded that the difference between a formal defense pact and a small solution was perhaps more of degree than of kind. In a secret letter to the Swedish government commenting on the negotiations, the Supreme Commander of the Swedish Defense Forces, Helge Jung, pointed to the need for good contacts with the Western nations, especially Britain and the United States.

The breakdown of the formal negotiations did not stop the ongoing discussions of the Scandinavian military leaders. In May 1949, three months after the end of the talks, the British Embassy reported to London the existence of an agreement between the Swedish and Norwegian air forces on a division of labor in case of a Soviet attack.[38] Sweden would deploy most of its air force to the south, repelling a Soviet attack from the southwest through Denmark; Norway would be responsible for the Northern defense.

The Chief of the Swedish Defense Staff, Nils Swedlund, suggested to Prime Minister Tage Erlander that some of the mutual defense planning that had been discussed during the talks should be allowed to continue—informally, of course, and concealed from the Swedish public and the world.[39] Swedlund produced a memorandum proposing that twelve detailed plans be drawn up to deal with matters such as radio and wire communications between military agencies in the three countries, coordination of air surveillance and air defense command and control, and the establishment of standardized procedures for reporting and identification processes. Furthermore, Swedlund suggested plans for army cooperation in the area around Treriksröset (the point where Norway, Sweden, and Finland meet in the North) and along the railway between Kiruna in Sweden and Narvik on the Norwegian coast. He also suggested plans for naval cooperation with Denmark in the Southern Baltic Sea, and with

Norway concerning Kattegat and Skagerak in the North Sea. Erlander made Swedlund understand that he was positively inclined toward at least some of the plans, namely those he termed technical. In his diary, Erlander noted that Swedlund's memorandum concerned coordination on military telephony. "It presumably concerned harmless matters," he added.[40] The notes in Swedlund's diary regarding the same meeting and its context shows that he had discussed the coordination issue with his Scandinavian colleagues. In these discussions, technical matters were basic to any cooperation. Without a removal of specific technological differences, there could be no cooperation to speak of. What Erlander saw as "harmless matters" represented for Swedlund and his colleges the basis for cooperation.[41]

In June 1949, Norway's Minister of Defense, Jens-Christian Hauge, met his Swedish counterpart, Allan Vougt, and expressed his (and, he claimed, the Danish government's) wish for combined planning. At a September meeting with the Swedish government, Swedlund again briefed them on the plans for cooperation; afterward, he was given instructions to go ahead with communications, coordinated air surveillance, air defense command and control, air force search and rescue, and military weather service. The army cooperation was delayed because Erlander was hesitant, although he was generally in favor. Minister of Foreign Affairs Östen Undén informed Swedlund of the decision of the government, and also of an earlier meeting in Copenhagen at which Scandinavian foreign ministers had agreed to exchange "technical results" and to prepare technical cooperation—the latter, of course, Undén stated, "without engagement."[42] It is hard judge this on the basis of the sparse material at hand. One interpretation that fits well with Swedlund's further actions is that he understood the latter to be a supplementary instruction—i.e, that he could set up communications, arrange air force cooperation, and, more generally, make further technical preparations.

Swedlund soon developed rather substantial cooperation with Norway, well beyond the instructions he had been given in 1949, but not necessarily without governmental acceptance. Norway's joining NATO in 1949 did not in any way deter the close Scandinavian connections; on the contrary, it provided Sweden with a potential channel to England and the United States. All the cooperation was conducted with strictest secrecy.

The "Dropshot" War Plan

The evolving division of labor between Norway and Sweden coincided with developments in US military planning. In 1949, a committee of the Joint Chiefs of Staff prepared an extensive plan for a possible war against the Soviet Union. This plan, called Dropshot, was only one in a series of war plans. It was, in the eyes of both Norwegians and Swedes, most disturbing because of its implications for their fate in a superpower conflict.

The overall objective of Dropshot was to destroy the USSR's will and capacity to resist. The starting point of the scenarios described in the plan was that the Soviet Union had launched an attack on continental Europe, going through the western parts of Germany, over the Rhine, and into France, Belgium, and the Netherlands. In connection with Scandinavia, the planners in Washington estimated a Soviet invasion of Denmark from the south using "5 line divisions and 400 combat aircraft," defeating and occupying Denmark within 14 days, and then making a "subsequent attack against southern Sweden."[43] The invasion of Sweden and Norway would commence about 90 days after the beginning of the attack on continental Europe, the Soviets using approximately "13 line divisions and 600–900 tactical aircraft" and having the capacity to "triple the initial number of divisions and build up to more than 2000 combat aircraft if required."[44] Contemporary Swedish war planning envisioned four alternatives for a Soviet invasion of Sweden: one in the southern province of Skåne, one in the north through Finland and Norway, a combination of the preceding two alternatives, and an invasion across the Baltic Sea toward Stockholm and central Sweden. In all these cases, Sweden's objective was to buy time while awaiting assistance from Britain and the United States.[45] At the end of 1947, the Supreme Commander of the Swedish Defense Forces ordered a separate study for two versions of the first alternative (a Soviet invasion of southern Sweden), focusing on an invasion from Danish territory at Öresund and/or Gothenburg.[46]

In a section on military alternatives, the Dropshot planners discussed Scandinavia in two separate scenarios, one concerning Norway and Sweden and the other concerning Denmark. The reason for this was that Sweden's neutrality precluded a combined defense of Scandinavia. Denmark, because of its topography, was said to be "virtually indefensible"—a conclusion much in line with what the Swedish military had come to conclude in the SDP negotiations, and a view expressed in a secret appendix by the Swedish members of the committee and delivered to the Swedish defense minister in connection with the negotiations.[47] Norway and Sweden were seen as strategically significant from both offensive and defensive standpoints. They could provide air bases 500–700 miles closer to the Soviet Union than those in the British Isles. In addition, holding Norway and Sweden would deny the Soviets free use of the Baltic Sea. Maintaining control over Swedish and Norwegian territory would deny the Soviet Union a base area for air and naval operations over the North Sea. If the Soviet Union occupied Norway and Sweden, it would be able to establish bases for air and naval operations against England and the North Atlantic.[48] Sweden's uranium, iron ore, and industrial products also figured in these plans.[49] The United States and the United Kingdom had tried in 1945 (before Hiroshima) to get some kind of control over Swedish uranium by demanding exclusive first option use—a request that was denied by the Swedish government, which instead formally assured the US and the UK that it would exercise absolute control over the uranium and informally agreed to keep them informed on the matter of Swedish use.[50] The Dropshot plan estimated that the equivalent of 20 US divisions and 1800 combat aircraft would be needed to hold the Scandinavian peninsula. The available Swedish and Norwegian forces (estimated forces available 45 days after the continental invasion) totaled 13 divisions and 1400 combat aircraft.[51]

Another important aspect of Scandinavia's military situation was the imbalance of military power. Though formally Norway could mobilize six divisions, these were characterized as "skeletonized." The report estimated that Norway would be able to mobilize two divisions 90 days after the continental invasion. The air force of Norway was estimated to contain about 100 combat aircraft. Sweden, on the other hand, was predicted to be able to mobilize 18 divisions 90 days after the continental invasion, and its air force, with 1300 combat aircraft, was described as "one of the best air forces in Western Europe." And Sweden would soon have the capacity to manufacture its own jet fighters.[52]

In a section titled "selection of allied courses of action," defending Central Europe was assigned the utmost importance. The alternative of holding Norway and Sweden was briefly touched upon, but Sweden's neutrality was seen as a factor hindering any effective help. The planners concluded:

The provision of the required aid would be justified only in the event it develops that there is a reasonable assurance that Sweden would join with Norway and Denmark in a concerted defense of Scandinavia in the event that the Soviets elect to exercise their capabilities against any of the three Scandinavian countries. . . . Based upon the above considerations, the course of action "Hold Norway and Sweden" is rejected.[53]

During the SDP negotiations, the United States had opposed the formation of such a pact. Thus, it seems rather surprising that the US military argued so forcefully for a concerted defense of Scandinavia. The Dropshot plan clearly indicates that the fate of Norway and that of Sweden were, in the eyes of the military planners in Washington, linked. If the results of the planning process of which Dropshot was a part were communicated to Norway (by then a member of NATO) or to Sweden, these considerations surely must have provided a strong incentive for closer military cooperation between those two countries.[54]

A Shift of Balance

Cooperation among Denmark, Norway, and Sweden developed further during the early 1950s.[55] But at the same, those countries gradually built up their relations with Britain and the United States. In 1949, in the above-mentioned special letter in 1949 to the Swedish defense minister in connection with the failure of the SDP negotiations, Supreme Commander Helge Jung provided a clear rational for this: ". . . while isolation from Denmark and Norway would be negative for our country, isolation from the Western powers would seriously jeopardize our defense possibilities and provision. . . . A policy that jeopardizes provision from the West must therefore, from a military point of view, be strongly rejected."[56] This was entirely consistent with Swedish war planning, which, as I have mentioned, was based on the assumption of Western help in the event of Soviet aggression.

A decisive step toward closer Norwegian relations with England and the United States was taken in September 1951, when NATO's Northern Command (AFNORTH) was established at Kolsås, near Oslo.[57] This temporarily affected Sweden's relations with Norway, since the Swedish military was not certain whether previous agreements between Norway and

Sweden were still valid. Nils Swedlund, who in 1951 became Supreme Commander, replacing Jung, did not know who was actually in charge of the military forces in Norway. At a meeting with Swedlund, Norwegian Prime Minister Oscar Torp assured him that the plans would remain in force, and also promised to instruct the Norwegian supreme commander to share the Atlantic Pact's defense plan for Scandinavia with him.

Contacts between Sweden and England soon increased in intensity. Interactions between military personnel in England and Sweden soon led to extensive exchanges of information. Sweden's Air Force and Navy both had strong traditions of affinity to their British counterparts. In March 1951, Swedish Defense Minister Allan Vougt sent a memorandum concerning Sweden's armed forces to the British Air Minister Henderson, who had asked Sweden for information on war planning in the case of a Soviet attack. Vougt's comprehensive memorandum contained top-secret information on Swedish war planning and capacity.

In mid 1952, the United States and Sweden exchanged diplomatic notes concerning Sweden's purchase of military equipment and services from the US government. On the request of the Swedes, the agreement was kept secret for a period of two years. Discussions concerning the conditions of such an agreement had been going on since the spring of 1951 and had concerned mainly the United States' desire that Sweden join the international system of export restrictions as implemented by the Coordinating Committee for East-West Trade (CoCom). In February 1952, following a new report by the National Security Council on the position of the United States with respect to Scandinavia and Finland, President Harry Truman declared that Sweden was eligible to purchase military technology in accordance with the Mutual Defense Assistance Act of 1949. The report of the NSC was a revaluation of the United States' position with respect to Sweden. After a prolonged period of trying to get Sweden "off the neutrality flagpole," the US now seemed to accept Sweden's neutrality:

Although on balance, and primarily because of the advantage to the organization of Scandinavian defense, it would be to our interest to have Sweden in NATO, we must for the predictable future accept as a political fact Sweden's policy of avoiding great power military alliances and calculate accordingly those means and methods best designed to increase Sweden's contribution to Western defense.[58]

The report concluded that the United States should make Sweden eligible to receive military assistance "on the same basis as other nations whose ability to defend themselves is important to the security of the United States,"[59] and that the Swedish military establishment should further be strengthened by exchanging military information and by having Swedish officers attend American technical schools. During an informal trip to the United States in April 1952, Prime Minister Tage Erlander met with President Truman and discussed the matter.[60] Sweden agreed not to re-export classified technology without prior permission from the United States, and also to develop internal security measures to prevent transfers of technology or knowledge to the Soviet bloc.[61]

Technological Dependence

But all the apparent logic of the above explanation is not sufficient for an understanding of either the change in US diplomacy toward Sweden or the reason for the increase in military cooperation. The British had sold Sweden military hardware, undermining Ambassador Matthews's strategy of curing Sweden's isolationism with isolation. But that in itself does not explain the change. So far, the story has focused on high politics and military considerations in connection with the foreign policy of Sweden (and Norway). A missing link is hidden in the considerations of the military, both in the United States and Sweden, and in the exchanges between these countries that started as soon as World War II ended. Thus, the story of the SDP negotiations and of the strategic considerations of how to defend Scandinavia against a Soviet attack must be complemented by observations of the ongoing cooperation between Sweden and the United States.

After World War II, the US War Department issued a Post-Hostilities Mapping Plan (the "Casey Jones" Plan) according to which the western parts of Europe and Africa would be scanned from the air to gather basic data for various maps and charts. One objective was to obtain data of sufficient quality for the compilation of strategic and tactical maps of, among other areas, the coastline of Northern Africa, the Middle East, Greece, Italy, Austria, and the western portions of Germany, Sweden, and Norway. Though most countries approved of the aerial survey, Sweden declined on the basis of its neutrality policy. This caused the War Department some concern, since the area was considered to be of strategic importance. As a way of getting around Sweden's neutrality, Lieutenant Colonel G. S. Smith of the Military Intelligence Division of the War Department suggested that the United States offer "to loan material and personnel as required to enable the Swedish government to conduct its own survey and to furnish the United States a copy of data obtained."[62] The Swedish government declined, on the ground that this arrangement too would formally violate Sweden's neutrality policy. Nevertheless, the United States sold Sweden six F-9 airplanes (a modified version of the B-17 bomber, stripped of all armament) and lent Sweden cameras and related equipment; in addition, a limited number of Swedes would be trained in the United States to carry out the survey project "independently." All this would of course be handled with the utmost discretion. Should it become known by the Soviets, Sweden could always claim that all the work was done with Swedish materiel and personnel, primarily for Swedish use. Selling the F-9s to Sweden avoided political complications for the Swedish government. The systemic military strategic requirements could be achieved without Sweden's having to depart formally from its self-declared neutrality. The subtext was quite different, though: the provision of maps to the US War Department indicated that Swedish airspace would be used by US forces in the event of war, and that Sweden was prepared to facilitate US military preparations well in advance. Obviously, the War Department considered the project highly important.

In April 1945, the United States sold 50 P51-D20 Mustang fighter planes to the Swedish Air Force. In March 1946, the Swedes wanted to buy 100 more Mustangs. In the United States, the sale was generally favored by both the State Department and the War Department, but it contradicted Senate regulations on the handling of surplus tactical aircraft. The State Department had earlier assured Congress that it would sell such planes to only a few specified countries.[63] The sale to Sweden received preliminary approval, but then Secretary of State Dean Acheson decided against it.[64] In a letter to Acheson, Secretary of War Robert Patterson argued in favor of the sale, referring to the joint mapping project.[65] In October the State Department finally approved the sale of 90 Mustangs, to the satisfaction of the Swedish Air Force. The deal was a chit linked to the ongoing mapping project. Both were steps, however small, toward Sweden's convergence with the West.

The mapping project was in itself only one of a series of preparations for possible air warfare against the Soviet Union. Sweden's geostrategic position was emphasized in various ways, one of them when the Strategic Air Command's Director of Intelligence, a Colonel Bently, received an invitation "to go hunting in Sweden" and "talk things over with his opposite number."[66] The trip was of course supposed to be completely unofficial, and publicity was to be avoided. The Strategic Air Command had also prepared war plans that involved crossing Norwegian and Swedish airspace on the way to the Soviet Union.[67]

Although the mapping project was carried out at an early date, the formation of strategy for atomic warfare developed over a long period. In 1950 President Truman issued an decision to allow secret air missions over Soviet territory for an aerial photographic survey, radar detection, and signal intelligence.[68] Aerial photographic and radar-detection missions were carried out with unmarked British and American aircraft from bases in Britain and West Germany, which deeply penetrated Soviet territory. The purposes of those penetrations were to collect data for specific approaches to bomb targets and to produce a sequence of radar pictures as guides for future bombing missions. Signals intelligence concerning the Soviet air defense system did not require penetration; it could in general be done along the Soviet Union's periphery. Still, aircraft sometimes crossed the Soviet border in order to investigate the reactions of the air defense system.

Sweden's intelligence organization had nurtured close relations to its British and American counterparts in the postwar period. Because of Sweden's proximity to the Soviet Union, US intelligence organizations approached their Swedish counterparts and offered their cooperation. Sweden bought signal-intelligence equipment from the United States at substantially reduced prices and used it for reconnaissance in the Baltic Sea.[69] At the beginning of the 1950s, Swedish officers started to take part in courses at American and British intelligence schools. One of the students, Stig Synnergren, eventually served as Sweden's Supreme Commander (1970–1978).[70]

We still know relatively little about the intelligence cooperation between Sweden and Western countries, since most of the material is classified. However, we do know of close cooperation between US military attachés in Stockholm and the Swedish Defense Staff. The flow of information from

Stockholm to Washington was impressive in volume.[71] From the available material it is clear that information concerning Sweden's infrastructure (airfields, railways, harbors, highways, industries, power plants, etc.) was combined with information on the status of the armed forces (men and materiel) and technological information on specific development projects (for, e.g., jet engines and fighter planes). The fact that the infrastructure reports often included aerial photos (positives and negatives) indicates that the American personnel in Sweden either had access to aircraft with which to take such photos or were given the photos by the Swedish military. The number of reports runs to thousands. Altogether, the information acquired in Sweden must have provided the US military with a very good picture of Sweden's military capacity and thinking.

In addition to the reports covering Swedish conditions, there were a number of reports that contained information on the Baltic Sea, Finland, the Soviet Union, Estonia, Latvia, Lithuania, Poland, and East Germany. Swedish reports on the Soviet merchant marine and navy were combined with aerial photos of Soviet naval ships and subsequent analysis of their gunnery and other vital information. Swedish intelligence concerning Soviet activities was communicated through the defense staff to the American attachés, as were police interrogation reports concerning refugees from the Eastern Bloc. For a number of years, the American embassy also collected "POLSAM" (Petroleum, Oil, Lubrication Samples) from Soviet ships in Sweden's harbors and sent them for analysis to Washington. The metallurgy of a Polish airplane of Soviet origin was made available through a small sample of aluminum. Most of the material from Sweden was widely distributed within the American defense and intelligence establishments. This military information was most likely collected from various sources and in various ways, some legal and some illegal. But it is also clear that the relationship between the Swedish and the US military was intimate. Such evidence, in and of itself, might not necessarily have compromised neutrality had the Swedish military treated all foreign military attachés the same. But there is no evidence that Soviet officers enjoyed similar access. At one point, the Soviet diplomatic delegation in Stockholm expressed its displeasure with the close contacts between the Swedish and the American military. The Swedes' answered such complaints by arranging larger meetings of all foreign attachés and by staging demonstrations of Swedish military facilities. After one demonstration, the American contingent reported to Washington with satisfaction that the Swedish military had hidden substantial amounts of equipment in order not to disclose too much to the Soviets.[72]

Systems and Strategic Missions

Seen from a strategic viewpoint, access to Sweden's airspace for Western forces in the event of war proved to be of vital importance. If Sweden wished to remain neutral in the event of war, its air force would have to counter any violation of Swedish airspace. Asked by the press, Swedish government ministers would answer that Sweden would defend its airspace against any penetration, regardless of origin. But this was generally an area of much ambiguity. American diplomats monitored and commented upon these public statements. At the same time, Swedish diplomats in Washington and military leaders in Stockholm reassured them that those statements should not be taken too seriously.[73] The foreign "policy" of Sweden was not decided on; the policy instead was a rather bewildering outcome of different signals at different levels. Thus, Donald MacKenzie's observation that policy is not a decision but an outcome is fully exemplified by the case of Swedish neutrality.

From 1947 to 1951, US war planning regarded the possibility of defending the European mainland with pessimism.[74] (This gradually gave way to a more optimistic view.[75]) Instead, US military plans emphasized maintaining some bridgeheads into Europe for an eventual counteroffensive. Britain and to some extent Scandinavia were judged crucial, for Scandinavia offered air bases more than 1000 kilometers closer than Britain to the industrial regions of the Soviet Union.[76] At least as important was keeping the Soviets from having access to Scandinavia. Although it is hard to find material on actual war planning concerning the Scandinavian peninsula in the 1950s and later, we do know that British war plans later included overflight of Sweden, with Gothenburg in Western Sweden designated the point of no return.[77] Extrapolating from the strategic bombing missions of World War II, one imagines that such missions would have been planned for bombers with fighter escorts. If Sweden were to be given so-called indirect support (for example, US bombing of Soviet harbors from which an attack against Sweden was anticipated), the Swedish Air Force would have to provide escorts or would have to put Swedish bases at the disposal of British and American forces.

The cooperation was partly channeled through an air force search and rescue agreement with Norway and Denmark, called SVENORDA. This cooperation involved air surveillance, air defense command and control standardization, and military weather service in case of war, as well as a net of wire communications between air guidance centers in Norway, Denmark, and Sweden. The cooperation involved exercises and the standardization of command terminology to correspond to NATO terminology, which included setting specific radio frequencies for communications between aircraft and air guidance. As part of the agreement, officers from the three countries met frequently and regularly.[78] This agreement was kept secret until 1960 because it was seen as sensitive. Through this cooperation, Swedish officers learned about the terrain, facilities, airfields, and communication routines of the two Scandinavian NATO countries—and, of course, vice versa.[79]

Norway and Sweden coordinated the installation of new radar stations to maximize their coverage. After Sweden had established a radio relay network as the backbone of a combat control system, Norway and Denmark connected to the system. The possibility of transferring communication on this net to NATO's European net existed, not least through AFNORTH in Kolsås, Norway. Swedish Air Force officers thus gained good insight into facilities, bases, and communication routines in Norway and Denmark, and, subsequently, into NATO routines. The 1994 report *Om kriget kommit . . .* discussed whether the arrangement made possible exchange of radar images, but concluded that the capacity of the communication probably was too low.[80]

Identification of aircraft is important in peacetime air search and rescue and in wartime air guidance and coordination. During World War II, England developed the first radar aircraft-recognition technology, which made it possible to determine the identity of an aircraft carrying a passive transponder. The Swedish Air Force bought the British Mark III IFF (identification friend or foe) system early on, but it also started a project to develop a Swedish IFF technology. Still, Swedish IFF technology was, from the start, based on Western contacts and designs.[81] The person responsible for the development at the Research Defence Establishment (FOA) traveled to the United States and Britain to gather information. The ground was partly prepared for him by Martin Fehrm, head of section 3 at FOA and later director-general. Fehrm and a military officer traveled to the United States in 1946 to study IFF. In a memorandum concluding the trip, they suggested procurement and testing of "foreign IFF-systems."[82] They listed about 15 different IFF systems, and they outlined the technological principles in detail. The person responsible for IFF development at FOA spent half a year at Marconi College in the UK in late 1947.[83] With the help of Fehrm's contacts and the Swedish military attachés in Washington, he also traveled to universities, research laboratories, and defense industries in the United States. Between late 1947 and early 1949, FOA developed a prototype intended for use by all the military services; later in 1949, it received grants from the services to continue design work.[84] A more nearly complete system was tested in 1957. It seems probable that Sweden's air defense system was made compatible with the Mark X system used by NATO at the time. At least one document exists in which the Chief of the Swedish Air Staff makes clear that any deployment of the system would be made dependent on the ability to install IFF Mark X stations at the same time.[85] That document clearly spells out the Swedish Air Force's desire to distinguish between Swedish aircraft and foreign aircraft. But the Swedish Air Force also wished to identify NATO aircraft—something not strictly necessary if Soviet and NATO planes counted equally as intruders. At the same time, Swedish military engineers were being trained in use of the Mark X system.[86] Only a few years later, NATO introduced yet another version, Mark XII. It is not clear whether Sweden's air defense system incorporated it. Military engineers claim that it would have been fairly easy to make the Mark XII system compatible with the Swedish IFF system, had Mark XII been supplied.[87]

At the end of the 1950s, the United States introduced LORAN C, a longwave, long-distance navigation system for strategic submarines and aircraft. In the 1950s, a precursor of this system, LORAN A, was used by NATO forces. During the buildup phase of Loran C, the United States asked Sweden to provide permission to calibrate the system in Stockholm and Gothenburg. After some investigations by a preparatory committee the government granted the request. The National Defense Research Establishment

(FOA) had previously studied LORAN A by obtaining technical data from the United States and was well aware of the potential strategic use of LORAN C by the US Air Force.[88] From a systems perspective, the calibration of LORAN C was of great importance, since calibration as close to the Soviet border as possible would make the system more reliable and navigation more precise. Furthermore, the Strategic Air Command needed good accuracy over Swedish territory for mission plans into the Soviet Union that included the use of Swedish airspace. Calibration might traditionally be considered a technical problem, but in this context it was what Thomas Hughes would label a "critical problem." The problems were not technical; rather, they had to do with access to Swedish airfields and airspace. NATO, the US Air Force, and Sweden solved them by political means.

Cooperation extended into much more mundane details as well. In January 1953, Secretary of State Dean Acheson sent a top-secret telegram informing the US embassy in Stockholm that the Department of Defense was "willing [to] consider additional stocks [of] AVGAS required by Swedish Airforce."[89] Acheson wanted to know the existing stock of gasoline, the peacetime monthly consumption, the minimum stock desired for war reserve, and Sweden's storage capacity. Acheson also wanted to know how many days the Swedes thought such a reserve would last in "wartime operations." There are almost no further sources to establish what happened subsequently, but we do know that the US ambassador briefed Prime Minister Erlander on the issue a few days later.[90] Acheson's telegram in itself indicates that the United States had been selling gasoline to the Swedish Air Force, and that there was a reserve. The level of handling and the classification surely indicate that this was a rather sensitive issue, and the available Swedish information (which touches on the subject only briefly) clearly states that Sweden switched to NATO-standard aviation fuel in 1960.[91]

Sweden's air defense strategy was based on the concept of decentralized air bases. Swedish aircraft were designed to be able to use short runways, and Swedish war planning centered on using reinforced highways as runways. NATO's strategic bombers and escort fighters were heavier and thus required longer runways and load capacity than the Swedish aircraft. Supreme Commander Helge Jung had discussed this subject with some in the Swedish government as early as 1948, and he had been instructed to analyze the question of US bases in Sweden. During the 1950s a small number of runways, primarily located along the eastern coast, were reinforced and extended to be able to accommodate NATO aircraft.[92] The exact reasons for building these runways remain unclear. They may have included accommodating damaged SAC aircraft returning from bombing missions over the Soviet Union, or they may have been intended for a more systematic reception.[93]

Conclusion

What all these preparations add up to is perhaps best understood by means of the concept of the operational scenario. Taken one by one, they might not seem to constitute more than fractured preparations, but taken together they all facilitated a strategic mission of the US Air Force across Swedish air territory with an infrastructure well prepared for warfare against the Soviet Union. The exact scenario is hard to discern, since the plans of the Strategic Air Command remain hidden in the archives (with other information). But it would be wrong to assume that all these technological preparations were carried out without a reason. It is also worth pointing out that the strategic importance of Scandinavia increased with the buildup of the Soviet Union's northern fleet during the 1960s and the 1970s.[94]

Through technological preparations, Sweden became a hidden member of NATO—something the Soviets, at times, complained about. As early as 1948, US Assistant Naval Attaché R. A. Winston had a conversation with Bengt Nordenskiöld, Chief of the Swedish Air Force, on the matter of US assistance. Nordenskiöld saw the possibility of channeling assistance through Swedish attachés in Washington and American attachés in Stockholm. In his report, Winston wrote: "This opens up the possibility of standardizing Swedish and US tactics and equipment, particularly in the air forces, since Sweden is using a considerable number of surplus US aircraft such as Mustang fighters, Harvard trainers etc. . . . In effect, it would build up a valuable backlog of highly trained combat units using US equipment and tactics, led by officers known to be strongly anti-Russian, who might be able to influence the Swedish government to abandon its policy of strict neutrality in the event of war."[95] During this period, the main objective of the United States was still to convince Sweden of the benefits of joining NATO.

In April 1954, the US ambassadors to Denmark, Norway, Sweden, Finland, Switzerland, and Iceland met in Copenhagen to discuss various aspects of US policy toward the Nordic countries. One issue was the "special cases of Sweden and Ireland," including the "military cooperation of Norway and Denmark with Sweden."[96] The meeting was chaired by Hayden Raynor of the US State Department. In the minutes of the meeting, the trend from 1949 on was described as "increasing *de facto* cooperation with the US."[97] Convincing Sweden to join NATO was no longer the prime objective. Now the ambassadors saw closer military cooperation with Denmark and Norway and dependence on US supplies as a way of reaching the "real gain represented by Swedish membership in NATO." The goal "should be the practical one of increasing the usefulness of Sweden to the defense of the Scandinavian area."[98] They concluded as follows:

The Swedish Government, while continuing its overt attachment to an independent course, has clearly shown its desire for closer relationships in fact. The question of Swedish membership in NATO thus becomes somewhat academic compared to certain practical questions:

1. The possibility of providing Sweden with certain information concerning NATO planning.
2. The ability to release to Sweden certain classified information and equipment which would directly strengthen its defense forces.
3. The arrangement of procedures for the exchange of technical information and intelligence with Sweden.
4. Encouragement of closer defense cooperation between Sweden and its NATO neighbors.

Attainment of objectives such as these would be just as effective as Swedish membership in NATO as a means of strengthening the West. . . . If, as is understood, important military advantages would flow from their achievement, we should consider whether it is not now necessary to realign our sights on the substantial objectives, without reference to the question of formal Swedish membership in NATO.[99]

In the subsequent summary, the ambassadors agreed to "live with the non-alliance policy."[100]

In conclusion, then, referring to the overall argument of my chapter and all the technological preparations and adjustments made in Sweden, this, for the US ambassadors, represented the "real gain," not formal Swedish membership in NATO. It might also be worth pointing out that, even though the strong and clear-cut wording of the ambassadors can hardly be misunderstood, much of the cooperation on the military level took place directly between the Swedish military and their foreign equivalents, not at the diplomatic level.

As is evident, the ambassadors clearly acknowledged the impossibility of changing Sweden's official policy of neutrality, but they also saw how military cooperation, intelligence exchange, and technological orientation could turn the neutrality policy into a rather empty concept. In this way, they reversed the argument that had been put forward by H. Freeman Matthews: isolationism was to be cured not by isolation but by cooperation.

Swedish debates have traditionally portrayed the technological and military cooperation between Sweden and member states of NATO as a way for Sweden to obtain liberty of action in the event of war. I would argue that this traditional portrayal is misleading. Technology is not made up of dispersed artifacts alone; it is interconnected, systemic, and social. Technology includes rules, career training, routines, and standardization, and it facilitates some possible actions while prohibiting others. Technology is a silent but very convincing "actor." Making those aspects of Sweden's defense system visible helps us to judge the content of Sweden's neutrality during the Cold War. The argument that technology opens specific avenues into the future and closes others may seem to be based on a deterministic understanding of technology's role in society; however, if the above case of Sweden's neutrality policy, written as history of technology, is portrayed as technological determinism, it is a socially and politically constructed determinism.

Notes

1. Walter LaFeber, "Presidential Address: Technology and US Foreign Relations," *Diplomatic History* 24 (2000), p. 3.

2. Ibid.

3. Michael J. Hogan, ed., *America in the World* (Cambridge University Press, 1995).

4. Jörg Martin Gabriel, *The American Conception of Neutrality after 1941* (Macmillan, 1988), p. 3.

5. Ibid., p. 1.

6. Marc Trachtenberg, *A Constructed Peace* (Princeton University Press, 1999).

7. Wilhelm Agrell, "Sweden and the Cold War: The Structure of a Neglected Field of Research," *Scandinavian Journal of History* 10 (1985): 239–253.

8. Ann-Marie Ekengren, Sverige under kalla kriget, 1945–1969. Report, Gothenburg University, 1995.

9. Among them the historian Wilhelm Agrell stands out as the leading scholar. He has published a series of books dealing with different aspects of Swedish neutrality and defense policy, including books on the Swedish nuclear weapons program, Swedish cooperation with the West, and the overall structure of the defense sector in relation to science, technology and doctrine. See *Alliansfrihet och atombomber* (Liber, 1985); *Den stora lögnen* (Ordfront, 1991); *Det välorganiserade nederlaget* (Ordfront, 1990); *Vetenskapen i försvarets tjänst* (Lund University Press, 1989).

10. See Agrell, *Den stora lögnen*; see also Paul Marion Cole, Neutralité du Jour: The Conduct of Swedish Security Policy since 1945, dissertation, Johns Hopkins University, 1990.

11. See, for example, Michael H. Hunt's three categories of (1) "realists" or "postrevisionists," (2) historians who focus on the interaction of policy and the domestic sphere, and (3) those who place foreign policy in an international context ("The Long Crisis in US Diplomatic History: Coming to Closure," in *America in the World*, ed. Hogan).

12. David Holloway, *Stalin and the Bomb* (Yale University Press, 1994), p. 364.

13. Walter A. McDougall, *The Heavens and the Earth* (Johns Hopkins University Press, 1985); Stuart W. Leslie, *The Cold War and American Science* (Columbia University Press, 1993); Paul Forman, "'Swords into ploughshares': Breaking new ground with radar hardware and technique in physical research after World War II," *Reviews of Modern Physics* 67 (1985): 297–455; Michael Aaron Dennis, "'Our First Line of Defence': Two University Laboratories in the Postwar American State," *Isis* 85 (1994): 427–455; David Hounshell, "The Cold War, RAND, and the Generation of Knowledge, 1946–1962," *Historical Studies in the Physical and Biological* 27 (1997): 237–267.

14. Paul N. Edwards, *The Closed World* (MIT Press, 1996).

15. Donald MacKenzie, *Inventing Accuracy* (MIT Press, 1990), pp. 412–413.

16. Ibid., pp. 395–409.

17. Ibid., p. 400.

18. Thomas P. Hughes, *Networks of Power* (Johns Hopkins University Press, 1983); Hughes, "The Evolution of Large Technological Systems," in *The Social Construction of Technological Systems*, ed. W. Bijker et al. (MIT Press 1987).

19. *Om kriget kommit . . .* , Betänkande av Neutralitetspolitikkommissionen (Statens offentliga utredningar, 1994). There is also a separate appendix with selected source material. All the following references are to the Swedish report, since the English translation (*Had There Been a War . . .* , 1994) omits the notes.

20. Nils Andrén, "On the Meaning and Uses of Neutrality," *Cooperation and Conflict* 26 (1991): 67–83.

21. H.W. Brands, *The Devil We Know* (Oxford University Press, 1993), p. 5.

22. Vladisav Zubok and Constantine Pleshakov, *Inside the Kremlin's Cold War* (Harvard University Press, 1996), p. 34.

23. See Geir Lundestad, *America, Scandinavia, and the Cold War, 1945–1949* (Columbia University Press, 1980), pp. 329–358. "With the Cold War temperature increasing and America's involvement in European affairs growing," Lundestad writes, "Washington stepped up its demands of loyalty from the Scandinavian countries" (ibid., p. 351). On the gradual change of the US position toward Scandinavia, see ibid., pp. 51, 62, 77, 94–108, 230–234, 238–249. Howard Jones and Randall B. Woods ("Origins of the Cold War in Europe and the Near East: Recent Historiography and National Security Imperatives," in *America in the World*, ed. M. Hogan, Cambridge University Press, 1995, p. 242) write that Lundestad showed that the US "was invited" by Norway; however, a close reading of Lundestad suggests a far more complex story. Cf. Gabriel, *The American Conception of Neutrality after 1941*, pp. 92–112; Jussi Hanhimäki, *Containing Coexistence* (Kent State University Press, 1997), pp. 104–105.

24. Karl Molin, *Omstridd neutralitet* (Tiden, 1991), p. 20.

25. Hanhimäki, *Containing Coexistence*, p. 105; Charles Silva, Keep Them Strong, Keep Them Friendly, dissertation, Stokholm University, 1999, pp. 58–79.

26. Bengt Wallerfelt, "Den svenska krigsplanläggningen i det kalla krigets inledande fas 1945–1958. Militära och politiska aspekter," in *Hotet från Öster*, ed. K. Zetterberg (Förvarshögskolan, 1997).

27. Matthews to Marshall, January 27, 1948, NARA, RG 59, Dec files 1945–49, 758.00/1-2748.

28. Matthews to Under Secretary of State, March 23, 1948, NARA, RG 59, Dec files 1945–49, 711.58/1-1547. The twelve fallacies were drawn up in a telegram from Matthews to Marshall (February 16, 1948, 758.00/2-1648).

29. Silva, Keep Them Strong, pp. 70–71; Nils Andrén and Åke Landqvist, *Svensk utrikespolitik efter 1945* (Almqvist & Wiksell, 1965), p. 3; Jacques Freymond, "The European Neutrals and the Atlantic Community," *International Organisation* 17 (1963), p. 603; Arne O. Brundtland, "The Nordic Balance—Past and Present," *Cooperation and conflict* 2 (1966): 30–63; Nils Örvik, "Scandinavia, NATO, and Northern Security," *International Organisation* 20 (1966): 380–396; Bo Huldt, "The Strategic North," *Washington Quarterly* (summer 1985): 99–109; Rodney Kennedy-Minott, Lonely Path to Follow, report, Hoover Institution, Stanford University, 1990.

30. Telegram, Matthews to Marshall, February 18, 1948, NARA, RG 59, Dec. files 1945–49, 758.00/2-1848.

31. Molin, *Omstridd neutralitet*.

32. Alf W. Johansson, *Herbert Tingsten och det kalla kriget* (Tiden, 1995).

33. Ambassador H. Freeman Matthews, American Embassy in Stockholm, to John D. Hickerson, Director of Office of European Affairs, State Department, Washington, October 6, 1948, National Archives and Record Administration

(NARA), RG 59, Records of the Office of British Commonwealth and Western European Affairs, 1941–1953, Lot files, Lot 54D224.

34. Krister Wahlbäck, "USA i Skandinavien, 1948–1949, del II," *Internationella studier* no. 1 (1977), p. 28.

35. Molin, *Omstridd neutralitet*, p. 29; Gabriel, *The American Conception of Neutrality after 1941*, pp. 93–112.

36. C. M. Ravndal to Secretary of State, July 9, 1946, NARA, RG 59, Dec files 1945–1949, 758.00/7-946.

37. Jussi M. Hanhimäki, "The First Line of Defence or a Springboard for Disintegration? European Neutrals in American Foreign Security Policy, 1945–1961," *Diplomacy and Statecraft* 7 (1996): 378–403.

38. *Om kriget kommit . . .*, p. 155.

39. Ibid.

40. Ibid., p. 286. See also Hans Weinberger, "'Det rörde sig väl om ofarliga ting': Svensk neutralitetspolitik och synen på teknik," *Arbetarhistoria* 23 (1999), no. 92, p. 34.

41. Weinberger, "'Det rörde sig väl om ofarliga ting.'"

42. *Om kriget kommit . . .*, p. 150.

43. A. Cave Brown, ed., *Dropshot* (Dial, 1978), p. 26. For an overview of US war planning during the Cold War, with an emphasis on target lists and strategic considerations, see David Allan Rosenberg, "A Smoking, Radiating Ruin at the End of Two Hours: Documents on American Plans for Nuclear War with the Soviet Union, 1954–1955," *International Security* 6 (1981–82), no. 3: 3–38; Rosenberg, "The Origins of Overkill: Nuclear Weapons and American Strategy, 1945–1960," *International Security* 7 (1983), no. 3: 3–71; Marc Trachtenberg, *A Constructed Peace* (Princeton, 1999), pp. 147–247. These works do not deal with specifics concerning the Scandinavian area.

44. Brown, *Dropshot*, p. 128.

45. Wallerfelt, "Den svenska krigsplanläggningen," p. 152.

46. Ibid., p. 161.

47. *Om kriget kommit . . .*, appendix, pp. 85–87.

48. Brown, *Dropshot*, p. 150.

49. Ibid.

50. Gunnar Skogmar, De nya malmfälten, Gothenburg University, 1997.

51. Brown, *Dropshot*, pp. 150–151.

52. Ibid., pp. 115–121.

53. Ibid., p. 180.

54. Ibid., p. 151.

55. *Om kriget kommit . . .*, appendix, pp. 149–159.

56. Ibid., p. 93.

57. Ibid., p. 153.

58. A report to the National Seecurity Council by the Executive Secretary on the Position of the United States with Respect to Scandinavia and Finland, January 8, 1952, NARA, RG 273, Box 16.

59. Ibid.

60. NARA, RG 59, Dec files 1950–1954, 758.13/1-1752, 758.13/1-2352, 758.5/3-1452, 758.13/4-152, 758.13/4-1152.

61. *Om kriget kommit* . . ., p. 130.

62. Lieutenant Colonel C.B. Smith, Military Intelligence Division, G-2, War Department General Staff, "Memorandum for Mr. E.L. McGinnis, Jr., Division of Aviation, Department of State," July 18, 1949, NARA, RG 59, Dec files 1945–1949, 858.014/7-1946.

63. Lundestad, *America, Scandinavia, and the Cold War*, p. 55.

64. NARA, RG 59, Dec files 1945–1949. Documents: 858.24/3-1346, 858.24/3-1546, 858.24/5-1746, 858.24/6-646.

65. Lundestad, *America, Scandinavia, and the Cold War*, p. 55.

66. NARA, RG 59, subject files 1941–1953. Records of the Office of British Commonwealth and Northern European Affairs, 1941–1953. Box 18. Telephone memorandum, Sept. 20, 1948. Colonel Shepley and Benjamin H. Hulley, Chief, NOE, State Department.

67. NARA, RG 341, Strategic Air Command Testimony Before the Vinson Committee, DS-49-782, undated, Office of the Secretary of the Air Force, 1946–1956, Box 1.

68. Agrell, *Den stora lögnen*, p. 132 (according to Agrell, based on Duncan Campbell, *The Unsinkable Aircraft Carrier*, 1984, and John Prados, *The Soviet Estimate*, 1982).

69. Jan Ottosson and Lars Magnusson, *Hemliga makter* (Tiden, 1991), p. 180; Agrell, *Den stora lögnen*, pp. 121–150.

70. Ottosson and Magnusson, *Hemliga makter*, p. 187.

71. Hans Weinberger, "Kartläggningen av Sverige," *Svenska Dagbladet*, January 29, 2000; NARA, Records of the Headquarter of the US Air Force (Air Staff), RG 341, Assistant Chief of Staff, Intelligence, Data Reference Branch, Intelligence Report Card Index, 1939–1962, Box 80-82.

72. NARA, RG 38, Chief of Naval Operations, Naval Intelligence Reports, Confidential 1949, Box 84. Serial 11-49, 12.1.1949. "A.N.A. visits Eighth Swedish Army at Uppsala Officially on 27 September 1948." Cf. Weinberger, "Kartläggningen."

73. *Om kriget kommit* . . ., p. 160. See also NARA, RG 341, Director of Operations, Air Force Plans, Project Decimal Files, 1942–1954, box 410, American Air Attaché in Stockholm to Washington, 8 July 1949: "I visited General Nordenskiöld

1st July at his invitation. He expressed concern at a discussion I had had with his Chief of Air Staff with regard to the possible danger to aircraft of the US Air Force in case of strategic situation which might demand our overflight of Sweden. He hastened to assure that Swedish interference with such flights was utterly inconceivable. Public statements by Defense Minister Vought which might be interpreted to the contrary were merely political eyewash. He, Nordenskiöld, had shown to Vought the absolute necessity for such overflights by our aircraft and Vougt had agreed that Sweden would not only permit but encourage such flights."

74. Agrell, *Den stora lögnen*, pp. 110–111.

75. For an overview on US strategic planning and its changes, see Trachtenberg, *A Constructed Peace*.

76. Agrell, *Den stora lögnen*, p. 111. See also Brown, *Dropshot*.

77. Duncan Cambell, "The 'deterrent' goes to war," *New Statesman*, 1.5.1980.

78. *Om kriget kommit . . .* , pp. 133-135.

79. Ibid., p. 135.

80. Ibid., pp. 242–245.

81. FOA Archives, Redogörelse för verksamheten vid försvarets forsknings, Stockholm 10.9.1946. Dnr H 192.

82. FOA Archives, Protokoll fört vid sammanträde den 26 juni 1946 angående apparater för igenkänningssignalering vid ekoradio, 26.6.1946, Dnr H 157.

83. FOA Archives, Redogörelse för verksamheten vid FOA 3 under budgetåret 1947/48 jämte förslag till program för den fortsatta verksamheten, 11.9.1948, Dnr H 158.

84. FOA Archives, Letter from Martin Fehrm, FOA, May 9, 1949, to the Ordnance Agencies of the Swedish Military Services, Dnr H 3043-478.

85. *Om kriget kommit . . .* , pp. 246–247.

86. Ibid., p. 247.

87. Ibid., p. 248.

88. Ibid., p. 252.

89. Telegram from Acheson to American Embassy, Stockholm, January 16, 1953, NARA, RG 59, Dec files 1950–54, 758.56/1-1653.

90. *Om kriget kommit . . .* , p. 269.

91. Ibid., p. 258.

92. Ibid., pp. 253–259.

93. Interview with former Prime Minister Carl Bildt by Peter Bratt, *Dagens Nyheter*, January 10, 1999.

94. Kennedy-Minott, *Lonely Path to Follow*, p. 15.

95. R.A. Winston, Commander, USN. Asst. Naval Attaché, Aviation: Military Cooperation, NARA, RG 38, Records of the Office of the Chief of Naval

Operations, Naval Intelligence Reports, "Secret" 1948, Box 24. Serial 5-S-48, 17.3.1948.

96. NARA, RG 59, Lot File 59D233, Miscellaneous Office Files of the Assistant Secretaries of State for European Affairs, 1943–1957, Box 32. Folder "Ambassadors Mtg Copenhagen, April 1954 (chaired by Hayden Raynor, State Department)."

97. Ibid., "Copenhagen Conference, Item # 6."

98. Both quotes: ibid.

99. Ibid.

100. Ibid.

About the Authors

Janet Abbate received her Ph.D. from the University of Pennsylvania in 1994 and is now a member of the history department at the University of Maryland. The author of *Inventing the Internet* (MIT Press, 1999) and a co-editor of *Standards Policy for Information Infrastructure* (MIT Press, 1995), she has also published numerous articles on the history, culture, and politics of the Internet. Her present research projects concern the adoption of the Internet in Eastern Europe and the history of women in computing.

Michael Thad Allen completed his Ph.D. with Thomas Parke Hughes at the University of Pennsylvania in 1995. During the period 1992–1995 he also worked with Wolfgang Scheffler at the Zentralinstitut für Anti-Semitismusforschung in Berlin. Since 1996 he has taught European history and the history of technology at the Georgia Institute of Technology. He has published articles in *Central European History*, in *Technology and Culture*, and in *Past and Present*. His first book is *The Business of Genocide: SS Business Administration, Slavery, and the Concentration Camps* (University of North Carolina Press, 2001).

W. Bernard Carlson studied with Thomas Parke Hughes at the University of Pennsylvania in the late 1970s and the early 1980s. He now teaches the history of technology in the Division of Technology, Culture, and Communication at the University of Virginia. Much of his writing has focused on the organizational, cultural, and cognitive dimensions of invention. He is the author of *Innovation as a Social Process: Elihu Thomson and the Rise of General Electric, 1870–1900* (Cambridge University Press, 1991) and an editor of the series Inside Technology (MIT Press). With support from the Sloan Foundation, he is writing a biography of the flamboyant inventor Nikola Tesla, tentatively titled *Ideal and Illusion.*

Gabrielle Hecht is an associate professor at the University of Michigan, jointly appointed in the Department of History and the Residential College. Her first book, *The Radiance of France: Nuclear Power and National Identity after World War II* (MIT Press, 1998), was awarded the American Historical Association's Herbert Baxter Adams Prize. She is working on a history of colonial and postcolonial uranium mining in Africa, Australia, and North America.

Erik P. Rau is a visiting assistant professor in Drexel University's Department of History and Politics. His contribution to this volume is part of a book project on the growth and spread of operations research in the United States since World War II. His research interests include the emergence and proliferation of new technical and engineering practices during World War II and the Cold War, the relationship between technical expertise and the state in postwar political culture, and the participation of American technical experts in decolonization and international development.

Eric Schatzberg is an associate professor in the Department of the History of Science at the University of Wisconsin in Madison, where he teaches the history of technology. He has published a number of works on the history of urban transit and airplane design. His book *Wings of Wood, Wings of Metal: Culture and Technical Choice in American Airplane Materials, 1914–1945* was published by the Princeton University Press in 1999. He is currently studying postwar American critiques of technology.

Amy Slaton received her Ph.D. from the Department of History and Sociology of Science at the University of Pennsylvania in 1995 and is now an assistant professor in the Department of History and Politics at Drexel University. Her book *Reinforced Concrete and the Modernization of American Building, 1900–1930* (Johns Hopkins University Press, forthcoming) was inspired in part by Thomas Hughes's interest in the encroachment of mass production on traditional building methods and craft-based occupations. Her current research addresses measurement and standardization in later-twentieth-century engineering contexts and the social displacements that such regularizing technologies continue to bring to the American workplace.

John Staudenmaier, S.J. has been the editor of *Technology and Culture* since 1995. A professor of history at the University of Detroit Mercy, he has served as a visiting faculty member at MIT, at Santa Clara University, and at Boston College. He has authored several historiographical studies of the history of technology, most notably *Technology's Storytellers: Reweaving the Human Fabric* (MIT Press, 1985). He also writes and lectures extensively to more popular audiences and consults for technology-oriented television programs. Typically, he interprets personal and communal behavior with an eye to the importance of core technologies in shaping behavior within an individual's host society.

Edmund N. Todd studied modern European history at the University of Florida and the history and sociology of science at the University of Pennsylvania. He has taught at the University of Maryland, at the State University of New York College at Potsdam, and the University of New Haven. His most recent publication is "Von Essen zur regionalen Stromversorgung, 1890–1920. Das Rheinisch-Westfälische Elektrizitätswerk," in *Elektrizitätswirtschaft zwischen Umwelt, Technik und Politik*, edited by Helmut Maier (Technische Universität Bergakademie Freiberg, 1999).

Hans Weinberger is an acting associate professor in the Department of History of Science and Technology at the Royal Institute of Technology in Stockholm. He is the editor of *Polhem: The Swedish Journal for the History of Technology*. His books include *Nätverksentreprenören* (Royal Institute of Technology, 1997), a study of Swedish science and technology policy during the Cold War. He is now doing research on Sweden's neutrality policy during the Cold War from a history-of-technology perspective.

Index

Printed in the United States
by Baker & Taylor Publisher Services